T0205939

Locations of regional case studies:–

① Mississippi Delta – Chapter 17.
② East coast of South America – Chapter 21.
③ Tees estuary and The Wash – Chapter 13.
④ The North Norfolk Coast – Chapter 19.
⑤ The Netherlands – Chapters 18 and 20.
⑥ Egypt – Chapter 16.
⑦ Bangladesh – Chapters 14, 15 and 16.
⑧ Hong Kong – Chapter 22.

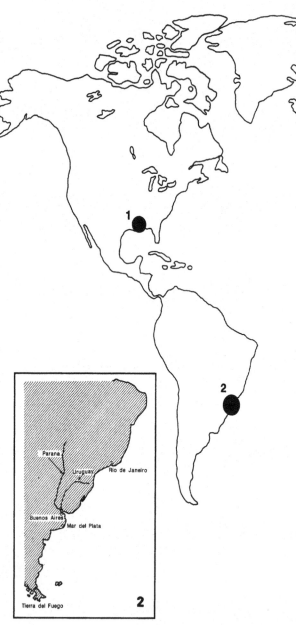

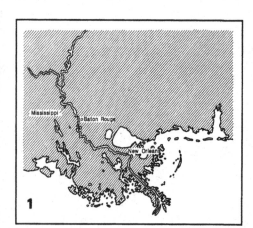

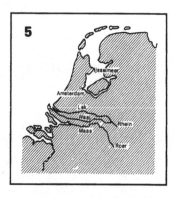

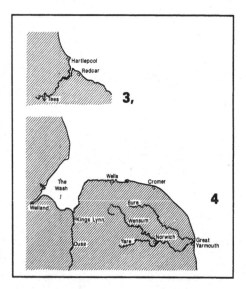

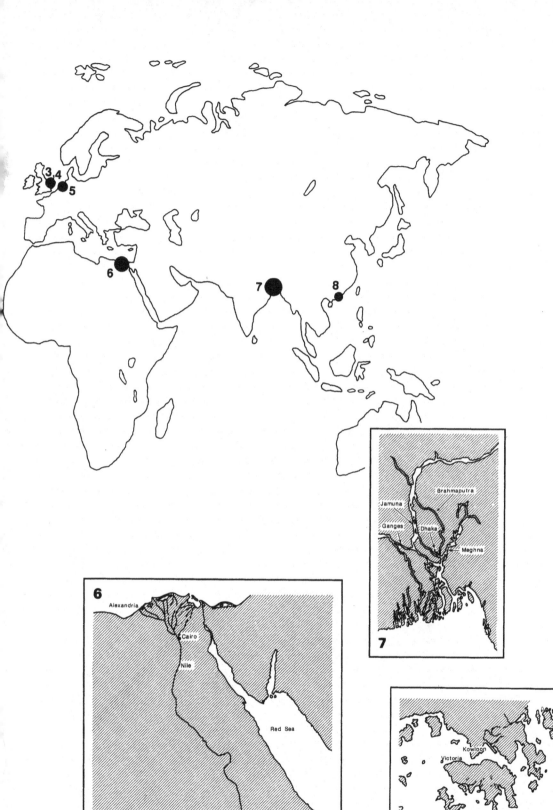

This book is primarily concerned with the prediction of future sea level rise and its possible impacts. It reviews past changes in sea level and their geological, oceanographic and climatological causes. New predictions are given of future rises due to oceanic thermal expansion and the melting of glaciers and ice sheets under the influence of increasing greenhouse gas concentrations. Whilst most emphasis is on the next 100 years, long-term changes (centuries to millennia) are also considered. Specific case studies explore the possible impacts of future rises in sea level in the UK, Bangladesh, The Netherlands, Hong Kong, Egypt, South America and the Mississippi delta in the USA.

Climate and sea level change: observations, projections and implications

Climate and sea level change: observations, projections and implications

EDITED BY

R. A. Warrick
Centre for Environmental Resource Studies
University of Waikato, Hamilton, New Zealand

E. M. Barrow and T. M. L. Wigley
Climatic Research Unit
University of East Anglia
Norwich, UK

CAMBRIDGE
UNIVERSITY PRESS

CAMBRIDGE UNIVERSITY PRESS
Cambridge, New York, Melbourne, Madrid, Cape Town, Singapore, São Paulo, Delhi

Cambridge University Press
The Edinburgh Building, Cambridge CB2 8RU, UK

Published in the United States of America by Cambridge University Press, New York

www.cambridge.org
Information on this title: www.cambridge.org/9780521115032

© Cambridge University Press 1993

This publication is in copyright. Subject to statutory exception
and to the provisions of relevant collective licensing agreements,
no reproduction of any part may take place without the written
permission of Cambridge University Press.

First published 1993
Reprinted 1994
This digitally printed version 2009

A catalogue record for this publication is available from the British Library

Library of Congress Cataloguing in Publication data

Warrick, R.A.
Climate and sea level change : observations, projections, and
implications / R.A. Warrick, E.M. Barrow, and T.M.L. Wigley.
p.   cm.
ISBN 0-521-39516-X (hardback)
1. Climatic changes.   2. Climatic changes–Case studies.   3. Sea
level.   4. Sea level–Case studies.   I. Barrow, E.M.   II. Wigley,
T.M.L.   III. Title.
QC981.8.C5W37   1993
551.4′58–dc20   92-7594 CIP

ISBN 978-0-521-39516-8 hardback
ISBN 978-0-521-11503-2 paperback

Additional resources for this publication at www.cambridge.org/9780521115032

Contents

Contributors

D.G. Aubrey, Woods Hole Oceanographic Institution, Woods Hole, MA 02543, USA

H. Brammer, 37 Kingsway Court, Hove, East Sussex, BN3 2LP, UK.

J.M. Broadus, Marine Policy and Ocean Management Center, Woods Hole Oceanographic Institution, Woods Hole, MA 02543, USA.

W.E. Carter, Geodetic Research and Development Laboratory, National Oceanic and Atmospheric Administration, Office of Charting and Geodetic Services, Rockville, Maryland 20852, USA.

K.M. Clayton, School of Environmental Sciences, University of East Anglia, Norwich, NR4 7TJ, UK.

W.H. Conner, Baruch Forest Science Institute, Clemson University, Box 596, Georgetown, SC 29442, USA.

R. Costanza, Chesapeake Biological Laboratory, University of Maryland, Box 38, Solomons, MD 20688, USA.

J.W. Day, Jr., Department of Oceanography and Coastal Sciences and Coastal Ecology Institute, Louisiana State University, Baton Rouge, LA 70803, USA.

J.G. de Ronde, Rijkswaterstaat, Postbus 20904, 2500 EX's-Gravenhage, The Netherlands.

J.M. Diamante, National Oceanic and Atmospheric Administration, R/CAR, 6010 Executive Boulevard, Rockville, MD 20852, USA.

C.S.M. Doake, British Antarctic Survey, Natural Environment Research Council, High Cross, Madingley Road, Cambridge, CB3 0ET, UK.

K.O. Emery, Woods Hole Oceanographic Institution, Woods Hole, MA 02543, USA.

R.A. Flather, Proudman Oceanographic Laboratory, Bidston Observatory, Birkenhead, Merseyside, L43 7RA, UK.

V. Gornitz, Goddard Institute for Space Studies, 2880 Broadway, New York, NY 10025, USA.

G.P. Kemp, Woodward–Clyde Consultants, Box 66317, Baton Rouge, LA 70896, USA.

H. Khandker, Department of Statistics and Computational Mathematics, The University of Liverpool, P.O. Box 147, Liverpool, L69 3BX, UK.
(Permanent address: Bangladesh Inland Water Transport Authority, 141–143 Motijheel Commercial Area, Dhaka-2, Bangladesh).

M. Kuhn, Institute of Meteorology, University of Innsbruck, Innrain 52, A6020 Innsbruck, Austria.

I.A. Mendelssohn, Department of Oceanography and Coastal Sciences and Laboratory for Wetland Soils and Sediments, Louisiana State University, Baton Rouge, LA 70803, USA.

J. Oerlemans, Institute of Meteorology and Oceanography, University of Utrecht, Princetonplein 5, Utrecht, The Netherlands.

J.G. Paren, British Antarctic Survey, Natural Environment Research Council, High Cross, Madingley Road, Cambridge, CB3 0ET, UK.

D.A. Peel, British Antarctic Survey, Natural Environment Research Council, High Cross, Madingley Road, Cambridge, CB3 0ET, UK.

E.B. Peerbolte, Delft Hydraulics, P.O. Box 152, 8300 Emmeloord, The Netherlands.

A.B. Pittock, CSIRO, Division of Atmospheric Physics, P.O. Box 77, Mordialloc, Victoria 3195, Australia.

D.T. Pugh, Institute of Oceanographic Sciences, Deacon Laboratory, Brook Road, Wormley, Godalming, Surrey, GU8 5UB, UK.

S.C.B. Raper, Climatic Research Unit, School of Environmental Sciences, University of East Anglia, Norwich, NR4 7TJ, UK.

W. Scherer, National Oceanic and Atmospheric Administration, Rockville, Maryland 20852, USA.

M.E. Schlesinger, Department of Atmospheric Sciences, University of Illinois at Urbana-Champaign, 105 South Gregory Avenue, Urbana, IL 61801–3070, USA.

E.J. Schnack, Arcan Ventures S.A., Rivadavia 611, 8°B, Buenos Aires, Argentina.

I. Shennan, Department of Geography, University of Durham, Science Site, South Road, Durham, DH1 3LE, UK.

J.G. Titus, (PM-220B), Strategic Studies Staff, Environmental Protection Agency, Washington DC 20460, USA.

M.J. Tooley, Department of Geography, University of Durham, Science Site, South Road, Durham, DH1 3LE, UK.

R.A. Warrick, Centre for Environmental and Resource Studies (CEARS), The University of Waikato, Hamilton, New Zealand.

T.M.L. Wigley, Climatic Research Unit, School of Environmental Sciences, University of East Anglia, Norwich, NR4 7TJ, UK.

H.G. Wind, University of Twente, P.O. Box 217, 7500 AE Enschede, The Netherlands.

P. Woodworth, Proudman Oceanographic Laboratory, Bidston Observatory, Birkenhead, Merseyside, L43 7RA, UK.

W.W.-S. Yim, Department of Geography and Geology, University of Hong Kong, Pokfulam Road, Hong Kong.

Preface

On 29 April 1991, a severe tropical cyclone swept up the Bay of Bengal and, like many other such storms in the region's turbulent past, caused the sea to rise rapidly and destructively. The driving wind and storm surge combined to wreak havoc in the massive, low-lying Ganges–Brahmaputra delta of Bangladesh. Although the death toll will never be known with precision, fatalities were estimated to be well over 100,000. Only two decades earlier a human tragedy of similar magnitude befell Bangladesh. Storm surge disasters of lesser proportions, but no less tragic for the many thousands of people affected, are a common occurrence in the Bay of Bengal. When they occur, they highlight dramatically the connections between climate and sea level variations – in this case on a time scale of hours to days – and the vulnerability of coastal inhabitants to extreme events.

In part, it was the chronic plight of Bangladesh and other low-lying coastal nations of the world, in addition to the landmark International Council for Scientific Union/ United Nations Environment Programme/World Meteorological Organisation (ICSU/UNEP/WMO) international conference on the greenhouse effect in Villach in 1985 (WMO, 1986; Bolin *et al.*, 1986), that prompted the United Nations Environment Programme to direct its attention to the longer-term issue of possible future sea level rise and its socio-economic consequences. How might global warming from increasing atmospheric concentrations of carbon dioxide and other greenhouse gases affect sea level and exacerbate problems of severe tropical storms and sea level rise in places like Bangladesh? Correctly, it was perceived that the answer to this question is not an easy one. The problem is multi-faceted and requires a synthesis of many strands of research within an array of disciplines – climatology, oceanography, meteorology, glaciology, geography, geology, economics. Furthermore, it was realised that there was a need to assess, from an interdisciplinary, international perspective, the current state of knowledge regarding the sea level rise issue as a basis for moving forward with research and policy initiatives.

Consequently, in September 1987, the United Nations Environment Programme (UNEP) and Directorate General XII of the European Commission (with additional support from the US Environmental Protection Agency and the Water Resources Centre in the UK) sponsored an International Workshop on Climatic Change, Sea

Level, Severe Tropical Storms and Associated Impacts, held in Norwich, UK and hosted by the Climatic Research Unit, University of East Anglia. The Workshop delegates were asked to prepare specific presentations which, when compiled, would serve as the basis for a 'state-of-the-art' assessment. Most of the Workshop contributions are presented in this book, after undergoing peer review. Additional contributions were solicited to fill obvious gaps.

Because of the scientific reviews of the greenhouse effect made by the Intergovernmental Panel on Climate Change (IPCC: Houghton *et al.*, 1990; Tegart *et al.*, 1990), publication of this book has been delayed, allowing the various contributions to be updated where appropriate and so take into account the IPCC work and recent research findings. However, in selected areas (e.g., climate modelling, global observing networks) progress has been so rapid that some chapters are slightly out of date; for these we apologise. The unifying theme of the book is sea level change, and the major subject areas addressed are observations, projections and implications.

R.A. Warrick
E.M. Barrow
T.M.L. Wigley

References

Bolin, B., Döös, B., Jäger, J. and Warrick, R., (eds.) (1986). *The Greenhouse Effect, Climate Change and Ecosystems.* John Wiley and Sons, Chichester.

Houghton, J.T., Jenkins, G.J. & Ephraums, J.J. (eds.) (1990). *Climate Change: The IPCC Scientific Assessment.* Cambridge University Press, Cambridge.

Tegart, W.J.McG., Sheldon, G.W. & Griffiths, D.C. (eds.) (1990). *Climate Change: The IPCC Impacts Assessment.*

WMO (1986). Report of the international conference on *The Assessment of the Role of Carbon Dioxide and of Other Greenhouse Gases in Climate Variations and Associated Impacts.* WMO-No.661, World Meteorological Organization, Geneva.

Abbreviations

AEC	adenylate energy charge
AGCM	atmospheric general circulation model
ASLR	apparent sea level rise
ATP	adenine triphosphate
BAU	Business as Usual
C.D.	chart datum
CFC	chlorofluorocarbons
CI	confidence interval
COADS	Comprehensive Ocean-Atmosphere Data Set
DMA	Defence Mapping Agency
EBM	energy-balance model
ECMWF	European Centre for Medium-Range Weather Forecasting
ENSO	El Niño – Southern Oscillation
EOF	empirical orthogonal function
EPA	Environmental Protection Agency
GCM	general circulation model
GFDL	Geophysical Fluid Dynamics Laboratory
DOE	Department of Energy
FAO	Food and Agriculture Organization
GDP	gross domestic product
GHGs	greenhouse gases
GIS	Geographic Information System
GISS	Goddard Institute for Space Studies
GLOSS	Global Level of the Sea Surface
GMT	Greenwich Mean Time
GNP	gross national product
GPS	Global Positioning System
HMSO	Her Majesty's Stationery Office
IAPSO	International Association for the Physical Science of the Ocean
ICSU	International Council for Scientific Union
IECo	International Engineering Co. Inc.

IERS	International Earth Rotation Service
IGCP	International Geological Correlation Programme
IGU	International Geophysical Union
IOC	Intergovernmental Oceanographic Commission
IPCC	Intergovernmental Panel on Climate Change
ISOS	Impact of Sea Level Rise on Society
ITCZ	Intertropical Convergence Zone
IUCN	World Conservation Union
JILA	Joint Institute for Laboratory Astrophysics
MSL	mean sea level
NAG	Navy Astronautics Group
NAS	National Academy of Sciences
NASA	National Aeronautics and Space Administration
NCAR	National Center for Atmospheric Research
NERC	Natural Environment Research Council
NGWLMS	Next generation water level measurement system
NOAA	National Oceanic and Atmospheric Administration
NRC	National Research Council
NRL	Naval Research Laboratories
OAS	Organization of American States
OD	ordnance datum
OGCM	ocean general circulation model
OSU	Oregon State University
P.D.	principal datum
PD	pure-diffusion
PRB	Polar Research Board
PSMSL	Permanent Service for Mean Sea Level
QBO	Quasi-Biennial Oscillation
RCM	radiative convective model
SL	sea level
SLP	sea level pressure
SPLASH	Special Programme for Listing Amplitude of Surge Heights
SST	sea surface temperature
TGBM	Tide-Gauge Bench Mark
TM	thematic mapper
TOGA	Tropical Ocean and Global Atmosphere programme
UD	upwelling-diffusion
UKMO	United Kingdom Meteorological Office
UNEP	United Nations Environment Programme
UNESCO	United Nations Educational, Scientific and Cultural Organization
USGS	US Geological Survey
VLBI	Very Long Baseline Interferometry
WAIS	West Antarctic ice sheet
WHOI	Woods Hole Oceanographic Institute
WMO	World Meteorological Organization
WOCE	World Ocean Circulation Experiment
WRM	water resources management

I

OVERVIEW

1

Climate and sea level change: a synthesis

R.A. Warrick

Change in climate and sea level is the rule, not the exception. Natural variations in sea level are clearly evident over a large range of time and space scales, from the pulse of diurnal tides to globally coherent variations in sea level occurring over many millenia (Fig. 1.1, from Pugh, Chapter 4). Climate varies similarly and has causal connections with the geological, biological and hydrological processes that affect sea level. The Earth is naturally a strongly interactive, dynamic system.

This fact is often overlooked in the recent excitement over the issue of 'global change'. The alarming rate at which the atmospheric concentrations of carbon dioxide, methane, chlorofluorocarbons, nitrous oxide and other greenhouse gases (GHGs) have been increasing raises the spectre of major global environmental changes within a century – a blink of the eye in geological time. In order to be properly assessed, such GHG-induced changes must be viewed in the context of their natural variations – past, present and future – and the causes of these variations.

But if global climate and sea level vary naturally, why the worry? It is not so much the prospect of global change *per se*, but rather the potential rates and magnitudes of the GHG-induced change that give rise to legitimate concerns about the future. These concerns include the following: firstly, that humankind may now be a potent factor in causing unidirectional global changes which could dominate over natural changes on the decade-to-century time-scale; secondly, that, in terms of recent human experience, changes in climate and sea level could accelerate to unprecedented rates; thirdly, that human tinkering with the global climate system could have unforeseen catastrophic consequences (e.g. 'runaway' warming or sea level rise from strong positive feedbacks); and finally, that the quickened rates of change could exceed the capacity of natural and human systems to adapt without undue disruption or cost.

In order to discover whether there is indeed reason to be concerned on these grounds, the connections between climate and sea level change, and their implications for environmental and societal systems, need to be examined. That is the purpose of this book. The objective of the present chapter is to provide an interpretive synthesis of the contributions to this volume. Let us begin by considering the observational data.

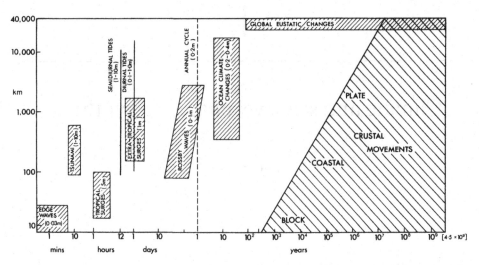

Fig. 1.1 Some of the physical and geological processes responsible for sea level changes, with indications of the length (km) and time-scales involved. Global processes have a length scale of 40,000 km. (From Pugh, Chapter 4).

Observations: past, present and future

Sea level change on the geological time-scale

On the geological time-scale, evidence from coral reefs, oxygen isotopes and other records leaves little doubt that there are globally coherent connections between sea level and climate. The fluctuation of sea level in association with the transitions between cold glacial and warm interglacial periods is the most obvious manifestation, one that is evident over many glacial–interglacial cycles spanning many millions of years (My). For example, from visual inspection of Fig. 1.2, taken from Tooley (Chapter 6), the temperature inferred from ice core measurements in Antarctica (Jouzel et al., 1987) is correlated with sea level estimates for the Huon Peninsula (which has been interpreted as a global curve; Chappell and Shackleton, 1986). During the last interglacial (120,000 years before present, BP), the global mean temperature was probably slightly warmer than today. Global mean sea level may have been 5–6 m higher than today, perhaps due to a disintegration of the West Antarctic ice sheet (Mercer, 1978) or a lowering of the surface of the East Antarctic ice sheet (Robin, 1986). During the last glacial maximum (18,000 BP) when the global temperature was about 4–5 °C colder than today, sea level was over 100 m lower.

Throughout the world, Late Holocene sea level indicators document the rate of rise in sea level since the last glacial maximum. There is strong corroborating evidence that the rise was swift between about 15,000 and 7,000 BP as the global temperature rose quickly and the large continental ice sheets disintegrated and increased the mass of the oceans. Thereafter, sea level continued to rise but at a decreasing rate (Gornitz, Chapter 2; Aubrey and Emery, Chapter 3; Tooley, Chapter 6). Over the last 1,000 years, the rate of rise has been of order 0.1–0.2 mm/yr, probably due not to the addition of mass to the oceans, but rather to lingering isostatic crustal adjustment (Peltier, 1986).

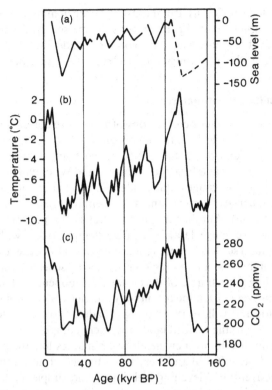

Fig. 1.2(*a*) Sea level changes during the past 160,000 years for the Huon Peninsula, recalculated after detailed correlation with the ^{18}O record from Pacific core V19–30 (Chappell and Shackleton, 1986. Reprinted by permission of *Nature*, **324**, copyright © 1986 Macmillan Magazines Ltd). (*b*) Variation with time of the Vostock isotope temperature record as a difference from the modern surface temperature value (Jouzel *et al.*, 1987. Reprinted by permission of *Nature*, **329**, copyright © 1987 Macmillan Magazines Ltd). (*c*) CO_2 concentrations plotted against age in the Vostock ice core record (Barnola *et al.*, 1987. Reprinted by permission of *Nature*, **329**, copyright © 1987 Macmillan Magazines Ltd). (From Tooley, Chapter 6).

That sea level is still adjusting to the effects of the last glaciation attests to the long response times involved, particularly when dealing with large ice sheets and their geophysical effects on the Earth's crust.

At finer spatial and temporal resolutions, the concept of a globally coherent, real (eustatic) change in the level of the ocean may not be entirely appropriate. As argued by Tooley (Chapter 6), because of deformations of the geoid and, particularly, the redistribution of ice mass during glacial–interglacial cycles, not all oceanic regions of the world will be affected identically by global climate change. Thus, strictly speaking, 'regional eustasy' is preferable to the more elusive concept of 'global eustasy'. Nonetheless, the existence of a globally coherent *component* of sea level change, which may be associated with climate change, cannot be denied.

On the geological time-scale, it is valid to speculate about an 'imminent' rise in global sea level of a few metres (Tooley, Chapter 6) due to the long-term effects of global warming of several degrees Celsius. But it must be borne in mind that such speculation is in the context of long-term equilibrium response – time-scales of thousands of years – not the time-dependent response within the coming decades (see, for example, Wigley and Raper, Fig. 7.9, in which the long-term sea level rise 'commitment' is illustrated).

Sea level changes on the secular time-scale

At the secular time-scale, direct observations of *relative* sea level changes have been derived mainly from tide-gauge data. In the conventional tide-gauge system, sea level is measured relative to a land-based tide-gauge bench mark. Hence, as discussed by Gornitz (Chapter 2), Aubrey and Emery (Chapter 3) and Woodworth (Chapter 23), the major problem in using tide-gauge data for identifying changes in eustatic sea level is the contaminating influence of vertical land movements (e.g. from isostatic effects, neo-tectonism, sedimentation). Another major problem is the geographical bias in the distribution of usable tide-gauge data. Even after attempts to deal with these problems by subtracting long-term sea level trends (e.g., using Late Holocene sea level indicators or geophysical Earth models) and by regional weighting schemes, there are still large uncertainties regarding the estimate of the eustatic component of change.

Considerable improvements in the technology and networks for observations of sea level have been made and will, in the coming decades, allow the uncertainties to be narrowed. With respect to the vertical land movement component of relative sea level measurements, the application of new geodetic techniques like the phase-differenced Global Positioning System (GPS) and Very Long Baseline Interferometry (VLBI), in conjunction with conventional levelling techniques and supplemented with absolute gravity measurements, is central to the goal of providing accurate local, regional and global geodetic control in the monitoring of temporal variations in the geocentric positions of tide gauges (Carter *et al.*, Chapter 5). Such methods can also be used to correct errors in direct measurements of the sea surface using satellite radar altimetry. Improving the quality and quantity of the global tide-gauge network through GLOSS (Global Level Of the Sea Surface) will address problems like the geographical distribution of measurements, and, by being tied in to the GPS/VLBI system, contribute to the overall objective of accurately monitoring the rate of change of absolute sea level (Pugh, Chapter 4). But such innovations aim for the future and do not rectify the problems inherent in the tide-gauge data set for the past century.

Considerable attention has been focused on identifying the global component of sea level change and its possible link to climate change over the last 100 years. Identification of the rate of globally coherent sea level change on the decade-to-century time-scale is important for establishing the underlying historical trend for future projections, for validating and constraining models, and for detecting possible accelerations in sea level rise. Because of the problems with tide-gauge data, some scientists, like Aubrey and Emery (Chapter 3), are cautious about estimating the value of the rate of global mean sea level change over the last 100 years. Others, like Gornitz (Chapter 2) believe that a true eustatic rise in sea level can, in fact, be discerned in the data after careful extraction of long-term trends and data averaging.

For example, in Chapter 2, Gornitz presents the results of two methods of estimating the eustatic component of sea level rise. First, by taking the arithmetic mean of a set of 130 station trends (corrected data) the rate of sea level rise over the last 100 years is estimated to be 1.2 mm/yr, although as shown in Fig. 1.3(*a*) substantial geographical variation is evident. (Statistically, these spatial variations in trend are such that there are wide confidence limits for the global mean value, which do not exclude the possibility of a zero, or even negative, change in sea level). In the second approach, the corrected data are averaged annually into a composite global mean sea level curve Fig. 1.3(*b*) and the least-squares slope is estimated to be 1.0 ± 0.1 mm/yr (95% confidence interval; with this composite curve the spatial variability is, of course, obscured). Many other analyses of tide-gauge data have been performed, all of which show a positive change in sea level. On the basis of these studies, Gornitz (as well as Woodworth, Chapter 23; Warrick and Oerlemans, 1990) conclude that the average change over the last 100 years most likely lies in the range 1.0–2.0 mm/yr. There is no firm evidence of an acceleration in the rate of sea level rise over this time period (Woodworth, 1990).

Based on Late Holocene sea level indicators and other data, a rate of global sea level rise of 1.0–2.0 mm/yr during the present century is probably an order of magnitude higher than that of preceding centuries. This would suggest that vertical land movements or other long-term geological processes are not responsible for the current rate of sea level rise. Such a change in the rate of rise on the decade-to-century time-scale is most likely due to climatic factors. The most obvious are temperature-related: oceanic thermal expansion and melting of land ice. Understanding of the effects of these factors is crucial to making projections of sea level rise as a result of global warming.

Projections

Factors affecting global sea level on the decade-to-century time-scale

Oceanic expansion. Without changing the mass, the volume of the oceans, and hence sea level, can change due to changes in the density of sea water – a 'steric' effect. Steric changes occur as a result of changes in ocean temperature and salinity. As discussed by Wigley and Raper (Chapter 7), temperature is by far the more important factor on the decade-to-century time-scale considered in projections of global sea level rise due to global warming.

In the context of greenhouse gas forcing changes, there is some trade-off between surface warming and thermal expansion depending on the rate at which heat from the surface mixed layer is able to penetrate into the deeper layers of the ocean. As the climate system is forced with increasing concentrations of greenhouse gases, oceanic thermal inertia creates a lag effect, slowing the surface warming. The more effectively heat is mixed into the deeper layers of the ocean the slower the warming, but the greater the thermal expansion as more heat is dispersed throughout the ocean depths.

In order to be consistent in the calculation of time-dependent global warming and sea level rise, it is thus preferable to use an ocean–climate model that, first, can account for these trade-offs and, second, can be readily adjusted in order to quantify the large uncertainties involved. A fully coupled dynamic ocean–atmosphere general circulation model (GCM) fits the bill on the first account, but unfortunately is impractical on the second: large computing requirements preclude extensive sensitivity analyses. For this reason Wigley and Raper (Chapter 7) and others (e.g. Gornitz *et al.*, 1982; Frei *et al.*, 1988; Wigley and Raper, 1987) use relatively simple, highly parameterized upwelling-diffusion (UD) or pure diffusion (PD) energy-balance climate models for making both

(a)

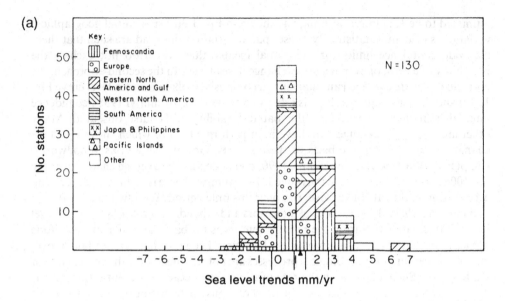

(b)

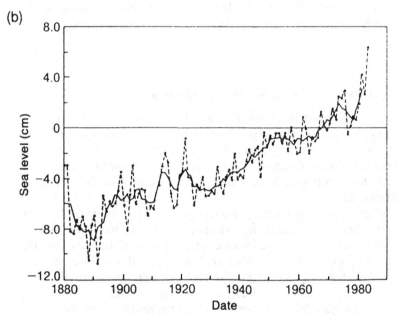

Fig. 1.3(*a*) Histogram of the number of tide-gauge stations *vs* sea level trends; long-range trends subtracted. Filled triangle ▲ indicates mean rate of sea level rise; lines indicate the bounds of the 95% confidence interval (C.I.) and ± 1σ, respectively. (From Gornitz, Chapter 2). (*b*) A composite global mean sea level change curve over the last century (reference period is 1951–70) with 5-year running mean represented by the solid line. (Adapted from Gornitz and Lebedeff, 1987).

past and future temperature and thermal expansion projections due to GHG-induced forcing. While such models cannot predict dynamic sea level effects arising from changes in ocean circulation, as an ocean–atmosphere GCM would do (see, e.g., Mikolajewicz *et al.*, 1990), they can easily be run many times under varying sets of assumptions. Using such a model, Wigley and Raper (Chapter 7) estimate that thermal expansion over the last 100 years (1880–1990) has been 3–6 cm, as constrained by a global temperature change of 0.5 °C. We shall summarize their results for future projections shortly.

Large ice sheets.   On the decade-to-century time-scale, the other most likely cause of global sea level rise is an increase in ocean mass from the melting of land ice. The potential sources are the large ice sheets of Greenland and Antarctica, and the smaller ice caps and mountain glaciers. As discussed by Oerlemans (Chapter 9), the large ice sheets comprise about 99% of the world's land ice, most of which is located in Antarctica. The Antarctic ice sheet is roughly equivalent to 65 m of sea level rise, if melted and distributed evenly over the world's oceans; the Greenland ice sheet would contribute an order of magnitude less. Although complete melting of the ice sheets over the next century is impossible, on longer time-scales relatively small changes in the mass balance of the large ice sheets could have very large effects on sea level.

For the large ice sheets, the contribution to sea level change is determined by imbalances between accumulation (precipitation of snow) and losses (due to melting/runoff and ice flow/calving, i.e., iceberg formation). The cold conditions of Antarctica are such that the accumulation rate is currently low, due to the low moisture content of the cold air; melting/runoff is negligible and would remain negligible even with substantial warming. The wastage of ice is controlled by the dynamics of ice flow and calving rates, which, in the context of climatic change, have a response time of the order of 10^4 to 10^5 years. Thus, on the decade-to-century time-scale a change in air temperature is likely to affect the surface mass balance only, i.e., the accumulation side of the equation. Warmer conditions would be expected to lead to a positive mass balance, that is, an effective *lowering* of sea level, as precipitation rates increase. In terms of sea level equivalent, Oerlemans (Chapter 9) estimates the sensitivity of the mass balance of Antarctica as a whole to be -0.5 mm/yr per degree warming. As a best estimate Wigley and Raper (Chapter 7) choose -0.2 mm/yr per degree after taking account of possible instability of the West Antarctic ice sheet, as discussed below.

In contrast, Schlesinger (Chapter 11), using the OSU (Oregon State University) GCM, finds both increases and decreases in the net accumulation rates over Antarctica in CO_2 doubling experiments, with an overall decrease suggesting a *rise* in sea level. Whether or not this is a reasonable estimate depends on a number of factors, including: the accuracy of the control run (if control run temperatures are significantly warmer than observed, too much melting will occur); the assumptions regarding runoff (if the proportion of meltwater that percolates and refreezes is underestimated, too much runoff will occur); and the high latitude enhancement of warming (the OSU equilibrium results show changes of up to 10 °C), which is much higher than results from more recent time-dependent GCM experiments (e.g., Stouffer *et al.*, 1989).

A major uncertainty is the possible instability of the West Antarctic ice sheet (WAIS). The sea level equivalent of the WAIS is about 5–6 m. The WAIS is composed of a grounded ice sheet, a marine ice sheet (grounded below sea level) and an extensive floating ice shelf. The latter is 'pinned' in several places where the ice shelf runs aground. The floating ice shelf exerts back pressure and has a buttressing effect, retarding the seaward flow of ice across the grounding line. The major issue, first raised by Mercer (1978), is whether warming will lead to a thinning and 'unpinning' of the ice shelf, causing an ice flow 'surge' and retreat of the grounding line, with eventual disintegration of the WAIS. Early model studies (e.g. Thomas *et al.*, 1979) suggested that this process could proceed rapidly. However, based on more recent model studies (e.g. Budd *et al.*, 1987), Oerlemans (Chapter 9) concludes that the rates of shelf thinning necessary for this to happen are unlikely to occur on the decade-to-century time-scale with projected rates of global warming. Nonetheless, the uncertainties are very large indeed.

Another special area of concern in Antarctica is the Antarctic Peninsula. The Peninsula contains the equivalent of about 0.5 m of sea level. It has sometimes been argued that, since the Peninsula is considerably warmer than the rest of Antarctica and the accumulation rates are larger, the Peninsula glaciers may be more akin to smaller mountain glaciers than large ice sheets and therefore more sensitive to climatic change. However, as Paren *et al.* (Chapter 10) point out, wastage is still dominated by calving with relatively minor melting/runoff. Thus, as in the case of the rest of Antarctica, on the decade-to-century time-scale the effect of warming will be primarily on surface mass balance through an increase in accumulation rates. Indeed, there is some evidence to suggest a temperature-related increase in accumulation rate over the last 30 years (Peel and Mulvaney, 1988), with no appreciable change in discharge rates (despite evidence of sizeable reductions of some ice shelves like the Wordie ice shelf; Doake and Vaughan, 1991). For the effect on future sea level, Paren *et al.* (Chapter 10) show that even with relatively large hypothetical changes in the mass balance components, the contribution to sea level change over 100 years is minor, several centimetres at most.

In comparison to Antarctica, Greenland has a warmer climate. As a consequence, the wastage of ice from the Greenland ice sheet occurs in nearly equal proportions from melting/runoff and calving. In the event of climate warming, both melting/runoff and accumulation rates should increase. In terms of surface mass balance, it is likely that melting/runoff will dominate despite the large interior area of accumulation (Oerlemans, Chapter 9). Calving rates are unlikely to be affected in the coming decades because the dynamic response, like that of the Antarctic ice sheet, occurs on time-scales of thousands of years. In terms of sea level equivalent, Oerlemans (Chapter 9) estimates that the sensitivity of the Greenland ice sheet is +0.5 mm/yr per degree warming. Wigley and Raper use a best guess value of 0.3 ± 0.2 mm/yr/°C.

Are the large ice sheets currently in balance? From direct, observational evidence of mass balance components there are no firm grounds for supposing that the mass balance of either ice sheet differs significantly from zero (Oerlemans, Chapter 9). However, there is circumstantial evidence to the contrary. From the few available long-term temperature records, it appears that Greenland has warmed over the last 100 years, with a peak warming of perhaps 2 °C occurring in the 1930s. Given the sensitivity value noted above, this implies a negative mass balance (assuming initial equilibrium conditions) and a positive contribution to sea level change, perhaps of the order of several centimetres (Warrick and Oerlemans, 1990). The rapid retreat of selected outlet glaciers in Greenland lends tentative support to this notion (Weidick, 1984).

With respect to Antarctica, recent analyses of temperature data suggest that Antarctica has also experienced a warming over the last 100 years of about 2 °C on average (Jones, 1990). As discussed by Wigley and Raper (Chapter 7), this implies a positive mass balance change (assuming initial equilibrium conditions) and a negative contribution to sea level change of the order of a few centimetres, depending on the sensitivity value used.

Mountain glaciers and small ice caps.   The smaller mountain glaciers comprise less than 1% of the world's land ice, but, in general, are relatively more sensitive to changes in climate than the large ice sheets. Complete melting of all such glaciers would be equivalent to a sea level rise of somewhere between 0.3 m and 0.6 m, depending on classification schemes for land-based ice and on assumptions about ice volumes, which are very uncertain (Kuhn, Chapter 8). Unlike the case of the large ice sheets, it *is* possible to substantially deplete glacier ice on the century time-scale under some scenarios of global warming.

It is likely that changes in glacier volumes, along with thermal expansion of the oceans, are responsible for most of the observed rise (10–20 cm) in global sea level over the last 100 years. A global survey shows that most of the world's glaciers retreated, on average, over that time period (Grove, 1988). However, it is difficult to estimate precisely the total change in global glacier volume because long-term mass balance data for individual glaciers are meagre. An estimate has been provided by Meier (1984). As reviewed by Kuhn (Chapter 8), Meier skirted the data problem by assuming that the magnitude of the long-term change is proportional to the seasonal mass balance amplitude. Since seasonal mass balance data are much more abundant, Meier used this proportion, derived from the few glaciers for which data are available, to extrapolate to a global scale. Based on Meier's work, it appears that glaciers may have contributed about 4 cm to global sea level over the last 100 years.

For future projections of small glacier contributions to sea level, the lack of glacier-specific data and models means that relatively simple global models have to be used. As discussed by Kuhn (Chapter 8) these include simple empirical balance–temperature relationships, hydrometeorological models and energy-balance models. For the latter, preferred type of model, Kuhn obtains about a 20 cm equivalent sea level rise over a 100 year period for a linear increase in temperature up to 4 °C. This compares favourably with estimates obtained from simple global glacier melt models used by Wigley and Raper (Chapter 7) and Oerlemans (1989) which explicitly take into account the dwindling amount of ice available for melting and the glacier response time.

Future global warming and sea level rise

Estimates of the total combined effects of GHG-induced oceanic thermal expansion and land ice changes on future global mean sea level are fraught with uncertainty. The major sources of uncertainty include: future GHG emission rates and consequent changes in atmospheric concentrations and radiative forcing; feedbacks in the climate system as they affect the climate sensitivity to a given change in radiative forcing (usually expressed as the equilibrium global mean temperature change for a CO_2 doubling); the penetration and transport of heat from the surface to the deeper layers of the ocean, which affects the time-dependent rate of ocean expansion and surface warming (and thus land ice mass balances); regional and seasonal changes in climate; and model uncertainties in the response of ice sheets and glaciers to climate changes.

Possible dynamic changes in ocean circulation could also affect sea level directly and indirectly, by compounding the uncertainties noted above, but have not yet been addressed adequately because fully coupled three-dimensional dynamic ocean–atmosphere models are still in the formative stages of development. In short, the full range of uncertainty surrounding any estimate of future sea level must be very large indeed. The only statement that can be made with near certainty based on the published literature is qualitative: global warming should cause sea level to rise. The pressing issue is, how high?

Various estimates of future sea level rise have been made over the last decade, as shown in Table 1.1. It is somewhat difficult to compare these estimates because of differences in the ways in which time dependence and uncertainties (often not explicitly stated) were handled. Nonetheless, perhaps with the exception of the early US Environmental Protection Agency (EPA) estimates (Hoffman *et al.*, 1983; 1986), there appears to be agreement that it is unlikely that sea level rise will exceed a metre before the end of the 21st Century.

The analysis by Wigley and Raper (Chapter 7) concentrates on giving time-dependent and internally consistent temperature and sea level changes, and on examining the effect of uncertainties in greenhouse gas forcing changes and model parameters on sea level rise estimates. In this analysis, the Intergovernmental Panel on Climate Change (IPCC) scenarios BaU ('Business as Usual'), B and C are considered to be the high, best-guess and low estimates of future forcing changes, respectively (Houghton *et al.*, 1990). For each forcing scenario, sets of high, best-guess and low values of model parameters are used in order to maximize the range of sea level rise estimates. As shown in Fig. 1.4, for the period 1990–2100 sea level rise is estimated to lie in the range 3–124 cm, with a best estimate of 46 cm (the global warming over the same period is 2.5 °C). In comparison to the results of IPCC (Warrick and Oerlemans, 1990), the range is larger because of the more explicit treatment of uncertainties, but the best estimates are nearly identical for comparable forcing scenarios (this is partly due to compensating differences in modelling procedures). It is interesting to note that these estimates are not dissimilar to the estimate of 61 cm by 2087 made in a more qualitative fashion by an assembled group of experts (Woodworth, Chapter 23).

Future changes in severe tropical storms and storm surges

In terms of impacts, an important issue is whether the intensity, frequency and/or area of occurrence of severe tropical storms (hurricanes, typhoons, cyclones) will be changed as a consequence of GHG-induced global warming. The energy for tropical storm development and intensification is derived from ocean heat. Insofar as there is an observed close correspondence between sea surface temperature (SST) and tropical storm formation – a SST of 27 °C or greater is required – one might expect increased tropical storm activity with global warming. However, as pointed out by Raper (Chapter 12), a warm SST is a necessary, but not sufficient, condition for the intensification of tropical disturbances into severe storms. Other atmospheric conditions, such as weak vertical wind shear over the storm centre and strong low-level cyclonic vorticity with high mid-tropospheric humidity, appear instrumental in storm intensification (e.g., Gray, 1968). How might global warming affect these processes? Limited evidence is available from three sources.

Table 1.1 *Estimates of future global sea level rise (cm) (Modified from Raper et al., 1990)*

	Contributing factors				Total rise[a]		
	Thermal Expansion	Alpine	Greenland	Antarctica	Best Estimate	Range[f]	To (Year)
Gornitz *et al.* (1982)	20	20 (combined)			40		2050
Revelle (1983)	30	12	13		71[b]		2080
Hoffman *et al.* (1983)	28–115	28–230 (combined)				56–345	2100
						26–39	2025
PRB (1985)	[c]	10–30	10–30	− 10–100		10–160	2100
Hoffman *et al.* (1986)	28–83	12–37	6–27	12–220		58–367	2100
Robin (1986)[d]	30–60[d]	20 ± 12[d]	to + 10[d]	to − 10[d]	80[i]	25–165[i]	2080
Thomas (1986)	28–83	14–35	9–45	13–80	100	60–230	2100
Villach (1987) (Jaeger, 1988)[d]					30	− 2–51	2025
Raper *et al.* (1990)	4–18	2–19	1–4	− 2–3	21[g]	5–44[g]	2030
Oerlemans (1989)					20	0–40	2025
Van der Veen (1988)[h]	8–16	10–25	0–10	− 5–0		28–66	2085
Warrick & Oerlemans (1990)[j]	28–66	8–20	3–23	− 7–0	66	31–110	2100
Wigley & Raper (this volume)					46[k]	3–124[l]	2100
Woodworth (this volume)	31	15	0	0	61[m]		2087

Notes:

[a] from the 1980s

[b] total includes additional 17 cm for trend extrapolation

[c] not considered

[d] for global warming of 3.5 °C

[f] extreme ranges, not always directly comparable

[g] internally consistent synthesis of components

[h] for a global warming of 2–4 °C

[i] estimated from global sea level and temperature change from 1880–1980 and global warming of 3.5 ± 2.0 °C for 1980–2080.

[j] for IPCC 'Business-as-Usual' forcing scenario only

[k] for IPCC Policy Scenario B, best estimate model parameters

[l] for IPCC forcing scenarios A and C with high and low model parameters, respectively

[m] includes 15 cm baseline trend

Empirical studies of observational data.   Based on *empirical analyses* of observed interannual variations in severe tropical storm frequencies and SSTs, Raper (Chapter 12) finds that there are predominantly positive correlations in four of the six cyclone regions. In the regions with highest positive correlation, simple regression suggests an increase of 37–63% in severe tropical storm frequency per degree increase in SST. But this relationship should not be construed as having predictive skill for long-term climate change. Firstly, there is evidence to suggest that the relationship is not

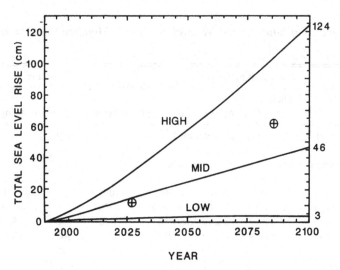

Fig. 1.4 Projections of sea level rise, 1990–2100. Projections reflect the full range of uncertainties in model parameters and radiative forcing changes due to increasing greenhouse gas concentrations (Wigley and Raper, Chapter 7). Crosses indicate best-guess estimates for 2027 and 2087, as reported by Woodworth (Chapter 23).

necessarily directly causal because the areas of correlation are not always coincident with the areas of cyclone origin and tracks. Thus, the effect of SSTs may be indirect, through the connection with the regional-scale atmospheric circulation. Secondly, two of the six regions (North Indian Ocean and West North Pacific) showed a predominantly *negative* correlation.

The picture is further complicated by the strong negative correlation in some regions between severe tropical storm frequency and El Niño occurrence (e.g. in the Australian region; Nicholls, 1984), either through SSTs or disruption of atmospheric circulation patterns favourable to tropical storm development and intensification (e.g. in the North Atlantic; Gray, 1984). As pointed out by Pittock and Flather (Chapter 24), severe tropical storm activity could be strongly affected by changes in the ocean circulation and the El Niño – Southern Oscillation (ENSO) system from global warming and its feedbacks, which emphasizes the need for further development of high resolution fully coupled ocean–atmosphere models.

Theoretical storm models.   As discussed by Schlesinger (Chapter 11), a *theoretical model* of tropical storms has been used to investigate the effect of increasing SST on maximum storm intensity (Emanuel, 1987). In these experiments the model was perturbed using changes predicted by the Goddard Institute for Space Studies (GISS) GCM. Central pressure decreased and windspeed increased non-linearly, leading to an increase in the destructive potential of storms by up to 60% in some places. However, this model result is not supported entirely by empirical evidence. While there *is* support for the notion of a threshold temperature for formation, additional temperature increases do not necessarily lead to further intensification (Merrill, 1988) due to other controlling factors (noted above; see Raper, Chapter 12).

General circulation model simulations.   The horizontal resolution of current *GCM simulations* of future climatic change is generally too coarse to resolve severe tropical storms explicitly (Mitchell *et al.*, 1990; Schlesinger, Chapter 11). Although several recent experiments (Broccoli and Manabe, 1990) for CO_2 doubling have been investigated for changes in storm days, the results are inconclusive (Mitchell *et al.*, 1990). In short, it is not yet possible to predict, based on either empirical or model results, how severe tropical storm intensity, frequency or areal occurrence will change in a warmer world.

One of the most life-threatening events associated with tropical storms is the extreme temporary increase in sea level and the resultant coastal flooding from storm surges. Higher mean sea level implies greater potential destruction from storm surges, since the initial sea surface 'platform' from which flooding occurs is raised. In addition, an important issue is whether higher mean sea level itself would affect the surge and tide components (and their interaction) of the rise. Flather and Khandker (Chapter 14) examined this issue using a storm surge model developed for the Bay of Bengal. The storm surge of 24–5 May 1985 was simulated with current mean sea level and with a 2 m rise. The difference was a slight *reduction* of the maximum flood elevations (above mean sea level). This was caused largely by a decrease in the ratio, wind stress/water depth, that occurred in the case of an elevated mean sea level, and is consistent with theory and other model studies (e.g., Heaps, 1967). This may be a compensating effect, but it is minor compared to the overall change in flood heights from the rise in mean sea level itself. A similar conclusion was reached by de Ronde (Chapter 20) using a model for the North Sea.

Impacts and case studies

The importance of changes in climate, sea level and severe tropical storms lies in how the physical environment is affected in ways which are judged detrimental to ecological and human social, economic and political systems. As summarized by Titus (Chapter 25), the principal physical effects of rising sea level include: land loss (through permanent inundation, erosion), temporary flooding (through storm surges, backwater effects) and salt-water intrusion (into groundwater, surface water, soils, wetlands). These, in turn, can preclude or reduce the potential use of land (for agriculture, recreation, residential, commercial or industrial purposes), result in occasional loss of life and property damage (to structures and infrastructure), and degrade environmental resources (water supplies, wetland ecosystems, fisheries). The seriousness of the threat of rising sea level, of course, depends on the particular coastal region and the characteristics of the environmental and socio-economic systems at risk and the valuation placed on them by society.

The approach taken for the present volume was to select a number of case studies which serve to illustrate a mix of the specific concerns noted above. As depicted in the endpapers, locations were selected in order to provide fairly comprehensive coverage – developing and developed countries, urban and rural uses, protected and unprotected coasts.

Three common themes

Three major themes emerge from these case studies. The first is that, in order to anticipate and respond effectively to the impacts of possible future sea level rise, *the dynamic processes involved in the relevant physical, environmental and socio-economic systems must be taken into account.* Too often, because of the uncertainties and for the sake of simplicity, the problem of assessing the impacts of sea level rise is approached statically: a scenario of X cm sea level rise is imposed on current conditions and the consequences are tallied. A typical product of this approach is a map with a new coastline redrawn to the contour that best matches the projected rise in global mean sea level! Environmental systems respond in much more complex ways.

For example, in the Mississippi delta region (Day *et al.*, Chapter 17), relative sea level change is governed by deltaic subsidence and global mean sea level increase on the one hand, and by vertical accretion due to sedimentation on the other. Relative sea level hangs in the balance. The Mississippi River and its distributaries feed the surrounding wetlands with sediments and freshwater and wetland vegetation plays an important role in the rate of sedimentation and therefore vertical land accretion. Global sea level rise itself can alter these processes and thereby affect relative sea level both directly and indirectly; the latter, for example, by causing vegetation stress through waterlogging and salt contamination. Understanding the dynamics of the coastal environmental system in the Mississippi delta region is crucial to predicting future sea level rise and to planning strategies for reducing its impacts.

The Ganges–Brahmaputra–Meghna delta of Bangladesh displays similar complexities. A simple static estimate of the potential area of the delta at risk from sea level rise can provide a rough approximation of the sectors of the economy (e.g. agriculture) at risk and the magnitude of impact (Broadus, Chapter 16; a similar analysis is conducted for Egypt). But, as described in detail by Brammer (Chapter 15), such an approach is insufficient for predictive purposes or planning given the complicated dynamic processes involved from one area of Bangladesh to the next. For example, in Bangladesh it is possible that the rise in eustatic sea level could be matched by a rise in river levees and, through natural sedimentation, tidal and estuarine floodplain land in the coastal region, with relative sea level remaining largely unchanged. Instead, the most likely adverse consequence of sea level rise, as argued by Brammer, is the increased incidence and depth of *inland* flooding due to higher river-bed channels and levees relative to the surrounding land, and to the impedance of drainage of inland areas as coastal lands accrete. For the same reasons, the chances of major shifts in river channels (as has happened in the not-so-distant past) could also increase, with major implications for the patterns of erosion, sedimentation and land use.

Similarly, in the low-lying Salado Basin depression in the Pampas region of Argentina, a major impact of sea level rise could be a drastic increase in inland flood frequencies as drainage is impeded due to rising water tables (Schnack, Chapter 21). Along the Norfolk coast of Britain, eustatic sea level rise does not necessarily mean that beaches will disappear and cliffs will retreat everywhere; rather, some areas will gain at the expense of others, as determined in part by the dynamic processes of erosion and littoral transport of sediment (Clayton, Chapter 19).

In other localities, biological factors are of fundamental importance. For example, in the Republic of the Maldives and other low-lying island states, eustatic sea level rise does not necessarily imply disappearance of land: coral reefs have the capacity to grow in response to sea level rise – up to a point - and mangroves play an important role in land maintenance. In such cases, it is critical to understand the dynamic response of coral and mangroves to continuous changes in water depth, temperature, salinity and pollution levels (Holdgate *et al.*, 1989).

The second theme to emphasize is that *human interference in these dynamic processes can cause, and has caused, changes in the rate of local sea level rise.* In some cases, these changes have been comparable to or greater than those projected as a result of future global warming. The Mississippi River delta is a good case in point. According to Day *et al.* (Chapter 17), the delta was, on average, in a state of dynamic balance, or slowly growing, for thousands of years prior to the present century. Then, as a result of human interventions in the natural systems, including levee construction along the banks of the Mississippi River, the construction of smaller-scale dykes, the damming of distributaries and the development of numerous canals and diversions, the wetlands began to be starved of needed freshwater and sediments. Relative sea level is now rising at around one metre per century in the Mississippi delta region, and up to 100 km² of wetlands are being lost each year (Gagliano *et al.*, 1981).

Similarly, at some locations along the Argentinian coast, particularly those being developed for tourism, the mining of beach sand, urbanisation of dune fields and ill-advised engineering works are encouraging beach erosion and cliff retreat (Schnack, Chapter 21). In Hong Kong, urbanisation on reclaimed land has caused land settlement, which has been increasing the risks of inundation from storm surges (Yim, Chapter 22). In the Republic of the Maldives, pollution and mining of coral for construction purposes are threatening to disrupt the islands' natural capacity to respond to rising sea level (Pernetta and Sestini, 1989).

In all of these examples, human interference has to be understood in the context of the prevailing cultural-historical circumstances under which decisions were made regarding the use of natural resources. In the Mississippi delta region, levee construction and dams, rather than, say, alternative settlement patterns, reflected a narrow-minded American preference for engineering solutions to flood problems, as guided by governmental, legislative and bureaucratic constraints of the time (White, 1969). In Hong Kong, acute population pressures, in part created by the political changes in the Far East after the Second World War, have forced intensive development of the coastal zone without proper consideration of longer-term consequences. Currently, mining of coral in the Maldives has been spurred on by a high demand for building material, largely for breakwaters for docks to support an expanding tourist industry and for construction of Western-style houses in preference to traditional styles (Pernetta and Sestini, 1989).

The third, related theme is that the dynamic processes can be manipulated to favourable advantage: *human adjustment can involve actions which nurture the ability of natural systems to respond so as to reduce relative sea level rise or the hazards arising therefrom.* This can be accomplished directly, or indirectly by altering the management contexts in which decisions are made or by broadening the range of perceived choices in resource use.

For example, eliminating the dykes and canals that are obstacles to freshwater and sediment flow could help re-establish land accretion processes in selected areas of the Mississippi delta region. Such actions could be promoted through incentives for less intensive land uses, e.g., wetland and wildlife reserves, that preclude the need for such obstacles (Day *et al.*, Chapter 17). In the Maldives, legislation to limit the mining of coral to a single (or perhaps several) uninhabited islands of the Maldives may well be preferable to random mining on populated islands close to construction sites. In the United Kingdom, changes in land use policies and innovative schemes to compensate landowners for mandated land use changes in threatened locations may well prove less disruptive and more economically effective than costly sea defences in the face of future sea level rise (Clayton, Chapter 19).

Improving impact assessments

Based on the previous discussion, it emerges that *there is a need to develop integrated modelling approaches for assessing the impacts of climate and sea level change in the coastal zone*. As emphasized above, there must be a move away from studies that involve the static imposition of fixed changes in sea level on the coast and that focus on sea level in isolation from other important regional factors. Even the current IPCC guide-lines for case studies suggest, in the interests of expediency for meeting short-term goals, a simple 0.5 m and 1.0 m sea level rise as a basis for approximating impacts (IPCC, 1991). Clearly, in the longer term, this is inadequate.

With the mounting international concern about global warming and sea level rise, a virtual explosion of case studies lies on the horizon. In order to derive the greatest benefit from such studies, some attention to methodological considerations is clearly warranted. It can be argued that there is strong merit in developing modelling approaches

> that are *time-dependent*. This is crucial in order to capture the dynamic processes, many of which may exhibit threshold and/or markedly non-linear effects.
>
> that include *climate changes*. The advantages of this are twofold: to assure consistency with scenarios of future greenhouse gas emissions; and to allow for the effects of interactions that can arise between climate change (e.g., temperature and precipitation) and sedimentation rates, vegetation changes, land use change and the like as they affect both relative sea level change itself and the vulnerability of environmental and human systems to the change.
>
> that include *detailed spatial information* on population, resources and land use. As demonstrated by Shennan (Chapter 13), one possibility is to organize and analyse such information within a Geographic Information System (GIS). As illustrated by Shennan for the Tees and the Fens in the United Kingdom, such information is critical as the baseline against which to assess impacts and policy changes.
>
> that are *model-based and flexible*. A major disadvantage of many sea level impact assessments is that they have used, for example, fixed scenarios of climate and sea level change. In such cases, if one wishes to examine the sensitivity to an alternative scenario, the analysis has to start again at 'square one'. This argues in favour of developing methodological frameworks that link global-scale to regional-scale changes, and that are computerized, model-based and modularized. This would allow updating as scientific information improves, and would provide a flexible tool for evaluating measures for adaptation or mitigation of greenhouse gas emissions.
>
> that focus on the *whole coastal zone*. The coastal zone, comprised of the various natural and human systems at the land–marine interface, should provide the logical, spatially coherent region of analysis.

Inclusion of physical, environmental and human systems as they interact to affect the coastal zone is crucial. In this regard, the importance of the presentation by Wind and

Peerbolte (Chapter 18) is that it spells out the reasoning behind the original development of such an integrated systems model for impact and policy analysis in The Netherlands. This model (ISOS, Impact of Sea Level Rise on Society) is a pioneering interdisciplinary approach which has subsequently been developed and tested in much greater detail and which has been applied to other sites (e.g. see Delft Hydraulics, 1989, 1991). The basic rationale behind the approach remains valid today.

There are signs of methodological developments in this direction. For example, in addition to the presentations in this volume mentioned above, the current work being supported by the World Conservation Union's Global Change Programme (Pernetta and Elder, 1990) and by the UNEP's Regional Seas Programme, particularly for the Mediterranean region (L. Jeftic, pers. comm.), endorse many of the points emphasized here. The way forward is to build on these efforts in order to develop methods that are credible, flexible and adaptable, and that will stand us in good stead in the years to come.

Acknowledgements

This review has been supported in part by grants from the Commission of the European Communities' Climatology and Natural Hazard Research Programme.

References

Broccoli, A. & Manabe, S. (1990). Can existing climate models be used to study anthropogenic changes in tropical cyclone climate? *Geophys. Res. Let.*, **17**(11), 1917–20.

Budd, W.F., McInnes, B.J., Jenssen, D. & Smith, I.N. (1987). Modelling the response of the West Antarctic ice sheet to a climatic warming. In *Dynamics of the West Antarctic ice sheet*, C.J. van der Veen & J. Oerlemans (eds.). Reidel, pp. 321–58.

Chappell, J. & Shackleton, N.J. (1986). Oxygen isotopes and sea level. *Nature*, **324**, 137–40.

Delft Hydraulics (1991). *Implications of Sea-level Rise on the Development of the Lower Nile Delta, Egypt: Pilot Study for a Quantitative Approach*. Final Report, H927, Delft Hydraulics, Delft.

Delft Hydraulics (1989). *Republic of Maldives, Implications of Sea-level Rise. Report on Identification Mission*. H926, Delft Hydraulics, Delft.

Doake, C.S.M. & Vaughan, D.G. (1991). Rapid disintegration of the Wordie Ice Shelf in response to atmospheric warming. *Nature*, **350**, 328–9.

Emanuel, K.A. (1987). The dependence of hurricane intensity on climate. *Nature*, **326**, 483–5.

Frei, A., MacCracken, M.C. & Hoffert, M.I. (1988). Eustatic sea level and CO_2. *Northeastern J. Environ. Sci.*, **7**, 91–6.

Gagliano, S.M., Meyer-Arendt, K.J. & Wicker, K.M. (1981). Land loss in the Mississippi River deltaic plain. *Trans. Gulf Coast Assoc. Geol. Soc.*, **31**, 295–300.

Gornitz, V. & Lebedeff, S. (1987). Global sea level changes during the past century. In *Sea Level Fluctuation and Coastal Evolution*, D. Nummedal, O.H. Pilkey and J.D. Howard (eds.). SEPM Special Publication No. 41.

Gornitz, V., Lebedeff, S. & Hansen, J. (1982). Global sea level trend in the past century. *Science*, **215**, 1611–14.

Gray, W.M. (1984). Atlantic seasonal hurricane frequency Part I: El Niño and the 30 mb Quasi-Biennial Oscillation influences. *Monthly Weather Review*, **112**, 1649–68.

Gray, W.M. (1968). Global view of the origin of tropical disturbances and storms. *Monthly Weather Review*, **96**, 669–700.

Grove, J.M. (1988). *The Little Ice Age*. Methuen, London.

Heaps, N.S. (1967). Storm surges. In *Oceanography and Marine Biology Annual Review, No. 5*, H. Barnes (ed.). Allen and Unwin, London, pp. 11–47.

Hoffman, J.S., Wells, J.B. & Titus, J.G. (1986). Future global warming and sea level rise. In *Iceland Coastal and River Symposium*, G. Sigbjarnason (ed.). Reykjavik, National Energy Authority, pp. 245–66.

Hoffman, J.S., Keyes, D. & Titus, J.G. (1983). *Projecting Future Sea Level Rise: Methodology, Estimates to the Year 2000, and Research Needs*, US GPO#055–000–0236–3, GPO, Washington, DC.

Holdgate, M.W. *et al.* (1989). *Climate Change: Meeting the Challenge. Report by a Commonwealth Group of Experts*. Commonwealth Secretariat, London.

Houghton, J.T., Jenkins G.J. & Ephraums, J.J. (eds.) (1990). *Climate Change: The IPCC Scientific Assessment*. Cambridge University Press, Cambridge

IPCC (1991). *The Seven Steps to the Vulnerability Assessment of Coastal Areas to Sea Level Rise: Guide-lines for Case Studies*. IPCC Response Strategies Working Group, Steering Committee for IPCC Sea Level Rise Vulnerability Assessment–Coastal Zone Management (draft report).

Jaeger, J. (1988). *Developing Policies for Responding to Climate Change: a Summary of the Discussions and Recommendations of the Workshops held in Villach 1987 and Bellagio 1987.* WMO/TD-No. 225.

Jones, P.D. (1990). Antarctic temperatures over the present century – a study of the early expedition record. *J. Clim.*, **3**, 1193–203.

Jouzel, J., Lorius, C., Petit, J.R., Genthon, C., Barkov, N.I., Kotlyakov, C.M. & Petrov, V.M. (1987). Vostok ice core: a continuous isotope temperature record over the last climatic cycle (160,000 yrs). *Nature*, **329**, 403–8.

Meier, M.F. (1984). Contribution of small glaciers to global sea level. *Science*, **226**, 1418–21.

Mercer, J.H. (1978). West Antarctic ice sheet and CO_2 greenhouse effect: a threat of disaster. *Nature*, **271**, 321–5.

Merrill, R.T. (1988). Environmental influences on hurricane intensification. *J. Atmos. Sci.*, **45**, 1678–87.

Mikolajewicz, U., Santer, B.D. & Maier-Reimer, E. (1990). Ocean response to greenhouse warming. *Nature*, **345**, 589–93.

Mitchell, J.F.B., Manabe, S., Meleshko, V. & Tokioka, T. (1990). Equilibrium climate change. In *Climate Change: The IPCC Scientific Assessment*, J.T. Houghton, G.J. Jenkins and J.J. Ephraums (eds.). Cambridge University Press, Cambridge, pp. 131–72

Nicholls, N. (1984). The Southern Oscillation, sea surface temperature, and interannual fluctuations in Australian tropical cyclone activity. *J. Clim.*, **4**, 661–70.

Oerlemans, J. (1989). A projection of future sea level. *Clim. Change*, **15**, 151–74.

Peel, D.A. & Mulvaney, R. (1988). Air temperature and snow accumulation in the Antarctic Peninsula during the past 50 years. *Annals of Glaciology*, **11**, 207.

Peltier, W.R. (1986). Deglaciation-induced vertical motion of the North American continent and transient lower mantle rheology. *J. Geophys. Res.*, **91**, 9099–123.

Pernetta, J.C. & Elder, D.L. (1990). Climate, sea level rise and the coastal zone: management and planning for global changes. In *Proceedings of the IUCN General Assembly Workshop on the Environmental Implications of Global Change, December, 1990, Perth, Australia*. IUCN, Gland.

Pernetta, J.C. & Sestini, G. (1989). The Maldives and the impact of expected climatic changes. *UNEP Regional Seas Reports and Studies No. 104*, UNEP, Nairobi.

Polar Research Board (PRB) (1985). *Glaciers, Ice Sheets and Sea Level: Effect of a CO_2-induced Climatic Change*. Report of a workshop held in Seattle, Washington, September 13–15, 1984. US DOE/ER/60235–1.

Raper, S.C.B., Warrick, R.A. & Wigley, T.M.L. (1990). Global sea level rise: past and future. In *Proceedings of the SCOPE Workshop in Rising Sea Level and Subsiding Coastal Areas, Bangkok 1988*, J.D. Milliman (ed.). John Wiley and Sons, Chichester (in press).

Revelle, R. (1983). Probable future changes in sea level resulting from increased atmospheric carbon dioxide. In *NAS Changing Climate*, pp. 433–47, NAS, Washington DC.

Robin, G. de Q. (1986). Changing sea level. In *The Greenhouse Effect, Climatic Change and Ecosystems*, B. Bolin, B.R. Döös, J. Jäger & R.A. Warrick (eds.). John Wiley and Sons, Chichester, pp. 323–59.

Stouffer, R.J., Manabe, S. & Bryan, K. (1989). Interhemispheric asymmetry in climate response to a gradual increase of atmospheric CO_2. *Nature*, **342**, 660–2.

Thomas, R.H. (1986). Future sea level rise and its early detection by satellite remote sensing. In *Effects of Changing Stratospheric Ozone and Global Climate, Vol. 4: Sea Level Rise*, J.G. Titus (ed.). UNEP and US EPA, pp. 19–36.

Thomas, R.H., Sanderson, T.J.D. & Rose, K.E. (1979). Effects of a climatic warming on the West Antarctic ice sheet. *Nature*, **227**, 355–8.

van der Veen, C.J. (1988). Projecting future sea level. *Surv. in Geophys.*, **9**, 389–418.

Warrick, R.A. & Oerlemans, J. (1990). Sea level rise. In *Climate Change: The IPCC Scientific Assessment*, J.T. Houghton, G.J. Jenkins and J.J. Ephraums (eds.). Cambridge University Press, Cambridge, pp. 257–81.

Weidick, A. (1984). Review of glacier changes in West Greenland. *Z. Gletscherk. Glazialgeol.*, **21**, 301–9.

White, G.F. (1969). *Strategies of American Water Management*. The MIT Press, Cambridge, Massachusetts.

Woodworth, P.L. (1990). A search for accelerations in records of European mean sea level. *Int. J. Clim.*, **10**, 129–43.

Wigley, T.M.L. & Raper, S.C.B. (1987). Thermal expansion of sea water associated with global warming. *Nature*, **330**, 127–31.

II

DATA

2

Mean sea level changes in the recent past

V. Gornitz

Abstract

Based on at least sixteen studies of tide-gauge data, estimates of global mean sea level change during the last 100 years indicate a rate of rise of 0.5–3.0 mm/yr, with most estimates within the range of 1.0–2.0 mm/yr. The agreement among these independent analyses, in spite of the diversity of sampling strategies and data processing techniques, lends support to the notion that the observed sea level rise represents a true eustatic change. Other supportive evidence includes: an apparent increase in the rate of rise over the last 100 years as compared to the preceding 100 years, as observed in long European tide-gauge records; and generally increased rates of rise over the last 100 years as compared to rates of change over the past few thousand years, as estimated from geological and other data.

However, estimates of mean sea level rise are subject to large uncertainties due, for example, to problems of data quality and geographical distribution, vertical land movements (resulting particularly from glacio-isostasy and neo-tectonics), atmospheric and oceanographic effects, and anthropogenic activity (water and sediment impoundment behind reservoirs and fluid withdrawal). Such effects vary widely in their relative importance; methods of filtering some of their contaminating signals are discussed in the paper.

Introduction

During the next 100 years, the increasing atmospheric concentrations of CO_2 and other greenhouse gases are expected to lead to a global warming of between 2–7 °C in the absence of determined steps to reduce emissions (Hansen *et al.*, 1988; Bretherton *et al.*, 1990; Wigley and Raper, this volume). Such warming could enhance melting/wastage of continental and alpine glaciers (Meier, 1984; Oerlemans, Kuhn, Paren *et al.*, this volume) and cause thermal expansion of the oceans (Gornitz *et al.*, 1982; Wigley and Raper, 1987, this volume), which, according to various estimates, could increase global mean sea level by up to 200 cm (National Academy of Sciences, NAS, 1987; Hoffman *et al.*, 1986; Thomas, 1986; Warrick and Oerlemans, 1990). The projected high rates of sea level rise exceed estimates of the current rate of rise by a factor of ten (e.g., Gornitz *et al.*, 1982; Barnett, 1983, 1984, 1988). Even at the current rates of sea level rise, erosion is prevalent along sandy beaches (Bird, 1985), and if future projected trends materialize, problems of erosion and inundation of low-lying coastal areas could be exacerbated.

Table 2.1. *Estimate of Sea Level Rise (SLR) (updated from Warrick and Oerlemans, 1990; Robin, 1986)*

Rate (mm/yr)	Comments	References
>0.5	Cryologic estimate	Thorarinsson (1940)
1.1 ± 0.8	Many stations, 1807–1939	Guetenburg (1941)
1.2–1.4	Combined methods	Kuenen (1950)
1.1 ± 0.4	Six stations, 1807–1943	Lisitzin (1958, in Lisitzin 1974)
1.2	Selected stations, 1900–1950	Fairbridge & Krebs (1962)
3.0	Many stations, 1935–1975	Emery (1980)
1.2 ± 0.1†	193 stations→14 regions, 1880–1980	Gornitz et al. (1982)
1.5	Many stations, 1900–1975	Klige (1982)
1.5 ± 0.15†	Selected stations, 1903–1969	Barnett (1983)
1.4 ± 0.14†	Many stations → regions, 1881–1980	Barnett (1984)
1.2 ± 0.3†	130 stations, 1880–1982	Gornitz & Lebedeff (1987)*
1.0 ± 0.1†	130 stations → 11 regions, 1880–1982	Gornitz & Lebedeff (1987)*
1.15	155 stations, 1880–1986	Barnett (1988)
2.4 ± 0.9§	40 stations, 1920–1970	Peltier & Tushingham (1989; 1991)**
1.75 ± 0.13§	84 stations, 1900–1986	Trupin and Wahr (1990)**
1.67 ± 0.33	69 stations, 1900–1986	Wahr and Trupin (1990)**
1.8 ± 0.1°	21 stations, 1880–1980	Douglas (1991)**

Notes:
† = Value plus 95% confidence interval
§ = Mean and standard deviation
° = Standard error
* = Long-term crustal motions removed
** = Glacio- and hydro-isostatic effects removed

Studies of global mean sea level change over the last 100 years based upon tide-gauge data yield values ranging between 0.5 mm/yr and 3 mm/yr; most estimates lie within 1.0–2.0 mm/yr (Table 2.1, modified from Warrick and Oerlemans, 1990). Because of a growing awareness of the complexity of factors which can affect relative sea level curves, and because of the high variability in the data due largely to crustal movements, some scientists have questioned whether a meaningful eustatic sea level trend can be determined simply by averaging tide-gauge measurements (Barnett, 1984; Aubrey, 1985, this volume; Pirazzoli, 1986; Peltier, 1986). Differences of up to 50% could arise just from varying averaging methods (Barnett, 1984).

An assumption common to many studies is that given a sufficiently large number of tide-gauge stations, crustal movements will be compensated over a large scale, such that the resultant global average sea level trend will mainly represent changes in water volume (i.e., the eustatic contribution). However, Pirazzoli (1986) objects to this assumption, which although valid in principle, has drawbacks if applied only to continental shelves and margins where most of the Earth's tide gauges are located. Pirazzoli points out that these regions are subject to several processes, all of which lead to an apparent sea level rise. These processes include:

1. Subsidence due to sediment loading on the continental shelf.

2. Depression of the ocean floor and continental margins by water loading from glacial melting (hydro-isostasy).
3. Subsidence of the ocean floor because of oceanic lithosphere cooling as the ocean plates move away from spreading centres.
4. Subsidence of coastal localities due to water, oil, gas withdrawal.

However, as will be shown below, on 100-year time-scales, the first and third of these processes are relatively insignificant contributors to global sea level change, while the second can be largely accounted for by viscoelastic Earth models. The last process may be significant on a very local scale.

Given the importance of monitoring current sea level trends as a basis of comparison with any future changes, it is necessary to evaluate these issues carefully. The aims of this paper are: to review measurements of recent sea level change; to examine some of the causes of high variability in sea level data; to determine whether a meaningful sea level trend exists; and to offer suggestions for future research.

Measurements of recent sea level changes

As mentioned above, most estimates of the rate of global eustatic sea level (SL) change over the last 100 years, as derived from tide-gauge records, range between 1.0 mm/yr and 2.0 mm/yr (Table 2.1). While many studies use the same data source[1], they vary in sampling strategies and data analysis. Fairbridge and Krebs (1962) and Barnett (1983) used single, widely spaced stations from geologically stable areas, assuming that they were representative of large regions. In other studies, a broader geographical distribution of records was averaged, again avoiding areas of known glacial rebound and tectonic activity, such that the net contribution of residual land movements reduces to zero (Barnett, 1984, 1988).

Another alternative has been to filter out long-wavelength crustal motions by use of Late Holocene paleo-sea level indicators (Gornitz et al., 1982; Gornitz and Lebedeff, 1987). These indicators (Pardi and Newman, 1987) were used to compensate for the effects of glacio-isostasy and long-term tectonism (see below). An average sea level trend was calculated for 130 stations in 71 cells, weighting stations by distance from the cell centre and by relative coastal area and subtracting long-term trends. These cells were grouped into 11 geographical regions, for which composite average SL curves were obtained. A least-squares slope was then fitted to each regional average SL curve. Finally, a global SL trend was obtained by averaging the 11 regional curves, weighting each region equally and also by a regional data reliability factor. The global mean corrected sea level change, with all regions weighted equally is 1.0 ± 0.1 mm/yr (95% C.I.), and with weighting for data reliability is 0.9 ± 0.1 mm/yr, which does not differ significantly. In addition, a global sea level trend was computed directly from the arithmetic mean of the least-squares slopes on individual stations, from which long-term vertical movements had been subtracted (130 stations). The global arithmetic SL change is 1.2 ± 0.3 mm/yr (Gornitz and Lebedeff, 1987)[2].

[1] The Permanent Service for Mean Sea Level (PSMSL), Bidston Observatory, Birkenhead, England.
[2] These values may slightly underestimate the total eustatic SL rise, as some Late Holocene eustatic SL rise may have been subtracted from present-day trends data, along with the 'land movements'.

The tide-gauge data are summarized in Fig. 2.1. Fig. 2.1(*a*) presents a histogram of 286 stations with record lengths $\geqslant 20$ years, uncorrected for long-range trends (arithmetic mean 0.6 ± 0.4 mm/yr [95%C.I.]). Fig. 2.1(*b*) is a subset of Fig. 2.1(*a*), for 130 tide-gauge stations that have associated long-range data (1.1 ± 0.5 mm/yr). The effect of subtracting the Late Holocene trends from raw tide-gauge data is seen by comparing Figs. 2.1(*a*), (*b*) and (*c*). The wide scatter present in the raw data (Figs. 2.1(*a*), (*b*)) attributed to glacial rebound (Fennoscandia), the collapsing forebulge (US East Coast and the Low Countries), and active tectonism (Japan) has been cut nearly in half by removal of the long-term trends (Fig. 2.1(*c*), 1.2 ± 0.3 mm/yr).

Regional sea level curves are shown in Fig. 2.2. Sea level, corrected for long-term movements, is rising in all but two regions. The anomalous trend of the Pacific Islands may be caused by a small sample population, and neo-tectonic activity on several islands near plate boundaries (e.g., Truk). Neo-tectonism (Emery and Aubrey, 1986; Uchupi and Aubrey, 1988) and inadequate Late Holocene data coverage are probably responsible for the erratic trend of Western North America. Residual regional variations are discussed further below.

More recently, glacial rebound models have been applied to tide-gauge data to determine elevation changes associated with ice melting and water loading (Peltier and Tushingham, 1989, 1991; Trupin and Wahr, 1990; Douglas, 1991). These models are calibrated by reference to the Holocene sea level record. The latter studies utilize different versions of the same model and also differ in their criteria of station selection and averaging methods. These studies also emphasize the need for long record lengths to minimize low frequency variations in sea level, but the few surviving stations provide reduced geographical coverage.

Both geophysical and geological methods have inherent limitations. Glacial isostatic models are sensitive to the choice of lithospheric thickness, viscoelastic properties of the Earth's interior and ice melting history. A major uncertainty has been the role of Antarctic ice melting. Older models assumed that Antarctic melting began prior to 14,000 BP (Wu and Peltier, 1983). However, more recent studies suggest that Antarctic deglaciation commenced around 9,000 BP, continuing into the mid-Holocene (Nakada and Lambeck, 1988, 1989; Lambeck, 1990; Tushingham and Peltier, 1991). Lateral variations in viscoelastic structures also contribute to the uncertainty (Gasperini *et al.*, 1990). On the other hand, the validity of paleo-sea level indicators decreases for locations far from tide-gauge stations. Availability of data sets on both century and millenial time-scales sharply curtails geographical coverage. Neither method considers Holocene climate changes such as neo-glacial advances or retreats. Finally, neither method accounts for local neo-tectonic or ground subsidence effects, nor for atmospheric/ocean forcing (see below).

Sources of variability in sea level data

Factors contributing to the high variability in sea level data generally fall into three major categories:

1. Data quality and distribution.
2. Vertical land motions
   (a) Natural (glacio- and hydro-isostasy, neo-tectonics, sediment loading).

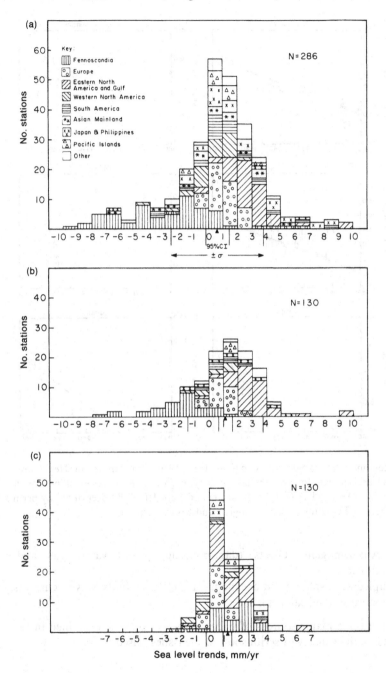

Fig. 2.1 Histogram of the number of tide-gauge stations *vs* sea level trends. Filled triangle ▲ indicates mean rate of sea level rise; bounding lines indicate the 95% C.I. and ±1σ, respectively.

(*a*) All tide-gauge stations with record length ≥ 20 years; raw data.

(*b*) Subset of tide-gauge stations; long-range trends included.

(*c*) Same subset of stations as (*b*); long-range trends subtracted.

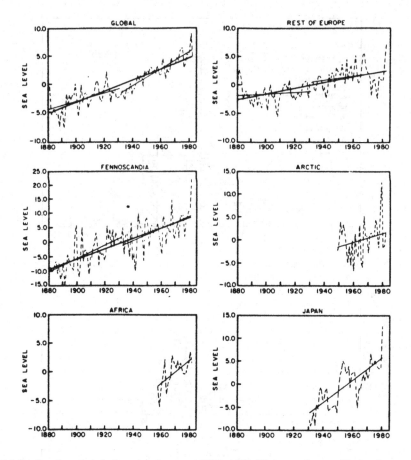

Fig. 2.2 Regional and global sea level curves, 1880–1980. Units are in cm. Heavy line represents least-squares slope for the last century. Lighter lines represent slopes for the intervals 1880–1931, and 1932–1980. (Gornitz and Lebedeff, 1987. Reprinted by permission of the Society of Economic Palaeontologists and Mineralogists).

    (b) Anthropogenic effects (dam-building, groundwater, gas and/or oil withdrawal).
3. Coupled atmospheric and oceanographic effects (winds, waves, currents, ocean temperatures and salinity).

Each of the three major sources of uncertainty will now be examined in turn, with comments on the validity or seriousness of the issues raised.

Data quality and distribution

Problems which affect sea level data quality include: tide-gauge records of too short or variable length, often with data gaps; abrupt changes in tide-gauge locations; and a pronounced geographical bias toward the Northern Hemisphere. The Permanent Service for Mean Sea Level (PSMSL) has compiled data from over 1,400 tide-gauge

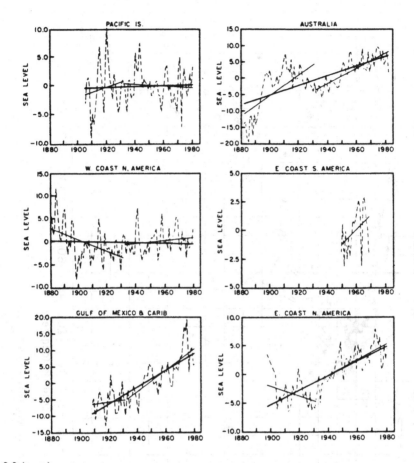

Fig. 2.2 (*cont.*)

stations world-wide. However, most tidal records are too short or too broken to be useful. A minimum record length of 20 years is advisable, because of interannual-to-decadal fluctuations in ocean temperature or currents (Namias and Huang, 1972). Periodic noise from the 18.6 year lunar nodal cycle (Lisitzin, 1974) is also more pronounced in short records. Limiting stations to those with records ≥ 20 years reduces the number of usable stations to ~ 400, which also reinforces the geographical bias (Fig. 2.3).

A marked geographical bias toward the Northern Hemisphere is unavoidable, given present data distribution. (Nearly 90% of the stations are located north of 23°N. Since the overwhelming majority of tide-gauge stations are distributed linearly along the coasts, and the shoreline north of 20°N represents around 70% of the world's total coast length, the sampling bias may not be as severe as implied by some authors). Geographical bias can be reduced by dividing the tide-gauge stations into a number of regions based on proximity and similarity in sea level behaviour, weighting each region equally, in order to obtain a 'global' average. A disadvantage is that regions with stations may be equated with those of a few (possibly poor or unrepresentative) stations (Aubrey and Emery, 1983). Therefore, one approach may be to weight the

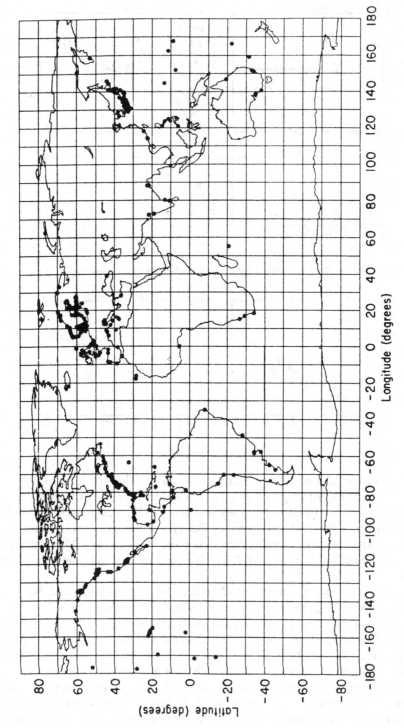

Fig. 2.3 Location of PSMSL tide-gauge stations with record lengths ≥ 20 years

regional averages by the area of ocean that they represent (Barnett, 1983), or alternatively to use weighting factors for regional reliability (see above). However, results using the latter approach are not significantly different from using equal weights (see above).

Vertical land motions

(a) Natural causes. Relative vertical movements of coastal regions can occur over a broad range of wavelengths. Warping of the lithosphere, with wavelengths of many hundreds or thousands of kilometres, can be caused by glacial and hydro-isostatic adjustments of the Earth's lithosphere to the removal of the Fennoscandian and Laurentide ice sheets (Peltier, 1986; Tushingham and Peltier, 1991), or by subsidence of the continental shelf due to sediment loading and cooling of the ocean lithosphere (Steckler and Watts, 1978; Heller *et al.*, 1982). Differential vertical movements with wavelengths shorter than 100 km are likely to be associated with tectonic deformation and fault displacements.

One method of removing the long-wavelength glacio- (and hydro-) isostatic components of relative sea level (RSL) change is to apply a correction factor based upon viscoelastic Earth models which explicitly describe gravitational interactions among ice sheets, land and ocean (Peltier, 1986; Tushingham and Peltier, 1991; Peltier and Tushingham, 1989, 1991). A second approach is to use Late Holocene ($\leqslant 6,000$ BP) sea level curves derived from ^{14}C-dated paleo-sea level indicators (Pardi and Newman, 1987; and Fig. 2.4). The two approaches are not completely independent, in that Peltier uses ^{14}C data to calibrate his models.

Since the completion of glacial melting by around 6,000 BP, the rate of sea level rise has decreased sharply, although the onset of this decrease is not globally synchronous. The Late Holocene curves largely reflect the continuing glacial isostatic response (Peltier, 1986; Tushingham and Peltier, 1991), although a small but ongoing Late Holocene eustatic trend and some short-wavelength neo-tectonic trends may still be included. Therefore, subtraction of the Late Holocene sea level trends from the tide-gauge trends should minimize these effects (Gornitz *et al.*, 1982; Gornitz and Lebedeff, 1987).

The Atlantic coast of North America can serve as a test case for a better understanding of the causes of variability in sea level data. Along this coast, nearly half of the recent observed regional mean sea level rise can be attributed to glacio-isostasy (Peltier, 1986; and Fig. 2.5). The average of the residuals (38 stations) is 1.26 ± 0.78 mm/yr as compared with 2.72 ± 0.71 mm/yr for the raw, uncorrected data (Gornitz and Seeber, 1990). The regional corrected mean lies within a few tenths mm/yr of the estimated eustatic sea level rise (~ 1.1 mm/yr). However, in this region, removal of the glacio-isostatic and/or long-wavelength tectonic signal has not completely eliminated the residual spatial variation in SL trends, nor substantially reduced the variance, which arises from such factors as short-wavelength neo-tectonism (Braatz and Aubrey, 1987; Uchupi and Aubrey, 1988; Gornitz and Seeber, 1990). Several anomalous areas have been identified, that are suggestive of gentle crustal warping.

Other factors, such as lithospheric cooling or sediment loading, are generally insignificant causes of global-scale sea level change over a 100-year period. For instance, subsidence rates have been calculated along the US East Coast continental

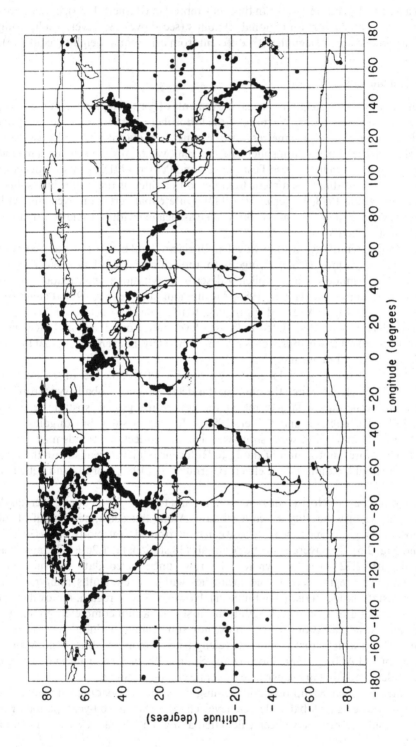

Fig. 2.4 Location of all ^{14}C-dated Holocene sea level indicators (Pardi and Newman, 1987. Reprinted by permission of the Society of Economic Palaeontologists and Mineralogists).

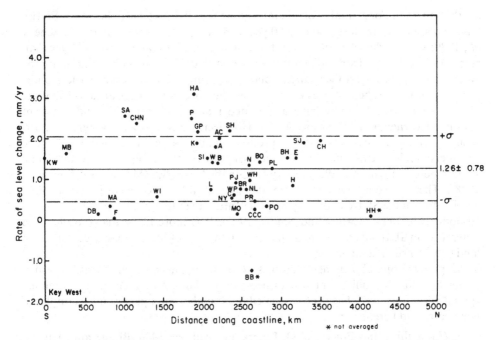

Fig. 2.5 Residual sea level change (tide-gauge trend minus ^{14}C-derived long-range trend for Eastern North America.

shelf and margin since the Cretaceous Period using the 'back-stripping' technique[3] (Heller *et al.*, 1982; Steckler and Watts, 1978). Total shelf subsidence is the sum of sediment and water loading (due to the original depth of deposition), and lithospheric cooling. The average total subsidence rate, over the last 135 My from all three processes is around 0.03 mm/yr (for the COST B-2 well, after Heller *et al.*, 1982). Over the last 12 My, average subsidence rates are also ~0.03 mm/yr. Holocene subsidence rates due to lithospheric cooling are not likely to differ substantially from the Neogene.

The global sea level rise attributable to the current deposition of suspended sediment load from the world's siltiest rivers at present (if averaged over the whole ocean area) is around 0.014–0.02 mm/yr (Milliman and Meade, 1983; Holeman, 1968). These rates are about 2% (or less) of the current estimated SL rise. These sedimentation rates represent a global average that may mask very high local rates, especially near major deltas (such as the Ganges–Brahmaputra, Amazon and Huangho deltas). Because of subsidence due to sediment loading and/or compaction of peats from tidal marshes, relative sea level curves near large deltaic areas may show a significant apparent rise: for example, rates of 9–12 mm/yr along the Louisiana coast (Day *et al.*, this volume; Penland and Ramsey, 1990), or of as much as 50 mm/yr in the Nile delta (Stanley, 1988).

Recent changes in sediment loading cannot easily explain differences between SL trends over the last few thousand years and those of the last 100 years (but see Day *et*

---

[3] 'Backstripping' involves subtracting sediment loads, and water loads due to changes in sea level, in order to obtain the depth to the bottom of the ocean.

al., this volume). For a number of localities along the US East Coast, the rate of SL rise has decreased over the last 2,000–3,000 years (Belknap *et al.*, 1989; van de Plassche *et al.*, 1989; van de Plassche, 1990; Kraft and Belknap, 1986), and is much lower than recent rates obtained from tide gauges. A similar pattern is also observed along the North Sea coast of West Germany (Rohde, 1980) and The Netherlands (de Ronde, pers. comm.). It has been suggested that forest clearing culminating around 150 years ago in the Eastern US may have accelerated marsh accretion rates and sediment loading (Kearney and Stevenson, 1991). However, not only is the weight of post-clearing sediment much too small to produce significant peat compaction, let alone subsidence under the weight of the sediments, but this explanation fails to account for a similar increase in the mid-19th century RSL in the North Sea (Mörner, 1973; Ekman, 1988). There is some evidence from the longest tide-gauge records for a slight acceleration in the SL trend during the past century over that of the preceding century (Woodworth, 1990; Gornitz and Solow, 1991). It is more likely that the observed differences in SL trends represent an increase in the rate of SL rise than acceleration of land subsidence or compaction.

As pointed out above, neo-tectonics form a significant component of short-wavelength SL variability. The role of neo-tectonic land movements in intraregional spatial variations in SL has been discussed in a number of papers, some examples of which are cited here:

1. *US*: Aubrey and Emery, 1983; Emery and Aubrey, 1986; Braatz and Aubrey, 1987; Gornitz and Seeber, 1990.
2. *Japan*: Aubrey and Emery, 1986a.
3. *Mediterranean*: Pirazzoli, 1987a; Emery *et al.*, 1988.
4. *Australia*: Aubrey and Emery, 1986b.

A global summary can be found in Emery and Aubrey (1991).

Other factors contributing to SL variability include long-term changes in river runoff (Meade and Emery, 1971), due to variations in rainfall, and changes in atmospheric pressure, ocean currents and wind stress (Chelton and Enfield, 1986).

(b) Anthropogenic effects.   Anthropogenic processes during the last 100 years have greatly altered the hydrologic regime in many places. For example, deforestation has increased runoff, soil erosion, flooding and downstream siltation in many parts of the world (Bird, 1985). However, forest regrowth, largely due to farm abandonment, has occurred in the Northern Hemisphere over the last 100 years (Clawson, 1979; Delcourt and Harris, 1980; Armentano and Ralston, 1980), thereby reducing sediment loads once more. Furthermore, dams and river channelization have decreased sediment supply to the world's coasts in many localities (for example, along the Mississippi drainage [Meade and Parker, 1985], and the Aswan dam on the Nile [Stanley, 1988]) thereby contributing to beach erosion (Bird, 1985, pp. 171–2).

Impoundment of water behind dams and reservoirs, and infiltration of water into aquifers, between 1932 and 1982, are equivalent to withholding a potential SL rise of ~ 0.75 mm/yr (Newman and Fairbridge, 1986). This effect could be nearly counteracted by groundwater mining (Korzun, 1978), which increases runoff to the oceans, tending to raise SL. However, Newman and Fairbridge believe that the figures for groundwater mining may have been overestimates, so that there may be a net reduction in SL rise due to impoundment.

Withdrawal of subsurface fluids (oil, gas, water) has exacerbated the natural rates of sea level rise at a number of coastal sites (Chi and Reilinger, 1984; Dolan and Goodell, 1986). Groundwater pumping can lead to lowering of the water table and compaction of fine-grained unconsolidated sediments, causing significant subsidence problems (Chi and Reilinger, 1984). However, this effect may be minimal at the shoreline, where the water table is essentially at sea level. Thus tide gauges located at the beach, as opposed to those located further upstream, may not be as seriously affected. This is borne out by the situation at Savannah, Georgia (US); between 1918 and 1975 land subsidence of up to 3 mm/yr occurred inland, whereas benchmarks near the tide gauge at Fort Pulaski (on the shore) showed no subsidence (Davis, 1987). In the Houston–Galveston area of Texas, maximum subsidence rates of up to 37.6 mm/yr have been measured between 1906 and 1987, falling to around 3.8 mm/yr at Galveston Island Pier 21 (Holdahl *et al.*, 1989), where the tide gauge is positioned. In this case, groundwater subsidence could account for as much as 60% of the RSL rise. In Venice, the human-induced subsidence accounted for approximately half of the RSL rise between 1872 and 1985 (Pirazzoli, 1987b). An extreme case is that of Bangkok, Thailand, where ground-water pumping has provoked local subsidence as great as 13 cm/yr (Milliman *et al.*, 1989).

A number of coastal cities are known to be affected by dewatering-induced subsidence (Dolan and Goodell, 1986). One might expect, therefore, that the mean RSL rise of cities located on unconsolidated, coastal plain sediments may be systematically higher than that for cities built on incompressible crystalline bed-rock. However, this is not supported by the data. Along the US East Coast, for example, the mean SL trend for 24 stations between Key West (Florida), Long Island (New York) and Cape Cod (Massachusetts) on the coastal plains is 2.90 ± 0.74 mm/yr, while that for 16 stations between Connecticut and St. John's (Newfoundland) on crystalline bed-rock is 2.46 ± 0.59 mm/yr; these means are not significantly different at the 95% confidence level (Gornitz and Seeber, 1990).

Atmospheric and oceanographic factors

Atmospheric and oceanographic processes produce a large fraction of the interannual variability in sea level records (Lisitzin, 1974). Sea level variability on time-scales of less than one year result from tides, atmospheric pressure changes and storm surges. Monthly and annual averages of tide-gauge measurements greatly reduce the < 1 year sources of variability. But annually averaged sea level curves still show fluctuations, 1–2 years to a decade in duration, coherent over long distances, which reflect steric changes (temperature or salinity; Namias and Huang, 1972), currents and combined oceanographic–atmospheric forcing, such as the El Niño-Southern Oscillation pheno-menon (ENSO) (Komar and Enfield, 1987).

Sea level varies inversely with atmospheric pressure (the so-called 'inverse barometer effect'). In other words, an increase in the downward force on the sea surface, due to the mass of the overlying atmosphere (an increase in atmospheric pressure) depresses the sea surface, and vice-versa. The inverse barometer response is -1.01 cm/mbar (Chelton and Enfield, 1986). The influence of atmospheric pressure variations can be removed by simple regression methods (Woodworth, 1987). Atmospheric pressure, however, is not a major contributor to interdecadal variations of global sea level. The effects of local wind stress on sea levels have been modelled for the North Atlantic (Thompson, 1990). Removing these effects has considerably reduced the variance in sea level records of the Eastern Atlantic.

Interdecadal changes in large-scale ocean circulation have a more pronounced effect on sea level. Shifts in ocean currents are recorded on tide-gauge records: for example, in South Florida and the Bahamas Bank, on opposite sides of the Florida Current, and on islands between the Pacific Equatorial Current and Counter Current (Komar and Enfield, 1987).

A major cause of interannual variability of sea level, particularly in the Pacific Ocean, is the ENSO phenomenon. ENSO events usually recur every 2-7 years, more commonly every 3-4 years. At the onset of an ENSO event, the pressure gradient that sustains the easterly trade winds weakens, reducing the E–W sea level gradient along the equator. The drop in SL in the western tropical Pacific propagates eastward as a long-wavelength equatorially-trapped wave (Kelvin wave), reaching the Eastern Pacific several months later. The eastward-propagating wave splits into N and S poleward-propagating waves, as it reaches the eastern margin of the Pacific (Komar and Enfield, 1987). This causes sea levels to rise along the East Pacific coasts of both hemispheres. The ENSO signal in sea level records can be recognized by the large-scale coherence and correlation with tropical interannual meteorological variability. Amplitudes of the ENSO signal in the tide-gauge records are between 10–50 cm (Table 2.2). Recently, eastward-propagating waves, associated with the 1986–87 El Niño were tracked, using GEOSAT altimetry (Miller *et al.*, 1988). The altimeter time series compares favourably with tide-gauge data from Christmas Island. These observations illustrate the future potential of satellite remote sensing for extending sea-surface measurements into ocean areas currently lacking tide gauges.

The various oceanic signals can be removed, or at least reduced, in several ways. Tides and seasonal cycles are largely suppressed by using annual means. Pressure effects and large-scale low frequency atmospheric forcing (winds, currents) can be removed by multivariate regression models. Empirical orthogonal function (EOF) analysis, as applied to sea level studies (Barnett, 1983; Aubrey and Emery, 1983), separates the tide-gauge data set into orthogonal spatial and temporal modes. This provides a concise description of the spatial and temporal structure of variability in the data. Spectral analysis of the temporal functions can be used to detect atmospheric or oceanic climatic cycles (Sturges, 1987; Trupin and Wahr, 1990). Chelton and Enfield (1986) recommend computation of EOF of residual monthly SL height anomalies over a large number of stations. Removal of the contribution from the most energetic mode eliminates large-scale coherent variability such as that associated with the ENSO.

Summary and conclusions

To date, most studies of secular, global mean sea level trends have used 'raw' mareograph records that are derived from tide-gauge stations of even geographical distribution, and that are contaminated by a variety of factors other than eustatic change. These factors, including vertical land movements, anthropogenic effects and oceanic–atmospheric phenomena, vary widely in their effects on recorded relative sea levels (as summarized in Table 2.2). Lithospheric cooling and shelf sedimentation rates (except near major deltas) are relatively minor on 100-year time-scales (Heller *et al.*,

Table 2.2. *Summary of processes affecting sea level changes*

Process			Rate (mm/yr)	Period (yr)
A	**Glacio-eustasy**		up to 10	Around 7,000 years following deglaciation
			~1	Last 100
B	**Vertical Land Movements**			
1	*Long-wavelength processes 100–1000 km*			
	a	Glacio-isostatic changes	±1–10	10^4
	b	Shelf subsidence due to oceanic lithosphere cooling and sediment/ water loading	0.03	10^7–10^8
	c	Shelf sediment accumulation, global	0.02–0.05	10–10^6
2	*Short-wavelength processes < 100 km*			
	a	Neo-tectonic uplift/subsidence	±1–5	10^2–10^4
	b	Shelf sediment accumulation, local – large river deltas	1–5	10–10^4
C	**Anthropogenic activity**			
	a	Water impoundment in dams, reservoirs	−0.75	<100
	b	Groundwater mining (to river runoff)	≤0.7 (?)	<100
	c	Subsidence due groundwater/oil/gas withdrawal (local)	3–5	<100
			Amplitude (cm)	**Period (yr)**
D	**Ocean-atmosphere effects**			
	Geostrophic currents		1–100	1–10
	Low-frequency atmospheric forcing		1–4	1–10
	El Niño		10–50	1–3

1982). It is possible that with respect to globally coherent secular changes in sea level, the effects of sediment and water impoundment, enhanced suspended sediment loads and groundwater mining may largely counteract each other, although local effects could still be pronounced. The net global anthropogenic impact on sea level change needs to be further quantified. On the other hand, glacio-isostasy, neo-tectonics and oceanic–atmospheric phenomena contribute significantly to the sea level signal.

In spite of diverse sampling and averaging strategies, data limitations and variability, one can recognize spatially and temporally coherent rises in sea level (Barnett, 1984, 1988; Gornitz and Lebedeff, 1987; Peltier and Tushingham, 1989; Trupin and Wahr, 1990). This is particularly evident for better-documented regions with large numbers of stations (Figs. 2.1 and 2.2). The fairly narrow range (1.0–2.0 mm/yr) of estimates of recent global sea level rise (Table 2.1), in spite of the numerous processes that contribute to the high variability, as outlined above, suggest a true eustatic change. It has been proposed that the apparent agreement in results is merely fortuitous, and reflects use of a common data set. However, rates of sea level rise over the last 100 years, in general, exceed those computed from Late Holocene (last 6,000 years) paleo-sea level indicators (Gornitz and Seeber, 1990; Shennan and Woodworth, 1992). More significantly, on both sides of the North Atlantic, rates obtained from tide gauges are much higher than rates derived from ^{14}C-dated materials in sediment cores during the last 2,000–3,000 years. Long records of the last few centuries suggest a possible acceleration of sea level rise within the last 150–200 years (Ekman, 1988; Mörner, 1973; Woodworth, 1990; Gornitz and Solow, 1991). Furthermore, this increase in sea level rise roughly corresponds to the end of the Little Ice Age (Grove, 1988) and onset of the recent climate warming (Folland *et al.*, 1990). It is therefore more likely that these changes represent an increase in the recent rate of sea level rise rather than an acceleration of land subsidence or sediment compaction.

In the near future, improved global geodetic networks and satellite monitoring of ocean topography will complement data obtained from tide-gauge measurements (Diamante *et al.*, 1987; Wyrtki, 1987; Bilham, 1991; Pugh, Carter *et al.*, this volume). Comparisons of SL positions from GEOSAT radar altimetry and tide gauges show a fairly good match over small areas of the equatorial Pacific Ocean, for periods of several months to over a year (Wyrtki, 1987; Miller *et al.*, 1988). These studies illustrate the potential of satellites, when 'calibrated' with tide gauges, to acquire long-term SL data over wide areas of ocean that are currently unsampled. This will help improve the spatial coverage and reduce the geographical bias. The tie-in between tide gauges and the Very Long Baseline Interferometry (VLBI) network (IAPSO, 1985; Diamante *et al.*, 1987; Pugh, Carter *et al.*, this volume) will determine absolute SL positions. This, in turn, will enable separation of vertical land motions from changes in ocean height, thus providing a more accurate estimate of the eustatic sea level rise.

Acknowledgements

Research was carried out under National Aeronautics and Space Administration Cooperative Agreement NCC-5-29 (Task A) and Martin Marietta Energy Systems MRETTA 19X-91348V. The author acknowledges prior collaborative work with the late Dr Sergej Lebedeff on two sea level papers. The author also wishes to express gratitude for the generosity of the late Prof. W. Newman, Queens College, New York, in making available his ^{14}C-dated Holocene paleo-sea level data base. Thanks are also extended to Dr D.T. Pugh and Dr P. Woodworth, Director, Permanent Service for Mean Sea Level, Bidston Observatory, Birkenhead, England and Dr S.D. Lyles, National Ocean Service, National Oceanographic and Atmospheric Administration, Rockville, Maryland, USA for providing updated tide-gauge records.

References

Armentano, T.V. & Ralston, C.W. (1980). The role of temperate forest zones in the global carbon cycle. *Can. J. Forest. Res.*, **10**, 53–60.

Aubrey, D.G. (1985). Recent sea levels from tide gauges: problems and prognosis. In *Glaciers, Ice Sheets and Sea Level: Effect of a CO_2-induced Climatic Change.* US Dept Energy DOE/ER/60235- 1, Attachment 1, pp. 73–91.

Aubrey, D.G. & Emery, K.O. (1986a). Relative sea levels of Japan from tide-gauge records. *Geol. Soc. Am. Bull.*, **97**, 194–205.

Aubrey, D.G. & Emery, K.O. (1986b). Australia – an unstable platform for tide-gauge measurements of changing sea levels. *J. Geol.*, **94**, 699–712.

Aubrey, D.G. & Emery, K.O. (1983). Eigenanalysis of recent U.S. sea levels. *Continental Shelf Res,.* **2**, 21–33.

Barnett, T.P. (1988). Global sea level. In *NCPO, Climate Variations Over the Past Century and the Greenhouse Effect.* A report based on the First Climate Trends Workshop, 7–9 Sept., 1988, Washington, DC, Nat'l Clim. Program Office/NOAA, Rockville, MD.

Barnett, T.P. (1984). The estimates of 'global' sea level change: a problem of uniqueness. *J. Geophys. Res.*, **89**, 7980–8.

Barnett, T.P. (1983). Recent changes in sea level and their possible causes. *Clim. Change*, **5**, 15–38.

Belknap, D.F., Shipp, R.C., Stuckenrath, R., Kelly, J.T. & Borns, H.W., Jr. (1989). Holocene sea level change in coastal Maine. In *Neotectonics of Maine*, W.A. Anderson and H.W. Borns, Jr. (eds.), Maine Geol. Surv. Dept. of Conserv., pp. 85–105.

Bilham, R. (1991). Earthquakes and sea level: space and terrestrial metrology on a changing planet. *Rev. Geophys.*, **29**(1), 1–29.

Bird, E.C.F. (1985). *Coastline Changes, A Global Review.* J. Wiley and Sons, 219 pp.

Braatz, B.V. & Aubrey, D.G. (1987). Recent sea level change in eastern North America. In *Sea Level Fluctuation and Coastal Evolution*, D. Nummedal, O.H. Pilkey and J.D. Howard (eds.). SEPM Special Publication No.41, pp. 29–46.

Bretherton, F.P., Bryan, K. & Woods, J.D. (1990). Time-dependent greenhouse gas-induced climate change. In *Climate Change: The IPCC Scientific Assessment*, J.T. Houghton, G.J. Jenkins and J.J. Ephraums (eds.), Cambridge University Press, Cambridge, pp. 173–93.

Chelton, D.B. & Enfield, D.B. (1986). Ocean signals in tide-gauge records. *J. Geophys. Res.*, **91**, 9081–98.

Chi, S.C. & Reilinger, R.E. (1984). Geodetic evidence for subsidence due to groundwater withdrawal in many parts of the USA. *J. Hydrology*, **67**, 155–82.

Clawson, M. (1979). Forests in the long sweep of American history. *Science*, **204**, 1168–74.

Davis, G.H. (1987). Land subsidence and sea level rise on the Atlantic Coastal Plain of the United States. *Environ. Geol. Water Sci.*, **10**(2), 67–80.

Delcourt, H.R. & Harris, W.F. (1980). Carbon budget of the southeastern US biota: analysis of historical change in trend from source to sink. *Science*, **210**, 321–3.

Diamante, J.M., Pyle, T.E., Carter, W.E. & Scherer, W. (1987). Global change and the measurement of absolute sea level. *Prog. Oceanogr.*, **18**, 1–21.

Dolan, R. & Goodell, H.G. (1986). Sinking cities. *Am. Sci.*, **74**, 38–47.

Douglas, B. (1991). Global sea level rise. *J. Geophys. Res.*, **96**, 6981–92.

Ekman, M. (1988). The world's longest continued series of sea level observations. *PAGEOPH*, **127**, 73–7.

Emery, K.O. (1980). Relative sea levels from tide-gauge records. *Proc. Natl. Acad. Sci., Washington, DC*, **77**(12), 6968–72.

Emery, K.O. & Aubrey, D.G. (1991). *Sea Levels, Land Levels, and Tide Gauges.* Springer-Verlag, Inc., New York, 237pp.

Emery, K.O. & Aubrey, D.G. (1986). Relative sea level change from tide-gauge records of Western North America. *J. Geophys. Res.*, **91**, 13941–53.

Emery, K.O., Aubrey, D.G. & Goldsmith, V. (1988). Coastal neo-tectonics of the Mediterranean from tide-gauge records. *Mar. Geol.*, **81**, 41–52.

Fairbridge, R.W. & Krebs, O.A. (1962). Sea level and the southern oscillation. *Geophys. J.*, **6**, 532–545.

Folland, C.K., Karl, T.R. & Vinnikov, K.Ya. (1990). Observed climate variations and change. In *Climate Change: The IPCC Scientific Assessment*, J.T. Houghton, G.J. Jenkins and J.J. Ephraums, (eds.). Cambridge University Press, Cambridge, pp. 195–238.

Gasperini, P., Yeun, D.A. & Sabadini, R. (1990). Effects of lateral viscosity variations on post glacial rebound: implications for recent sea level trends. *Geophys. Res. Lett.*, **17**(1), 5–8.

Gornitz, V., Lebedeff, S. & Hansen, J. (1982). Global sea level trend in the past century. *Science*, **215**, 1611–14.

Gornitz, V. & Lebedeff, S. (1987). Global sea level changes during the past century. In *Sea Level Fluctuation and Coastal Evolution*, D. Nummedal, O.H. Pilkey and J.D. Howard (eds.). SEPM Special Publication No. 41, pp. 3–16.

Gornitz, V. & Seeber, L. (1990). Vertical crustal movements along the East Coast, North America, from historic and Late Holocene sea level data. *Tectonophys.*, **178**, 127–50.

Gornitz, V. & Solow, A. (1991). Observations of long-term tide-gauge records for indicators of accelerated sea level rise. In *Greenhouse-Gas-Induced Climatic Change: A Critical Appraisal of Simulations and Observations*, M.E. Schlesinger, (ed.). Elsevier, Amsterdam, pp. 347–67.

Grove, J.M. (1988). *The Little Ice Age*. Methuen, London, 498pp.

Guetenberg, B. (1941). Changes in sea level, post-glacial uplift and mobility on the earth's interior. *Bull. Geol. Soc. Am.*, **52**, 721–72.

Hansen, J., Fung, I., Lacis, A., Rind, D., Lebedeff, S., Ruedy, R. & Russell, G. (1988). Global climate change as forecast by Goddard Institute for Space Studies three-dimensional model. *J. Geophys. Res.*, **93**, 9341–64.

Heller, P.L., Wentworth, C.M., & Poag, C.W. (1982). Episodic post-rift subsidence of the United States Atlantic continental margin. *Geol. Soc. Am. Bull.*, **93**, 379–90.

Hoffman, J.S., Wells, J.B. & Titus, J.G. (1986). Future global warming and sea level rise. In *Iceland Coastal and River Symposium '85*, G. Sigbjarnarson (ed.). Reykjavik, National Energy Authority.

Holdahl, S.R., Holzschuh, J.C. & Zilkoski, D.B. (1989). *Subsidence at Houston, Texas 1973–1987*. NOAA Tech. Rept. NOS 131 NGS 44, 21pp.

Holeman, J.N. (1968). The sediment yield of major rivers of the world. *Water Resources Res.*, **4**, 737–47.

IAPSO Advisory Committee (1985). Changes in relative mean sea level. *EOS*, **66**, 754–6.

Kearney, M.S. & Stevenson, J.C. (1991). Island land loss and marsh vertical accretion rate evidence for historical sea level changes in Chesapeake Bay. *J. Coast. Res.*, **7**, 403–15.

Klige, R.K. (1982). Oceanic level fluctuations in the history of the earth. In *Sea and Oceanic Level Fluctuations for 15,000 years*. Acad. Sc. USSR, Institute of Geography, Moscow, Nauka (in Russian), pp. 11–22.

Komar, P.D. & Enfield, D.B. (1987). Short-term sea level changes and shoreline erosion. In *Sea level Fluctuation and Coastal Evolution*, D. Nummedal *et al.* (eds.). SEPM Special Publication No. 41, pp. 17–27.

Korzun, V.I., Ed. (1978). *World Water Balance and Resources of the World*. UNESCO, Moscow.

Kraft, J.C. & Belknap (1986). Holocene epoch coastal geomorphologies based on local relative sea level data and stratigraphic interpretations of paralic sediments. *J. Coastal Res. Spec. Issue*, **1**, 53–9.

Kuenen, Ph. H. (1950). *Marine Geology*. Wiley, New York.

Lambeck, K. (1990). Glacial rebound, sea level change and mantle viscosity. *Q. J. Roy. Astron. Soc.*, **31**, 1–30.

Lisitzin, E. (1974). *Sea Level Changes*. Elsevier, New York.

Meade, R.H. & Emery, K.O. (1971). Sea level as affected by river runoff, Eastern United States. *Science*, **173**, 425–7.

Meade, R.H. & Parker, R.S. (1985). Sediment in rivers of the United States. In *National Water Summary 1984*, USGS Water-Supply Paper 2275, pp. 49–70.

Meier, M.F. (1984). The contribution of small glaciers to global sea level. *Science*, **226**, 1418–21.

Miller, L., Cheney, R.E. & Douglas, B.C. (1988). GEOSAT altimeter observations of Kelvin waves and the 1986–87 El Niño. *Science*, **239**, 52–4.

Milliman, J.D., Broadus, J.M. & Gable, F. (1989). Environmental and economic implications of rising sea level and subsiding deltas: the Nile and Bengal examples. *Ambio*, **18**, 340–5.

Milliman, J.D. & Meade, R.H. (1983). World-wide delivery of river sediment to the oceans. *J. Geol.*, **91**, 1–21.

Mörner, N.A. (1973). Eustatic changes during the last 300 years. *Palaeog., Palaeoclim., Palaeoecol.*, **13**, 1–14.

Nakada, M. & Lambeck, K. (1988). The melting history of the Late Pleistocene Antarctic ice sheet. *Nature*, **333**, 36–40.

Nakada, M. & Lambeck, K. (1989). Late Pleistocene and Holocene sea-level change in the Australian region and rheology. *Geophys. J.*, **96**, 497–517.

Namias, J. & Huang, J.C.K. (1972). Sea level at Southern California: a decadal fluctuation. *Science*, **177**, 351–3.

National Academy of Sciences (NAS), National Research Council (1987). *Responding to Changes in Sea Level; Engineering Implications*. National Academy Press, Washington, DC, 148 pp.

Newman, W.S. and Fairbridge, R.W. (1986). The management of sea level rise. *Nature*, **320**, 319–21.

Pardi, R.R. & Newman, W.S. (1987). Late Quaternary sea levels along the Atlantic coast of North America. *J. Coastal Res.*, **3**, 325–30.

Peltier, W.R. (1986). Deglaciation-induced vertical motion of the North American continent and transient lower mantle rheology. *J. Geophys. Res.*, **91**, 9099–123.

Peltier, W.R. & Tushingham, A.M. (1991). Influence of glacial isostatic adjustment on tide-gauge measurements of secular sea level change. *J. Geophys. Res.*, **96**(B4), 6779–96.

Peltier, W.R. & Tushingham, A.M. (1989). Global sea level rise and the greenhouse effect: might they be connected? *Science*, **244**, 806–10.

Penland, S. & Ramsey, K.E. (1990). Relative sea level rise in Louisiana and the Gulf of Mexico, 1908–1988. *J. Coast. Res.*, **6**(2), 323–42.

Pirazzoli, P.A. (1987a). Sea level changes in the Mediterranean. In *Sea Level Changes*, M.J. Tooley and I. Shennan (eds.), Basil Blackwell, Oxford, UK, pp. 152–81.

Pirazzoli, P.A. (1987b). Recent sea level changes and related engineering problems in the lagoon of Venice (Italy). *Prog. Oceanogr.*, **18**, 323–46.

Pirazzoli, P.A. (1986). Secular trends of relative sea level changes (RSL) indicated by tide-gauge records. *J. Coastal Res. Spec. Issue*, **1**, 1–26.

Rohde, H. (1980). Changes in sea level in the German Bight. *Geophys. J. R. Astron. Soc.*, **62**, 291–302.

Shennan, I. & Woodworth, P.L. (1992). A comparison of sea level trends from the UK-North Sea region during the twentieth century to those from the Late Holocene. *Geophys. J. Int.*, (in press).

Stanley, D.J. (1988). Subsidence in the Northeastern Nile delta: rapid rates, possible causes and consequences. *Science*, **240**, 497–500.

Steckler, M.S. & Watts, A.B. (1978). Subsidence of the Atlantic-type continental margin off New York. *Earth Planet. Sci. Lett.*, **41**, 1–13.

Sturges, W. (1987). Large-scale coherence of sea level at very low frequencies. *J. Phys. Oceanogr.*, **17**, 2084–94.

Thomas, R.H. (1986). Future sea level rise and its early detection by satellite remote sensing. In *Effects of Changing Stratospheric Ozone and Global Climate, Volume 4: Sea Level Rise*, J.G. Titus, (ed.). UNEP and US EPA, pp. 19–36.

Thompson, K.R. (1990). North-Atlantic sea level and circulation. In *Sea Level Change*. National Research Council, National Academy Press, Washington, DC, pp. 52–62.

Thorarinsson, S. (1940). Present glacier shrinkage and eustatic change in sea level. *Geogr. Ann.*, **22**, 131–59.

Trupin, A. & Wahr, J. (1990). Spectroscopic analysis of global tide-gauge sea level data. *Geophys. J. Int.*, **100**, 441–53.

Tushingham, A.M. & Peltier, W.R. (1991). ICE-3G: A new global model of late-Pleistocene deglaciation based upon geophysical predictions of post-glacial relative sea level change. *J.*

Geophys. Res., **96**(B3), 4497–523.

Uchupi, E. & Aubrey, D.G. (1988). Suspect terranes in the North American margins and relative sea levels. *J. Geol.*, **96**(1), 79–90.

Van de Plassche, O. (1990). Mid-Holocene sea level change on the eastern shore of Virginia. *Mar. Geol.*, **91**, 149–54.

Van de Plassche, O., Mook, W.G. & Bloom, A.L. (1989). Submergence of coastal Connecticut 6,000–3,000 (^{14}C) years BP. *Mar. Geol.*, **86**, 349–54.

Wahr, J.M. & Trupin, A.S. (1990). New computation of global sea level rise from 1990 tide-gauge data. *EOS*, **71**, 1267.

Warrick, R.A. & Oerlemans, J. (1990). Sea level rise. In *Climate Change: The IPCC Scientific Assessment*, J.T. Houghton, G.J. Jenkins and J.J. Ephraums (eds.). Cambridge University Press, Cambridge, pp. 257–81.

Wigley, T.M.L. & Raper, S.C.B. (1987). Thermal expansion of sea water associated with global warming. *Nature*, **330**, 127–31.

Woodworth, P.L. (1990). A search for acceleration in records of European mean sea level. *Int. J. Clim.*, **10**, 129–43.

Woodworth, P.L. (1987). Trends in U.K. mean sea level. *Marine Geod.*, **11**, 57–87.

Wu, P. & Peltier, W.R. (1983). Glacial isostatic adjustment and the free air gravity anomaly as a constraint on deep mantle viscosity. *Geophys. J.R. Astron. Soc.*, **74**, 377–449.

Wyrtki, K. (1987). Comparing GEOSAT altimetry and sea level. *EOS*, **68**, 731.

3

Recent global sea levels and land levels

D.G. Aubrey and K.O. Emery

Abstract

Tide-gauge records from around the world ambiguously document the rise and fall of relative sea levels during the past century. Analyses of these records, non-uniformly distributed in time and space, reveal that land levels as well as ocean levels are changing, complicating the estimation of the eustatic component of sea level change. Because most tide gauges are concentrated in the Northern Hemisphere instead of the Southern Hemisphere (where most ocean area is located), tide-gauge data reflect major continental motions due to glacio-isostasy and neo-tectonism, which suppress and mask the signal from eustatic sea level change.

Estimation of the magnitude of sea level change is important for many reasons, including the interpretation of possible effects of global climate change resulting from carbon dioxide and other trace gas loading of the atmosphere. Proper interpretation of past sea level changes is valuable for the calibration and evaluation of certain global climate, ocean and geophysical Earth models. This interpretation can be accomplished using tide-gauge records and other records of relative sea levels only if land level changes are sufficiently distinguished from actual sea level changes.

Introduction

Past and future climate changes drive various responses in the oceans. One such response is variation in sea levels as the oceans warm or cool, and as water is removed from or returned to the ocean basins and glaciers. During the past 15,000–18,000 years following the last glacial maximum, the Earth has experienced an average warming of about 4–5 °C. As a result, sea levels have risen between 60 and 150 metres (Milliman and Emery, 1968; Blackwelder *et al.*, 1979; Oldale and O'Hara, 1980). On the century time-scale, the Earth's climate has also undergone changes, including the climatic optimum of around AD 1000–1300 and the Little Ice Age of AD 1300–1850 (Thompson *et al.*, 1986). These short-term climate changes have produced global mean temperature fluctuations of perhaps as much as several degrees (e.g., Willett, 1950; Mörner, 1973; Wigley and Jones, 1981; Ellsaesser *et al.*, 1986), as well as unresolved variations in mean sea levels.

In addition to these historic changes in climate, concern has arisen over potential future global warming resulting from increasing atmospheric concentrations of trace gases such as carbon dioxide, methane, nitrous oxide and chlorofluorocarbons (NRC, 1983; Bolin *et al.*, 1986; Houghton *et al.*, 1990). One expected response from global warming is an increase in ocean levels. Based on observational data and on simple models of ocean response to climate change, estimates of sea level rise during the next 100 years range from a few tens of centimetres to several metres (Warrick and Oerlemans, 1990). Given the uncertainty in these calculations and the potential economic impact of increasing coastal inundation (Hoffman *et al.*, 1983; Broadus *et al.*, 1986), it is essential that research reduce this large range of uncertainty. For the sea level rise issue, improved knowledge of the response of sea levels to past climate change is needed, incorporating both the magnitude and phasing of such ocean response to climatic forcing.

Eustatic sea levels

Tide-gauge records have been used extensively during this century to estimate sea level changes (Gornitz, Pugh, this volume; Gornitz *et al.*, 1982; Emery, 1980; Barnett, 1984; Aubrey, 1985). Most estimates of the rate of sea level rise over the last 100 years are in the range of 1.0–1.5 mm/yr, despite the use of records from different geographic regions and the spanning of dissimilar time intervals (Woodworth, this volume). The similarity of these estimates gives false confidence to their validity, since the roles of glacio-isostasy and neo-tectonism on a global basis have only recently begun to be explored systematically (Emery and Aubrey, 1991).

Average sea level rise since deglaciation (due to both steric response and glacial meltwater input) was 6–12 mm/yr prior to about 6,000 years ago, depending on the speed of deglaciation and maximum depression of ocean levels during the Wisconsinan (Würm). Subsequently, the rate of rise of sea level has been less. Estimates of the ocean level at 7,000 and 5,000 BP are, respectively, 10 m and 5 m lower than today (Curray, 1964; Milliman and Emery, 1968; Pirazzoli, 1977). This suggests that the average rate of global sea level rise since then has been between 1.4 mm/yr and 1.0 mm/yr. Some radiocarbon dating indicates that the rate of rise has decreased continuously during the past 4,000 years (Redfield, 1967); however, sparse sampling and lack of discrimination between ocean levels and land levels present difficulties in quantifying the decreased rate of rise (Clark and Lingle, 1979).

Recent investigations of time series of sea levels over large areas suggest that an increase in the rate of rise of relative sea levels (RSL) may have occurred during the post-1930 era compared with previous records (Barnett, 1984; Braatz and Aubrey, 1987; see Fig. 3.1). Such increases have been found at all locations where sufficiently long record lengths exist, including North America, Northern Europe, and Eastern Asia. On average, rates of RSL increased by approximately 0.5 mm/yr following the decade of the 1930s. The source of this apparent increase in relative sea level rise is uncertain. Although it is tempting to relate this acceleration to human-induced climate change resulting from trace gas effects, there is no direct support for this view. Rather, the recent increase may reflect the delayed ocean response to climate warming following the Little Ice Age (that ended about AD 1850; Thompson *et al.*, 1986). As global mean atmospheric temperatures have risen by approximately 0.5 °C since the end of the Little Ice Age (Jones *et al.*, 1986a, 1986b), sea level would be expected to respond. If this increase in sea level is a response to global warming, then the delay

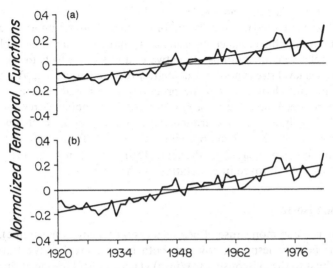

Fig. 3.1 Apparent change in rate of relative sea level rise along the US East Coast from 1920 to 1980, derived from tide-gauge data. (*a*) The first temporal eigenfunction computed from all 44 stations used in the study. (*b*) The first temporal eigenfunction computed from the 26 stations which were in operation before 1935.

There is almost no difference between the two results, indicating that the change in slope of the first temporal eigenfunction at 1934 is not an artifact of lower sampling density between 1920 and 1934, nor is it a bias resulting from the analysis. The reasons for this change in slope are uncertain, the changes having an origin perhaps in oceanographic variability or changes in volume of the world oceans (Braatz and Aubrey, 1987. Reprinted by permission of the Society of Economic Palaeontologists and Mineralogists).

between atmospheric warming and sea level response could be of the order of many decades, instead of the commonly cited 20-year lag (Cess and Goldenberg, 1981; Revelle, 1983). Alternatively, the apparent change in rate of sea level rise may be attributable to oceanographic factors (Sturges, 1987; Pugh, 1987). Finally, it is possible that the apparent mid-century acceleration in sea level rise is not statistically significant (Woodworth, 1990).

Two possible climate-related explanations for a globally coherent rise in sea level at the century time-scale are oceanic thermal expansion and increased glacial melt. Global ocean warming is thought to have occurred during the past century (Roemmich and Wunsch, 1984; Roemmich, 1985), although there are uncertainties about its existence and magnitude (Barnett, 1983b). Roemmich and Wunsch (1984) estimated that steric expansion due to warming in the North Atlantic between water depths of 1,000–3,000 m accounts for about 0.4 mm/yr during the interval 1957–72, assuming that the effects found in the North Atlantic ocean are characteristic of the global oceans as a whole. Available data are inadequate to resolve whether such an expansion is a global or local phenomenon. For global estimates, reliance has to be placed on model-based estimates (see Wigley and Raper, this volume).

Estimation of the contribution of meltwater input to oceans similarly is limited by inadequate data. Recent reviews of contributions of land ice to ocean levels (PRB,

1985; Robin, 1986; Warrick and Oerlemans, 1990) document the uncertainty in estimates of present mass balances of the major ice sheets and mountain glaciers. This makes the estimation of historical changes in mass balances even more problematical.

In short, the relative contributions of meltwater and steric effects to the increased rate of absolute sea level rise cannot be adequately resolved at present. An increase in rate is consistent with changes in climate processes as opposed to being tectonic in origin, because no mechanisms have been suggested for a global increase in tectonic processes occurring simultaneously around the globe in such a short time interval. Global warming may have reached approximately 0.25 °C from 1860 through 1940. Whether or not this is the cause of the increase in RSL can be resolved only by modelling studies and improved interpretation of observations.

Glacial rebound isostasy

Glacio-isostasy is a common cause of neo-tectonism (Tooley, this volume). Isostatic adjustment has been widespread, covering both hemispheres, as the Earth's crust continues to adjust to the Wisconsinan (Würm) glaciation and deglaciation (Walcott, 1972; Clark *et al.*, 1978; Peltier, 1984, 1986). Glacio-isostasy dominates the many tide-gauge records of Northern Europe (Emery and Aubrey, 1985; see Fig. 3.2). Centred over the Gulf of Bothnia and also over the Northern British Isles is an isostatic rebound that reflects the removal of large masses of glacial ice during the Late Pleistocene period. Rates of rebound reach 10 mm/yr, about equal in absolute value to the global mean rate of rise of sea level during the deglaciation. Of the available high quality global tide-gauge records, the 134 records from Fennoscandia are contaminated by isostatic adjustment which cannot be removed without using a rebound model. Although numerical models of isostatic effects have been developed (Peltier, 1984, 1986; Mörner, 1980), the approximations needed for the models make it difficult to verify their accuracy, particularly for the past 100 years.

Reconstructions of postglacial rebound in Fennoscandia, based on diverse geological data (Mörner, 1969; Emery and Aubrey, 1985), suggest a rebound of up to 700 m during the past 10,000–15,000 years or so. This implies an average rate of rebound in the central depressed area of about 50–70 mm/yr, much more rapid than the eustatic sea level change. Present rates of rebound, although still significantly higher than eustatic sea level change, are lower than previous rates of rebound by a factor of five. This finding is consistent with numerical model studies (Clark and Lingle, 1979).

Glacio-isostasy also may dominate records from the northeastern coast of North America (Aubrey and Emery, 1983; Braatz and Aubrey, 1987; see Fig. 3.3). Using numerical model estimates of glacial rebound (Peltier, 1986), the total relative sea level record can be reconstructed, as can the residual following removal of the estimated deglaciation effect (in the Eastern US isostatic submergence is taking place as the glacial forebulge that formerly covered most of this area relaxes, while the central glaciated area over Canada is rebounding, leading to belts of submergence and emergence). Removal of the glacio-isostatic effect suggests a residual relative sea level rise of between 1.0 and 1.5 mm/yr, certainly within the range of previous estimates of sea level rise. Superimposed on this mean rate is a variability that bears some relationship to the distribution of exotic terranes, although the relationship is qualitative (Uchupi and Aubrey, 1988). Thus the 44 most usable tide gauges of Eastern North America are contaminated by glacio-isostatic and tectonic processes. Although adjust-

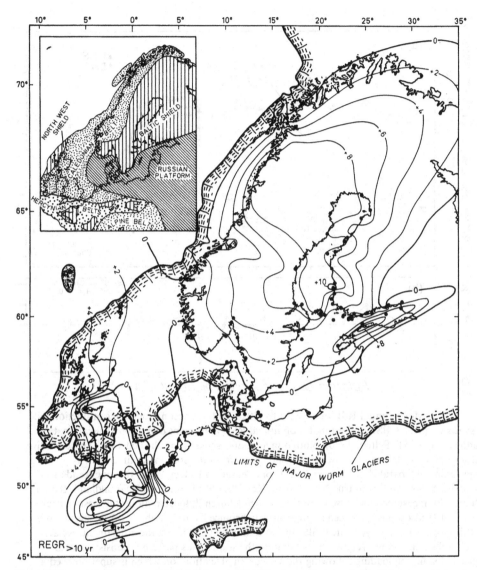

Fig. 3.2 Rates of relative sea level rise (mm/yr) in Northern Europe derived from analysis of tide-gauge records covering the time period 1900–85. Positive values indicate relative rise of land; negative values indicate relative sinking of land. The patterns of relative sea level change are consistent with ongoing crustal response to the latest glaciation/deglaciation cycle (Emery and Aubrey, 1985. Reprinted by permission of Elsevier Science Publishers).

ments have been made for these processes, they represent only approximations that are certain to be refined in the future.

Plate Movements

Tectonism affects tide-gauge records both by changing the volume of ocean basins and by altering the level of recording instruments. Ocean-basin volume changes can arise

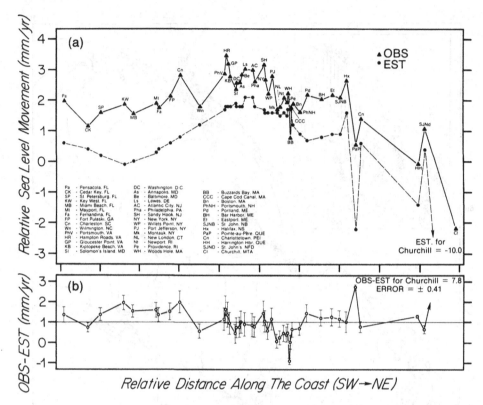

Fig. 3.3(a) Mean annual RSL movement for 1920–80 from reconstructed eigenfunction data (synthetic rates, OBS), and estimates of annual RSL movement due to postglacial isostatic adjustment (EST, Peltier, 1986), plotted in a relative sense along the coastline. These relative distances are obtained by drawing perpendiculars from the stations to lines drawn approximately parallel to the coastline. From Pensacola to Key West, this line trends 146° measured clockwise from true North; from Key West to St. John's, Newfoundland, the line trends 40° measured clockwise from true North. Churchill, located along the west central coast of Hudson Bay, is placed at an arbitrary distance from St. John's. (b) Residual annual RSL movement, i.e., synthetic (OBS) minus estimated (EST) isostatic adjustment. Relative sea level change shows relative rise of land to the south. This pattern is consistent with glacial loading/unloading following the Wisconsin glaciation, on which is superimposed a eustatic rise of poorly constrained magnitude. The convention for land rising/sinking is negative for relative rise of the land, positive for sinking of the land (Braatz and Aubrey, 1987. Reprinted by permission of the Society of Economic Palaeontologists and Mineralogists).

from changes in rates or directions of sea-floor spreading (for instance, Pitman, 1979; Kennett, 1982). During the Cretaceous Period, sea levels stood about 350 m above present levels, flooding nearly 35% of the present land surface; the rates of sea-floor spreading were higher than present rates. When spreading rates decreased, the average ocean depth increased and sea levels fell. Estimates for the maximum rates of sea level fall responding to such rate changes are approximately 0.01 mm/yr, clearly important on a geological time-scale if persistent, but not significant on historical time-scales.

Neo-tectonism also affects tide-gauge records by changing the level of the tide-gauge station. At present, changes in mean sea level measured at a single station cannot be separated into vertical land movement and eustatic sea level components, because they cover the same part of the spectrum (although geodetic techniques in development may achieve such a separation in the future; Diamante *et al.*, 1987; Carter *et al.*, this volume). Alternatively, analyses of regional arrays of tide gauges, in combination with geological studies, provide a promising means of estimating the various contributing factors.

Plate tectonics directly affect relative sea levels as the surfaces of plates continuously adjust to relative horizontal plate motion, deforming the continental margins on which tide gauges are located. Intra-plate earthquake activity (Sykes, 1963) concentrated about the margins of the Pacific reflects this plate interaction. Since the tide gauges are located along the oceanic margins where much tectonic activity exists, it is expected that their records would mirror these effects.

Using tide-gauge data from around the world (approximately 1243 stations, of which 563 exceed 10 years in length and 332 exceed 20 years in length; Pugh *et al.*, 1987; Pugh, 1987), recent investigations have clarified the extent of tectonic impact on relative sea levels (Hicks, 1972; Aubrey and Emery, 1983, 1986; Emery and Aubrey, 1985, 1986a, 1986b, 1989, 1991; Braatz and Aubrey, 1987; Aubrey *et al.*, 1988; Pirazzoli, 1986). The tide-gauge records of Western North America, long known to be tectonically active, also are affected by tectonism (Emery and Aubrey, 1986b; see Fig. 3.4). Similarly, the records of the many tide gauges of Japan and other island chains contain a broad-scale tectonic trend related to plate convergence processes that mask changes in ocean level itself (Aubrey and Emery, 1986; Emery and Aubrey, 1991; see Fig. 3.5). From none of these records can a change in eustatic sea level be extracted with confidence. Of the 200 tide gauges covering these locations, none is obviously free from tectonic impact. Thus, they must be used with care in deriving estimates of sea level rise.

Continental margins consist in large part of exotic or suspect terranes composed of oceanic and continental fragments from past plate interactions, creating a complex mélange of geology and hence variable structural strength (Coney *et al.*, 1980; Uchupi and Aubrey, 1988). As stress fields are set up due to deglaciation and plate interactions, one might expect the response of different suspect terranes to vary. The pervasiveness of such tectonic fabrics and stresses argues strongly for more local tectonic control on tide-gauge records than has been envisioned previously (Barnett, 1983a; Chelton and Davis, 1982).

Other studies have been made of South American, Mediterranean, East Asian, Indian, and Australian tide gauges (Aubrey *et al.*, 1988; Aubrey and Emery, 1986; Emery and Aubrey, 1986a; 1989; 1991; Emery *et al.*, 1988). These studies show similar results: that tectonism and isostatic processes dominate tide-gauge records and make interpretation of sea level rise difficult. Of the total number of tide gauges available for analysis, more than 90% of the Permanent Service for Mean Sea Level (PSMSL) stations exceeding ten years in length are located in the nine localities discussed above. Given the tectonic bias at these stations, estimation of absolute sea level rise is not unique. In addition, Man's direct impacts on tide gauges are severe and pervasive. River diversion, pore-fluid mining and sedimentation control negate data from dozens of tide gauges world-wide (Emery and Aubrey, 1991). However, much has been learned about relative land level changes and about the response of continental margins to plate

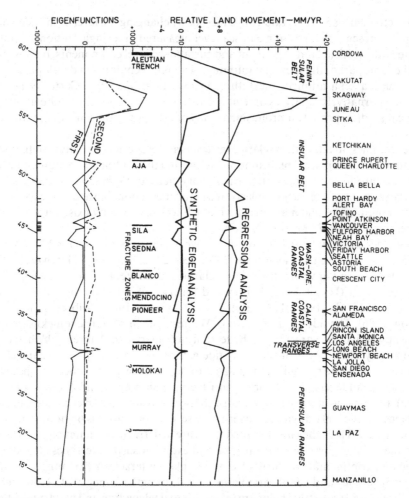

Fig. 3.4 Tide-gauge data from the Western North American region reflect the importance of tectonism on relative sea level. While some of the variability is due to differing record lengths of station data, larger scale trends are thought to reflect trends in relative sea level variability arising from tectonic causes (Emery and Aubrey, 1986b. Reprinted by permission of the American Geophysical Union).

motions. Because of the difficulty in estimating sea level changes, their use in calibrating models of ocean response to climate change is limited by available data.

Conclusions

The data relating historical climate change to relative sea levels are incomplete. Yet, judicious analysis of past observations in combination with modelling programmes should provide important information regarding climate–sea level relationships.

With respect to estimating globally coherent changes in sea level during the last 100 years, the biases arising from vertical land movements mean that estimates based on crude averages of tide-gauge data are subject to large uncertainties; significant elements

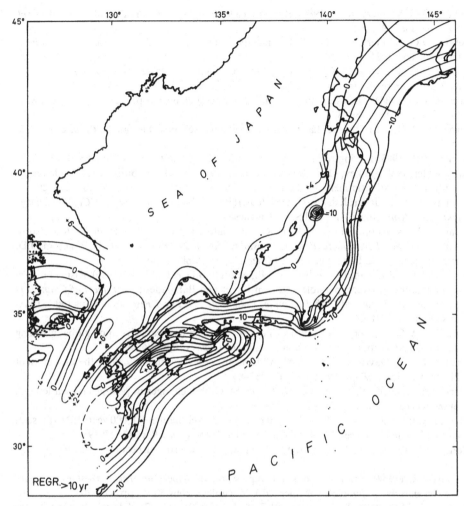

Fig. 3.5 Rates of relative sea level rise (mm/yr) along the coast of Japan as indicated by study of more than 100 tide gauges of that country. The strong geographic variability in this region is consistent with previous hypotheses of crustal deformation along convergent margins, notably along the trenches bordering Eastern and Southern Japan (Aubrey and Emery 1986. Reprinted by permission of the American Geophysical Union).

of the physics involved may be missed. New geodetic technologies such as Very Long Baseline Interferometry, differential Global Positioning System, and absolute gravity meters should allow for more precise discrimination between land motion and ocean volume change. This information, gathered over decades, should significantly reduce the uncertainties in estimates of past, present and future sea level behaviour.

References

Aubrey, D.G. (1985). Recent sea levels from tide gauges: problems and prognosis. In *Glaciers, Ice Sheets and Sea Level: Effect of a CO₂-induced Climatic Change*, (Polar Research Board), National Academy of Science Press, Washington DC, pp. 73–91.

Aubrey, D.G. & Emery, K.O. (1986). Relative sea levels of Japan from tide-gauge records. *Geol. Soc. Am. Bull.*, **97**, 194–205.

Aubrey, D.G. & Emery, K.O. (1983). Eigenanalysis of recent United States sea levels. *Continental Shelf Res.*, **2**, 21–33.

Aubrey, D.G., Emery, K.O. & Uchupi, E. (1988). Changing coastal levels of South America and the Caribbean region from tide-gauge records. *Tectonophys.*, **154**, 269–84.

Barnett, T.P. (1984). The estimation of 'global' sea level change: a problem of uniqueness. *J. Geophys. Res.*, **89**, 7980–8.

Barnett, T.P. (1983a). Recent changes in sea level and their possible causes. *Clim. Change*, **5**, 15–38.

Barnett, T.P. (1983b). Long-term changes in dynamic height. *J. Geophys. Res.*, **88**, 9547–52.

Blackwelder, B.W., Pilkey, O.H. & Howard, J.D. (1979). Late Wisconsinan sea levels on the Southeast US Atlantic shelf based on in-place shoreline indicators. *Science*, **204**, 618–20.

Bolin, B., Döös, B., Jäger, J. & Warrick, R., (eds.) (1986). *The Greenhouse Effect, Climate Change and Ecosystems*. John Wiley and Sons, Chichester.

Braatz, B.V. & Aubrey, D.G. (1987). Recent relative sea-level changes in Eastern North America. In *Sea-Level Fluctuation and Coastal Evolution*, D. Nummedal, O.H. Pilkey and J.D. Howard (eds.). Soc. Econ. Paleontol. Mineral. Spec. Publ. 41, pp. 29–46.

Broadus, J.M., Milliman, J.D., Edwards, S.F., Aubrey, D.G. & Gable, F. (1986). Rising sea level and damming of rivers: possible effects in Egypt and Bangladesh. In *Effects of Changes in Stratospheric Ozone and Global Climate, Volume 4: Sea Level Rise*, J.G. Titus (ed.). EPA/UNEP, pp. 165–189.

Cess, R.D. & Goldenberg, S.D. (1981). The effect of ocean heat capacity upon global warming due to increased atmospheric carbon dioxide. *J. Geophys. Res.*, **86**, 498.

Chelton, D.B. & Davis, R.E. (1982). Monthly mean sea level variability along the west coast of North America. *J. Phys. Oceanogr.*, **12**, 757–84.

Clark, J.A., Farrell, W.E. & Peltier, W.R. (1978). Global changes in post-glacial sea level: A numerical calculation. *Quat. Res.*, **9**, 265–87.

Clark, J.A. & Lingle, C.S. (1979). Predicted relative sea level changes (18,000 years BP to present) caused by late-glacial retreat of the Antarctic Ice Sheet. *Quat. Res.*, **11**, 279–98.

Coney, P.J., Jones, D.L. & Monger, J.W.H. (1980). Cordilleron suspect terranes. *Nature*, **288**, 329–33.

Curray, J.R. (1964). Transgressions and regression. In *Papers in Marine Geology, Shepard Commemorative Volume*, R.C. Miller (ed.). Macmillan, New York, pp. 175–203.

Diamante, J.M., Pyle, T.E., Carter, W.E. & Scherer, W. (1987). Global change and the measurement of absolute sea level. *Prog. Oceanogr.*, **18**, 1–21.

Ellsaesser, H.W., McCracken, M.C., Walton, J.J. & Grotch, S.L. (1986). Global climatic trends as revealed by the recorded data. *Rev. Geophys.*, **24**, 745–92.

Emery, K.O. (1980). Relative sea levels from tide-gauge records. *Proc. Nat. Acad. Sci.*, **77**, 6968–72.

Emery, K.O. & Aubrey, D.G. (1991). *Sea Levels, Land Levels and Tide Gauges*. Springer Verlag, New York, 237pp.

Emery, K.O. & Aubrey, D.G. (1989). Tide gauges of India. *J. Coastal Res.*, **5**, 489–501.

Emery, K.O. & Aubrey, D.G. (1986a). Relative sea level changes from tide-gauge records of Eastern Asia mainland. *Mar. Geol.*, **72**, 33–45.

Emery, K.O. & Aubrey, D.G. (1986b). Relative sea level changes from tide-gauge records of Western North America. *J. Geophys. Res.*, **91**, 13941–53.

Emery, K.O. & Aubrey, D.G. (1985). Glacial rebound and relative sea levels in Europe from tide-gauge records. *Tectonophys.*, **120**, 239–55.

Emery, K.O., Aubrey, D.G. & Goldsmith, V. (1988). Coastal neo-tectonics of the Mediterranean from tide-gauge records. *Mar. Geol.*, **81**, 41–52.

Gornitz, V., Lebedeff, S. & Hansen, J. (1982). Global sea level trend in the past century. *Science*, **215**, 1611–14.

Hicks, S.D. (1972). Vertical crustal movements from sea level measurements along the east coast of the United States. *J. Geophys. Res.*, **77**, 5930–4.

Hoffman, J.S., Keyes, D. & Titus, J.G. (1983). *Projecting Future Sea Level Rise, Methodology, Estimates to the Year 2100, and Research Needs.* US EPA Report no. 230–09–007, 121 pp.

Houghton, J.T., Jenkins, G.J. & Ephraums, J.J. (eds.) (1990). *Climate Change: The IPCC Scientific Assessment.* Cambridge University Press, Cambridge.

Jones, P.D., Raper, S.C.B., Bradley, R.S., Diaz, H.F., Kelly, P.M. & Wigley, T.M.L. (1986a). Northern hemisphere surface air temperature variations 1851–1984. *J. Climate Applied Met.*, **25**, 161–79.

Jones, P.D., Raper, S.C.B. & Wigley, T.M.L. (1986b). Southern hemisphere air temperature variations, 1851–1984. *J. Climate Applied Met.*, **25**, 1213–30.

Kennett, J.P. (1982). *Marine Geology.* Prentice-Hall, Inc., Englewood Cliffs, NJ, 813 pp.

Milliman, J.D. & Emery, K.O. (1968). Sea levels during the past 35,000 years. *Science*, **162**, 1121–3.

Mörner, N.A., Ed. (1980). *Earth Rheology, Isostasy and Eustasy.* Wiley, London–New York, 599 pp.

Mörner, N.A. (1973). Eustatic changes during the last 300 years. *Palaeog. Palaeoclim. Palaeoecol.*, **13**, 1–14.

Mörner, N.A., Ed. (1969). The late Quaternary history of the Kattegatt Sea and the Swedish West Coast, Deglaciation, Shore level Displacement, Chronology, Isostasy and Eustasy. *Sveriges Geologiska Undersörning 63*, Ser. C, No. 640, 404–53.

National Research Council (NRC) (1983). *Changing Climate, Report of the Carbon Dioxide Assessment Committee.* National Academy of Sciences Press, Washington, DC, 496 pp.

Oldale, R.N. & O'Hara, C.J. (1980). New radiocarbon dates from the inner continental shelf off Southeastern Massachusetts and a local sea level rise curve for the past 12,000 yr. *Geology*, **8**, 102–6.

Peltier, W.R. (1986). Deglaciation-induced vertical motion of the North American continent. *J. Geophys. Res.*, **91**, 9099–123.

Peltier, W.R. (1984). The thickness of the continental lithosphere. *J. Geophys. Res.*, **89**, 11303–16.

Pirazzoli, P.A. (1986). Secular trends of relative sea level (RSL) changes indicated by tide-gauge records. *J. Coastal Res.*, **1**, 1–26.

Pirazzoli, P.A. (1977). Sea level relative variations in the world during the last 2000 years. *Z. Geomorphol.*, **21**, 284–96.

Pitman, W.C., III (1979). The effect of eustatic sea level changes on stratigraphic sequences at Atlantic margins. *Am. Assoc. Pet. Geol. Mem.*, **29**, 453–60.

Polar Research Board (PRB) (1985). *Glaciers, Ice Sheets and Sea Level: Effects of a CO_2-induced Climatic Change.* National Academy of Sciences Press, Washington, D.C., 330 pp.

Pugh, D.T. (1987). Tides, *Surges and Mean Sea Level.* John Wiley and Sons, New York, 472 pp.

Pugh, D.T., Spencer, N.E. & Woodworth, P.L. (1987). *Data Holdings of the Permanent Service for Mean Sea Level.* Bidston Observatory, United Kingdom, 156 pp.

Redfield, A.C. (1967). Postglacial change in sea level in the Western North Atlantic Ocean. *Science*, **157**, 687–92.

Revelle, R.R. (1983). Probable future changes in sea level resulting from increased atmospheric carbon dioxide. In *Changing Climate, Report of the Carbon Dioxide Assessment Committee.* National Academy Press, Washington, DC, pp. 433–448.

Robin, G. de Q. (1986). Changing the sea level. In *The Greenhouse Effect, Climate Change and Ecosystems*, B. Bolin, B. Döös, J. Jäger and R. Warrick (eds.). John Wiley and Sons, Chichester, pp. 323–59.

Roemmich, D. (1985). Sea level and the thermal variability of the ocean. In *Glaciers, Ice Sheets and Sea Level: Effect of CO$_2$-induced Climatic Change*, (Polar Research Board), National Academy of Science Press, Washington, DC, pp. 104–15.

Roemmich, D. & Wunsch, C. (1984). Apparent changes in the climatic state of the deep North Atlantic Ocean. *Nature*, **307**, 447–50.

Sturges, W. (1987). Large-scale coherence of sea level at very low frequencies. *J. Phys. Oceanogr.*, **17**, 2084–94.

Sykes, L.R. (1963). Seismicity of the South Pacific Ocean. *J. Geophys. Res.*, **68**, 5999–6006.

Thompson, L.G., Mosley-Thompson, E., Dansgaard, W. & Grootes, P.M. (1986). The 'Little Ice Age' as recorded in the stratigraphy of the Quelccaya ice cap. *Science*, **234**, 361–4.

Uchupi, E. & Aubrey, D.G. (1988). Autochthonous/allochthonous terranes in the North American margins and sea level from tide gauges. *J. Geol.*, **96**, 79–90.

Warrick, R.A. & Oerlemans, J. (1990). Sea level rise. In *Climate Change: The IPCC Scientific Assessment*, J.T. Houghton, G.J. Jenkins and J.J. Ephraums (eds.). Cambridge University Press, Cambridge, pp. 257–81.

Walcott, R.I. (1972). Late Quaternary vertical movements in Eastern North America: Quantitative evidence of glacio-isostatic rebound. *Rev. Geophys. Space Phys.*, **10**, 849–84.

Wigley, T.M.L. & Jones, P.D. (1981). Detecting CO$_2$-induced climatic change. *Nature*, **292**, 205–8.

Willett, H.C. (1950). Temperature trends of the last century. *Proc. R. Meteorological Society*, Special Volume, 195–206.

Woodworth, P.L. (1990). A search for accelerations in records of European mean sea level. *Int. J. Clim.*, **10**, 129–43.

4

Improving sea level data

D.T. Pugh

Abstract

Direct measurements of sea level have a comparatively recent history compared with the period covered by geological indicators. However, they constitute the most reliable evidence over the past 100 years for sea level changes relative to local tide-gauge benchmarks. This account outlines the history of direct sea level measurements, describes the various techniques which are used and indicates the methods for converting these measurements to a mean sea level value. Improvements in the quality and quantity of sea level measurements are being made through the Intergovernmental Oceanographic Commission GLOSS Tide-Gauge Network in which many countries participate. Details are given of the operation and development of GLOSS (to measure the Global Level of the Sea Surface) and the role of new technology in improving sea level data.

Introduction

The space- and time-scales over which sea level variations extend are enormous. At one extreme, wind waves have periods of about 10 seconds and length-scales of tens of metres; at the other extreme, global changes of sea level due to changing rates of sea floor spreading have periods of hundreds of millions of years. Fig. 4.1 summarizes the various physical and geological processes which contribute to sea level changes. This figure is not intended to be a definitive representation; it may stimulate the reader to identify additional effects and to question the importance of those shown. Concern about possible increases of sea level due to global warming necessitates studies of sea level changes on a global scale over periods of decades and longer. However, these changes, which may proceed by as much as 0.01 m per year, must be identified against interannual ocean climate changes and long-term crustal movements.

Traditional technology allows measurements of sea level relative to a fixed coastal Tide-Gauge Bench Mark (TGBM) so that only relative land-sea vertical movements are detected. These levels are shown as A_1 and B_1 in Fig. 4.2. If the TGBM could be fixed in an absolute geometric framework, so that changes in A_2 and B_2 were monitored, it would be possible to distinguish between vertical movements of the land and of the sea. Oceanographers would also like the sea levels to be measured relative to the geoid, the equipotential surface which sea level would assume in the absence of

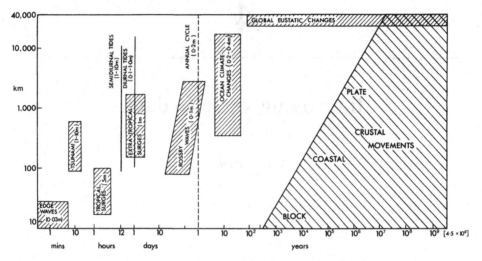

Fig. 4.1 Some of the physical and geological processes responsible for sea level changes, with indications of the length (km) and time-scales involved. Global processes have a length scale of 40,000 km.

disturbing forces such as currents, winds and air pressure. If these distances, A_3 and B_3 were known, then the absolute currents in the ocean between sites A and B could be computed by applying the difference A_3–B_3 in the geostrophic equation. However, A_3, B_3 cannot be measured at present, but A_2, B_2 can now be fixed to useful accuracy (Carter *et al.*, 1989).

This chapter begins with a summary of the history of direct sea level measurement, the instruments used and their limitations, and the methods for calculating monthly and annual mean sea levels (MSL) from these observations. The second part describes the Global Tide-Gauge Network, GLOSS, which is being established to monitor sea level variations for many purposes. Finally, some of the new technical developments which will overcome the limitations outlined above are discussed, to show that our ability to monitor the impact of global climate changes on sea level may improve dramatically over the next ten years.

Historical summary

The first detailed records of mean sea level were started in Brest in 1807, but earlier measurements of high and low water levels were made at Amsterdam and Stockholm from 1682 and from 1774 respectively (van Veen, 1954; Mörner, 1973). These early readings of fixed markers by eye would have been very difficult to continue over long periods. The first automatic recording tide gauge, using the stilling-well and float principle, was first deployed at Sheerness in the Thames Estuary (Palmer, 1831). Sir James Clark Ross (1847, p.43) reported that the '... fixing of solid and well-secured marks for the purpose of showing the mean level of the ocean at a given epoch, was suggested by Baron von Humboldt ... subsequent to the sailing of the expedition ...' (in 1839). He further reported that von Humboldt made the following observation: 'if similar measures had been taken in Cook and Bougainville's earliest voyages, we

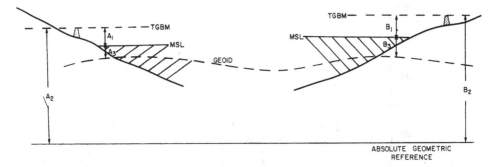

Fig. 4.2 Sea levels relative to Tide-Gauge Bench Marks (A_1, B_1), an absolute geometric system (A_2, B_2) and the geoid (A_3, B_3). The level differences ($A_3 - B_3$) are related to the geostrophic ocean currents across the section.

should now be in possession of necessary data for determining whether secular variation in the relative level of land and sea is a general or merely a local phenomenon, and whether any law is discoverable in the direction of the points which rise or sink simultaneously'. Ross made measurements of sea level in Tasmania and the Falkland Islands, and the original Tide-Gauge Bench Marks have been relocated in both cases (Hamon, 1985).

By this time, the relationship between mean sea level and accurate survey work seems to have been widely recognized. While Ross was completing his voyage in the Southern Ocean, a detailed survey of sea levels around Ireland was made in the summer of 1842 at 22 sites during an Ordnance Survey levelling operation (Airy, 1845; Dixon, 1979; Pugh, 1982). Since the mid-nineteenth century there has been a steady growth in the number of stations established for measuring long-term changes in sea level. Fig. 4.3, which shows this growth, is based on data held by the Permanent Service for Mean Sea Level (PSMSL; Pugh *et al.*, 1987).

Established in 1933, PSMSL is now supported under the auspices of the Federation of Astronomical and Geophysical Services and by the Intergovernmental Oceanographic Commission of UNESCO, and is based at the Proudman Oceanographic Laboratory, Bidston Observatory, United Kingdom. The PSMSL receives monthly and annual mean values of sea level from approximately 110 national authorities, distributed around the world, responsible for sea level monitoring in each country or region. Data from each station are entered directly as received from the authority into the PSMSL raw data file for that station (usually called the 'Metric' file). In order to construct time series of sea level measurements at each station, the monthly and annual means have to be reduced to a common datum where possible. This reduction is performed by the PSMSL, making use of the Tide-Gauge Datum history provided by the supplying authority. These homogeneous series are termed the 'Revised Local Reference' dataset. For scientific purposes, the RLR dataset is superior to the 'Metric', although all data held by PSMSL are available on request. At present, PSMSL holds only monthly and annual mean values: more detailed observations, from which these monthly values are determined, are held by the national authorities. However, it is possible that future applications of mean sea level will require more frequent values, for example, daily means, and if so, the activities of the PSMSL will be reviewed.

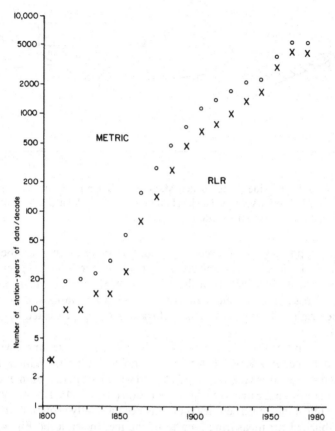

Fig. 4.3 The number of station-years of data per decade held by PSMSL in the 'Metric' and uniform 'RLR' series. The RLR series are distinguished by their having a homogeneous datum history.

Instruments

Methods for measuring sea level may be divided into two categories: those appropriate for coastal measurements, and those which can be used off-shore. Off-shore gauges cannot be levelled into an absolute reference system using present technology and so have not been used for identifying long-term trends. Coastal sea level measurements were first made using graduated poles and then by Palmer's self-recording gauge. The stilling-well and float gauge has been the standard instrument for many years and is still the most common. However, there are fundamental limitations to the accuracy of measurements made using stilling-wells, particularly due to their non-linear response to waves. Also, stilling-well systems are expensive and difficult to install, requiring a pier or similar vertical structure for mounting over deep water. Reading charts over long periods is a tedious procedure and prone to errors. As a result, new types of gauges which do not use charts have been developed and are being introduced gradually.

The two most widely used new systems are the Bubbler System and the Acoustic Pulse System. In the Bubbler System, compressed air from a cylinder is reduced in pressure through one or two valves so that there is a small steady flow down a connecting tube which escapes through an orifice in an underwater canister, called a pressure-point. At this underwater outlet, the low rates of gas escape means that the gas pressure is equal to the water pressure. This is also the pressure transmitted along the tube to the measuring and recording system, apart from small corrections for pressure gradients. If the water density and gravitational acceleration are known, then the water level relative to the pressure-point orifice datum may be calculated.

In the Acoustic Pulse System, gauges measure the travel time of a sound pulse returning after reflection at the sea surface. This system is being installed by the United States National Oceanic and Atmospheric Administration (Carter *et al.*, this volume). The NOAA System uses a relatively small diameter vertical tube which provides a minimal amount of stilling, guides the sound pulses, and contains an automatic calibration system that corrects the observations for the changes in the temperature, humidity and air pressure between the transmitter–receiver and the sea surface (Scherer, 1986).

Calculation of mean sea level

Fig. 4.4(*a*) shows a typical chart from the Lowestoft (UK) Tide Gauge. The tidal range here is relatively small but is still far bigger than the long-term trends which are being investigated. The small-scale oscillations on the tidal traces are due to local water oscillations known as seiches. Fig. 4.4(*b*) and 4.4(*c*) show the monthly and annual variations at selected stations held by PSMSL, and the natural variability against which trends must be detected. There are several ways of determining these monthly and annual mean sea levels from observations but it is now normal practice for authorities to calculate monthly and annual mean sea levels from digitized hourly values of observed sea level.

The most direct way of calculating monthly mean sea levels is to add together all the hourly values observed in the month, and then divide the total by the number of hours. The annual mean level can be calculated from the sum of the monthly mean levels, weighted for the number of days in each month. Any days for which hourly values are missing should be excluded, but only these days are lost in the analysis. Small errors may be introduced by the incomplete tidal cycles included at the end of the month. The tidal effects on monthly mean sea levels are best removed by applying a low-pass numerical filter to the hourly values to get a smooth daily noon value, before calculating the average of these values. Several filters have been developed for this (see for example, Pugh, 1987). The Doodson X_0 filter, which requires only 39 hourly values, is suitable for data containing occasional gaps, because not too much data is lost on either side of the gap. The monthly means differ insignificantly from the means calculated after applying longer filters. The standard deviation of the differences between the monthly means calculated at the Newlyn tide gauge over a year, applying a 168-hour filter, and the means calculated by the X_0 filter was 1.5 mm; for the differences between the 168-hour filter and the arithmetic means the differences had a standard deviation of 2.0 mm. Even if the data is free of gaps, there is little point in applying more elaborate filters for mean sea level studies alone, because the oceanographic variability is much greater than the differences due to applying different filters.

The mean tide level is the average of all the high and low water levels in a specified period (Pugh, 1987). To give an equal number of both high and low waters, the last maximum or minimum in a month may have to be omitted. Mean tide level is not the

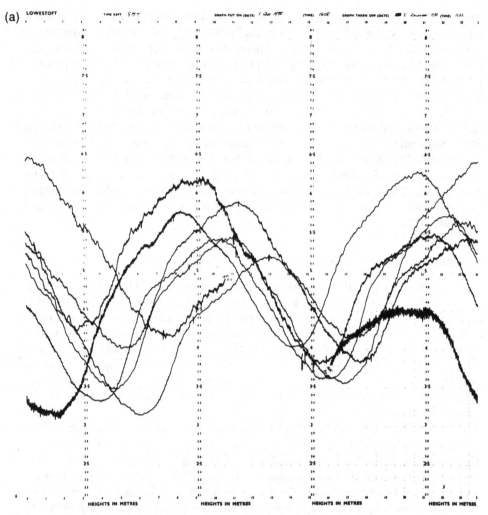

Fig. 4.4 The oceanic variability against which mean sea level trends must be identified (Pugh, 1987. Reprinted by permission of John Wiley and Sons, Ltd.). (*a*) An original tide-gauge chart from Lowestoft, UK, showing tidal variations and seiches. (*b*) Monthly mean values for 5 stations. (*c*) The annual means for 4 long-term stations held by PSMSL.

same as mean sea level because of the influence of shallow water tidal distortions, although variations in the two values are highly correlated; in extreme cases the difference may be 10% of the tidal range. Many of the very old estimates of average sea levels were computed as mean tide level because there were no continuously recording instruments available, but the practice is no longer acceptable.

The global sea level observing system (GLOSS)

Although measurements of sea level have been made regularly for over a century and have been collected and published by PSMSL, the recent concern about global

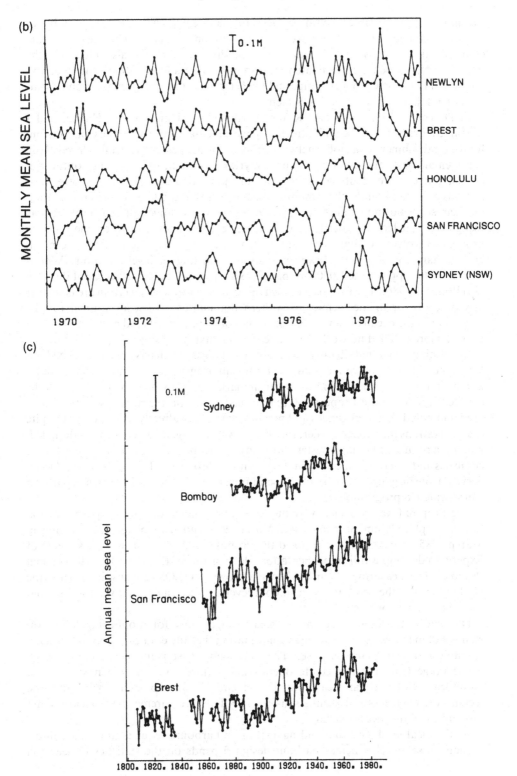

warming and the possible resulting sea level increases emphasizes the need for a better
co-ordinated network of gauges, working to common standards. The oceanographic
components of the World Climate Research Programme have also identified the
requirements for better co-ordinated measurements of sea level, as have the Intergovernmental Panel on Climate Change and the Intergovernmental Oceanographic Commission Global Ocean Observing System.

In December 1983, Professor Klaus Wyrtki of the University of Hawaii and I
prepared a Proposal, on behalf of the Intergovernmental Oceanographic Commission,
for the establishment of a global network of some 250 gauges, strategically placed along
continental coasts and on ocean islands (Wyrtki and Pugh, 1985). Along continental
coasts, gauges were identified at intervals of approximately 1,000 km, whereas for
stations on ocean islands, the minimum spacing was reduced to 500 km in recognition
of their strategic importance for monitoring global changes of sea level. It was
recognized that length-scales for coherent sea level changes depend on the physical
processes involved so that, for example, length-scales along the equator are much
greater than those across it. However, analysis of existing sea level variations (Woodworth, 1986) supports the 1,000 km spacing as a typical value for ocean variability (see
also Wunsch, 1986). The Proposal refined this coarse resolution in certain areas such as
across straits, or in the vicinity of western boundaries, which are critical for the
behaviour of the oceans as a whole. For studies of local vertical land movements at the
coast, a more detailed network may be required in many regions.

In selecting individual sites priority is given to gauges which have been installed for a
long time. All gauges are required to aim for an accuracy of 10 mm in level, and 1
minute in time. All must be linked to bench marks against which their datum is to be
checked regularly. The Proposal placed no restrictions on the instrumentation to be
used, but noted that float gauges and pressure gauges are already widely used. Despite
the reservations mentioned earlier, the stilling-well arrangement, if properly designed,
remains a robust and reliable system for many applications, particularly in developing
countries and where advanced technology is not appropriate. The US National Ocean
Service acoustic gauges have become operational at several coastal stations as part of a
world-wide US programme.

The Proposal was considered by the Executive Council of the Intergovernmental
Oceanographic Commission in Paris in March 1984, and again by the full Assembly in
March 1985. The Assembly approved the Proposal and established a Task Team of
Experts to develop a full Implementation Plan for the system, based on the original
Proposal. The measuring system has become known as GLOSS because it measures the
Global Level of the Sea Surface, a smooth level after averaging out waves, tides and
short-period meteorological events.

National authorities, reluctant to install gauges only for monitoring long-term
changes of mean sea level, welcomed confirmation that the data can be used for many
scientific and operational purposes. There are many reasons for measuring sea level.
These range from the immediate operational requirements of ship navigation to
monitoring global ocean heat content and circulation. However, hydrographers,
technicians, engineers and scientists may have different requirements for accuracy and
availability of the measurements.

At the local level, charting and navigation in harbours requires immediate information on sea level, whereas harbour design depends on the statistics of sea level

variations, measured over several years. Tidal prediction to facilitate ship movements depends on careful analysis of a long period of data, preferably at least a year. Coastal defences against flooding are also designed on the basis of long-term statistics. Datum definitions for both hydrographic charts and land surveys are based on analysis of long periods of sea level. Over long periods, planning for coastal zone management depends on long-term estimates of local sea level change.

On a regional scale, sea level measurements are required with minimum delay to give warnings of coastal flooding. The system established in the North Sea for forecasting flooding is a well-known example and similar systems exist on the East Coast of the United States. Numerical models, used in some cases to forecast flooding, need sea level measurements as an input. These models can also benefit from careful measurements of sea level gradients along their boundaries because these may allow water circulation in coastal seas to be determined. National and State Boundaries are defined in the coastal zone in terms of sea level statistics: in the USA 18.6 years (a nodal cycle in the tidal range) are often required for legal purposes and for National Ocean Service charts.

On an ocean scale, the Tsunami Warning System in the Pacific Ocean exists to warn against catastrophic local flooding, but the warning system, located in Hawaii, co-ordinates measurements from gauges throughout the whole ocean. In tropical latitudes the heat content of the upper layers of ocean, indirectly related to climate events such as El Niño, may be determined directly from sea levels; higher levels indicate high heat content. Lower sea levels in tropical regions indicate that the warm surface layer is relatively thin and that the colder deeper water, usually rich in nutrients, is near the surface. If so, there may be an upwelling of the nutrient-rich water and an enhancement of the fisheries stocks. The relationship between sea level gradients and ocean currents, in terms of geostrophic balance, is well-known and, properly applied, is a valuable tool for monitoring ocean circulation (see Fig. 4.2 and Introduction).

The Implementation Plan for GLOSS (IOC, 1990) emphasizes that the system provides high quality standardized data for international and regional research pro-grammes, as well as for the practical applications of national importance. The elements of GLOSS are:

1. The global network of permanent sea level stations to obtain standardized sea level observations; this forms the primary network to which regional and national sea levels can be related;
2. Data collection for international exchange with unified formats and standard procedures which may include near-real-time data collection;
3. Data analysis and product preparation required for scientific and/or practical applications;
4. Assistance and training for establishing and maintaining sea level stations as part of GLOSS and improving national sea level networks;
5. A selected set of GLOSS tide-gauge bench marks accurately connected to a global geodetic reference system (i.e., the conventional terrestrial frame estab-lished by the International Earth Rotation Service [IERS]).

The objective is to have an operational global network of permanent sea level stations reporting monthly mean averages to the PSMSL. This network will be the framework for other regional and scientific programmes such as the Tropical Ocean and Global Atmosphere (TOGA) Programme and the World Ocean Circulation Experiment

(WOCE). As programmes are developed and implemented, certain GLOSS stations will be upgraded to near-real-time transmission of data. The development of GLOSS is seen as a dynamic activity and the results of global oceanographic experiments, such as WOCE, will be used to adjust the number and location of gauges to meet operational and research requirements. The proposed distribution of gauges is shown in Fig. 4.5.

The first stage in the development of GLOSS is for participating countries to make a commitment to the Intergovernmental Oceanographic Commission to operate and maintain gauges in their country, as required by GLOSS in accordance with the Implementation Plan. Most of the major potential contributors have now made this commitment. For consistent measurements over long periods the important point is that the commitments are made by national authorities, such as a hydrographic department, a mapping agency, or a permanent research institution. Valuable data are being collected by individual research teams but without the guarantee of permanent national support. Many countries require training in measurement and analysis techniques. Annual courses have been run at the Proudman Oceanographic Labora-tory, UK, since 1983; numbers of participants are deliberately kept low, in order to maintain an emphasis on practical training. Successful courses were held in China in 1984 and in France in 1990. Further courses in Spanish and Russian are planned. The teaching material used at the UK course has been prepared in the form of a manual, published by the IOC, and translated by them from English into French, Russian and Spanish (IOC, 1985). This common training manual which will help to maintain common observational standards is to be updated in 1992.

One of the major difficulties has been to obtain new tide-gauge equipment for developing countries. Sweden has recently offered ten tide gauges and China has offered two. The United States NOAA has also made resources available for installing tide gauges in the Indian Ocean and elsewhere. Funding for measurements made as part of major scientific research programmes such as TOGA have allowed scientists to purchase and install equipment permanently at sites in developing countries in collab-oration with national authorities. For example, a scientist from Mauritius who attended the first UK course has participated in the installation of three gauges in the Indian Ocean and will be responsible for gathering and analysing the data: through this he will be a full participant in the TOGA and WOCE programmes.

The basic requirements for network gauges is that they supply monthly and annual sea level values to the PSMSL, but they are also required to make the hourly values of sea level available to interested scientists on request. A special centre for collecting and analysing observations has been established in Hawaii for TOGA and similar centres are being considered for WOCE. The data centres prepare several products for users; these include catalogues, newsletters, a chart of monthly anomalies in the Pacific and the Northern Atlantic and analyses of variability and sea level trends. Further products will be added as the data flow increases and the scientific requirements are defined. Although it is essential that close contact be maintained between all the data centres, authorities are encouraged to send their data to each centre individually.

Within the GLOSS system, an important component is the development of regional networks. Gauge distribution within these regional networks is more concentrated than in the global network, and they are expected to provide information which can be analysed in terms of local problems: for example, storm surges and statistics of sea level extremes. Regional components will allow for local exchange of data, ideas and equipment, and will also allow for training to be arranged to meet local needs.

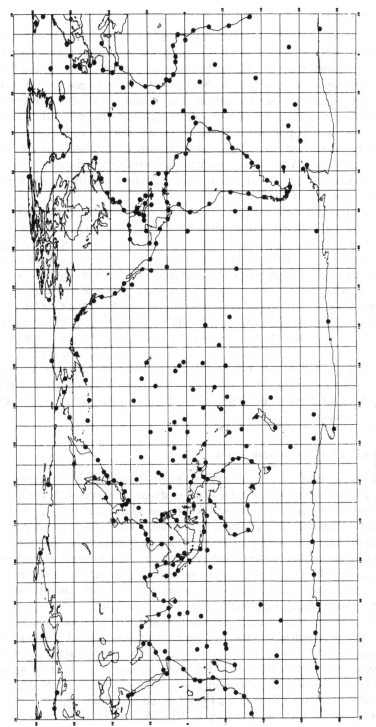

Fig. 4.5 Gauges proposed for inclusion in the GLOSS network.

Future developments

Any increase in mean sea level which has immediate practical consequences will be easily identified without recourse to detailed statistical analysis of the measurements. Nevertheless, it is important to consider how the data already collected and the data which will gradually accrue in future can be analysed in order to give early warning of increased rates of sea level rise. An early indication that the volume of water in the oceans is increasing would allow advance planning for coastal inundation, and may also trigger a social and political response which could avoid irreversible changes. Three possible areas of activity towards this goal are: consolidation of the GLOSS system; development of techniques for reducing the variability in the time series due to ocean dynamics; and application of new technology to fix tide-gauge bench marks in a global co-ordinate system.

Further development and consolidation of the GLOSS measuring network needs a commitment from a few key countries which have not already confirmed their interest, and a major effort from all participants to bring their tide gauges up to the required standard and to maintain them at that level. At present, it is not possible to say with certainty which of the gauges in the GLOSS network are representative of ocean conditions and which are severely influenced by coastal, estuarine or harbour characteristics. One of the advantages of collecting hourly data from the GLOSS network through the period of the World Ocean Circulation Experiment (1990–7) will be the ability to analyse the complete data set and to identify those gauges which are most representative of global conditions. While this would not exclude other countries whose gauges are part of the GLOSS activities, it would be sensible to give top priority to collecting and analysing the data from this subset of the network, probably enhancing the measurements of these sites with other meteorological and oceanographic measurements.

Reduction of variability in time series

One of the major difficulties in identifying trends is the high level of interannual variability. Part of this variability is due to ocean dynamic processes. At present, these processes can only be treated as statistical noise in the data. But if, through the WOCE studies and the development of global ocean numerical models, it becomes possible to identify and correct for the effects of ocean dynamics, then a much clearer signal will remain. This information is urgently needed. Some investigators (e.g. Barnett, 1984), who have examined the problem of identifying sea level trends against a background of year-to-year ocean variability, have even concluded that the existing data set is inadequate for determining a global rate of change of sea level.

One of the most direct adjustments for weather-related sea level variability is simply to add the mean atmospheric pressure to the mean sea level. Fig. 4.6 shows the high degree of correlation between annual mean sea levels at Newlyn and nearby atmospheric pressures. More elaborate models for removing monthly and annual variability which can be correlated with meteorological variables allow for both atmospheric pressure and wind effects (Thompson, 1980; Garrett and Toulany, 1982; Pugh and Thompson, 1986). Each of the gauges in the GLOSS network should be analysed so that these adjustments can be made to the long-term mean sea level series.

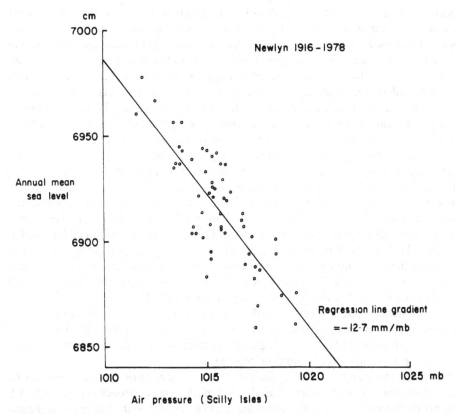

Fig. 4.6 The relationship between annual mean sea level and annual mean atmospheric pressure at Newlyn, UK. Adjusting for atmospheric effects can leave a clearer indication of long-term mean sea level changes (Pugh, 1987. Reprinted by permission of John Wiley and Sons, Ltd.).

Another method of removing, or at least reducing, variations of sea level due to ocean dynamics is to analyse the differences in the mean sea level from month-to-month or from year-to-year between pairs of stations. Fig. 4.4(*b*) shows a high degree of correlation between the monthly mean sea levels at Newlyn and at Brest. Wood-worth (1987) has used this simple but elegant technique to investigate possible different rates of vertical movement of the Tide-Gauge Bench Marks at paired stations. However, although this is a useful technique for looking at changes over length-scales which are short compared with the length-scales for ocean mean sea level variations, further examination of Fig. 4.4(*b*) shows that there is no coherence in the variations over separations comparable with ocean dimensions.

New technologies

One of the most promising developments for the next decade is the application of new instruments for fixing TGBMs in a global co-ordinate system. Carter *et al.* (1986; this volume) describe how Very Long Base Line Interferometry and the Global Positioning

System can be used to achieve this. Another independent check on crustal movements may be obtained from measurements using absolute gravimeters. A strategy for applying these new technologies has been considered by the International Association for the Physical Sciences of the Ocean. Now that the technology is capable of absolute measurements at the 2–3 cm level, general global-positioning surveys of TGBMs should be undertaken. If this is done, then future generations cannot criticize us as von Humboldt criticized Cook and Bougainville! The goal of locating TGBMs relative to the geoid is a more distant prospect, but precise gravity-measuring satellites are being considered which would make this a reality before the end of the century.

Whether the traditional methods of measuring sea level at the coast will eventually be superseded by new satellite technology remains an open question. At present satellite altimetry has the potential to measure sea levels to about 0.02 m after careful correction for atmospheric effects. However, these levels are relative measurements and the stability required for monitoring sea level changes of a few centimetres over several decades demands far greater absolute accuracy than is at present envisaged for satellite altimetry. Perhaps the first role for satellites is in detecting sea level change to identify changes in the total volume of polar ice; the surface area is approximately one thirtieth of the ocean surface area. Thus a change of 1.0 m in the average surface level of the ice would be equivalent to a change of 0.03 m in mean sea level.

Another new approach would be to develop permanent pressure measuring devices which could be located on the sea bed off-shore, away from local coastal effects on sea level; ideally these would be fixed in a global reference system and would transmit their data ashore, without further attention, over decades.

The basic requirement of measuring sea level in geocentric co-ordinates can be satisfied without reference to land-based bench marks. One approach is to mount GPS systems on moored buoys in the open sea (Rocken *et al.*, 1990). There are technical problems in terms of the mean buoy position relative to mean sea level, but in theory such systems could monitor sea level over long periods and transmit the information back for analysis. Movement of the buoy on its moorings within a circle of several hundred metres would be quite acceptable because changes in the geocentric co-ordinates of the geoid in the area would be small, and in any case could be mapped by monitoring the buoy position as it moved within the circle. Using this technology, measurements would be possible anywhere over more than 70% of the Earth's surface, rather than being confined, as at present, to the narrow coastal zone where potential problems of local sea level distortion exist.

These are formidable technical challenges, well beyond present capabilities. Nevertheless, the problem of identifying and responding to sea level changes will be with us for decades and even centuries ahead. We cannot be dogmatic about what will happen, nor about the scientific tools which will be available in the future.

References

Airy, G.B. (1845). On the laws of the tides on the coasts of Ireland, as inferred from an extensive series of observations made in connection with the Ordnance Survey of Ireland. *Phil. Trans. R. Soc. London*, **135**, 1–124.

Barnett, T.P. (1984). The estimation of 'global' sea level change: a problem of uniqueness. *J. Geophys. Res.*, **89**, 7980–8.

Carter, W.E., Aubrey, D.G., Baker, T.F., Boucher, C., Provost, C. Le., Pugh, D.T., Peltier,

W.R., Zumberge, M., Rapp, R.H., Schutz, R.E., Emery, K.O. & Enfield, D.B. (1989). *Geodetic Fixing of Tide-Gauge Bench Marks.* Woods Hole Oceanographic Institution Technical Report, WHOI-89-31, 44 pp.

Carter, W.E., Robertson, D.S., Pyle, T.E. & Diamante, J.M. (1986). The application of geodetic radio interferometric surveying to the monitoring of sea level. *Geophys. J. R. Astron. Soc.*, **87**, 3–13.

Dixon, J. (1979). Apparent sea level slopes – Ireland. *Chartered Land Surveyor, Chartered Minerals Surveyor*, **1**, 46–50.

Garrett, C.J. & Toulany, B. (1982). Sea level variability due to meteorological forcing in the Northeast Gulf of St. Lawrence. *J. Geophys. Res.*, **87**, 1968–78.

Hamon, B. (1985). Early mean sea levels and tides in Tasmania. *Search*, **16**, 274–7.

Intergovernmental Oceanographic Commission (1985). Manual on sea level measurement and interpretation. *IOC Manuals and Guides, No. 14.* UNESCO. (Available in English, French, Russian and Spanish).

Intergovernmental Oceanographic Commission (1990) *Global Sea Level Observing System Implementation plan* IOC Technical Series No. 35.

Mörner, N.A. (1973). Eustatic changes during the last 300 years. *Palaeog., Palaeoclim., Palaeoecol.*, **13**, 1–14.

Palmer, H.R. (1831). Description of graphical register of tides and winds. *Phil. Trans. R. Soc. London*, **121**, 209–13.

Pugh, D.T. (1987). *Tides, Surges and Mean Sea Level.* John Wiley, Chichester, 472 pp.

Pugh, D.T. (1982). A comparison of recent and historical tides and mean sea levels off Ireland. *Geophys. J. R. Astron. Soc.*, **71**, 809–15.

Pugh, D.T., Spencer, N.E. & Woodworth, P.L. (1987). *Data Holdings of the Permanent Service for Mean Sea Level.* PSMSL, Bidston, Birkenhead, 156 pp.

Pugh, D.T. & Thompson, K.R. (1986). The subtidal behaviour of the Celtic Sea – 1: Sea level and bottom pressures. *Continental Shelf Res.*, **5**, 293–319.

Rocken, C., Kelecy, T.M., Born, G.H., Young, L.E., Purcell, G.H. & Wolf, S.K. (1990). Measuring precise sea level from a buoy using the Global Positioning System. *Geophys. Res. Lett.*, **17**(12), 2145–8.

Ross, J.C. (1847). *A Voyage of Discovery and Research in the Southern and Antarctic Regions During the Years 1839–43.* John Murray, London, 447 pp.

Scherer, W.D. (1986). National Ocean Services' Next Generation Water Level Measurement System. FIG, International Congress of Surveyors, 1–11 June, Toronto, Canada, pp. 232–43.

Thompson, K.R. (1980). An analysis of British monthly mean sea level. *Geophys. J. R. Astron. Soc.*, **63**, 57–73.

van Veen, J. (1954). Tide-gauges, subsidence-gauges and flood-stones in the Netherlands. *Geologie en Mijnbouw*, **16**, 214–19.

Woodworth, P.L. (1987). Trends in U.K. mean sea level. *Marine Geod.*, **11**, 57–87.

Woodworth, P.L. (1986). A global sea level network: how many gauges are enough? *Tropical Ocean–Atmosphere Newsletter*, October 1986, 3–5.

Wunsch, C. (1986). Calibrating an altimeter: how many gauges is enough? *J. Atmospheric and Oceanic Technology*, **3**, 746–54.

Wyrtki, K. & Pugh, D.T. (1985). *Proposal for Development of an IOC Global Network of Sea Level Stations.* 13th Session of the Assembly, Intergovernmental Oceanographic Commission, Paris, 45pp. and 7 Annexes.

5

Global absolute sea level: the Hawaiian and US Atlantic coast-Bermuda regional networks

W.E. Carter, W. Scherer and J.M. Diamante

Abstract

The US National Oceanic and Atmospheric Administration (NOAA) and the Department of Energy (DOE) have jointly developed regional absolute sea level monitoring networks in the Hawaiian Islands and the US Atlantic Coast–Bermuda region. These networks were established to test and refine the instrumentation and methods for further developing a Global Absolute Sea Level Monitoring System. Conventional levelling, phase-differenced Global Positioning System (GPS) observations and Very Long Baseline Interferometry (VLBI) are being used to establish local, regional and global geodetic control to monitor temporal variations in the geocentric positions of tide-gauges with centimetre level accuracy. Absolute gravity measurements, precise to one microgal (equivalent to a 0.3 cm change in height) provide a relatively inexpensive independent check on vertical land motion measurements. The tide-gauge facilities are being upgraded to NOAA's newly developed Next Generation Water Level Measurement System (NGWLMS). This will improve the accuracy of the tidal records and collect important ancillary data, such as wind speed and direction, barometric pressure, salinity, and air and water temperatures. The ultimate goal is to assist in the establishment of an internationally operated system that is sufficiently accurate to determine conclusively the rate of change in absolute sea level.

Introduction

Existing tide-gauge records indicate a rise in global sea level of about 10–15 cm during the past 100 years (see Woodworth, this volume). There are, however, significant questions about the adequacy of the tide-gauge measurements and the sampling obtained with the current number and distribution of the stations (Pugh, this volume). Even in regions where numerous, well-maintained tide-gauge stations have been carefully operated for decades, such as along the Atlantic Coast of the United States, the apparent rate of rise in sea level varies widely among individual gauges and among groups of gauges (Barnett, 1984; Aubrey and Emery, this volume). At least some portion of the scatter in rates is almost certainly due to vertical land motions which, on global scales (the effects of glacial rebound actually causes crustal deformation of the entire globe), are comparable to the estimated long-term rise in sea level, and on local and regional scales are often an order of magnitude larger. If we are to refine our estimate of the rate of rise of sea level and determine if that rate is increasing with time, we must extract the effects of the vertical land motions from the tide-gauge records.

Recent advances in geodesy now make it possible to do this by connecting the tide gauges to a global geodetic reference frame with centimetre level accuracy. The most important advances have been in the development of VLBI and phase-differenced observations of the GPS satellites. Also, improved absolute gravity meters capable of microgal repeatability (one microgal is equivalent to approximately 0.3 cm difference in height) may provide a relatively inexpensive and independent method of checking the VLBI/GPS measurements.

Geodesy can also contribute to understanding changes in global sea level in a less direct manner. The long-term global scale deformations of the Earth associated with glaciation-deglaciation, which must be accounted for in analysing the tide-gauge records, may be treated in the immediate future most efficiently by use of a model such as that developed by Peltier (1986). The deformations assumed in such models imply rates of change in the length-of-day and secular motion of the rotation axis that must be consistent with the observed values. The introduction of VLBI has improved both the accuracy and temporal resolution of the Earth orientation measurements by more than an order of magnitude during the past few years, and these improved measurements will place increasingly tighter constraints on Earth models (for further discussion see Carter *et al.*, 1986).

The new geodetic techniques can make another contribution to the monitoring of global sea level. Satellite radar altimeters have demonstrated a capability for monitoring the global sea surface (see, for example, Douglas *et al.*, 1987). Sea surface measurements from conventional sea level gauges positioned on islands, combined with island-based satellite laser ranging, already have been successfully employed in calibrating satellite radar altimeter instruments in orbit. Now, VLBI/GPS positioning offers the prospect of removing the longest wavelength errors from satellite-derived sea surface topography and relating the time sequence of such altimetric surfaces to the same land-based VLBI/GPS geodetic system in which the absolute sea level gauges are referenced, even across multiple satellite missions. Hence, complementary satellite altimeter and sea level gauge measurements can provide long-term monitoring of global sea level changes, related to a common global geodetic reference system.

The purpose of this chapter is to report on the development of an Absolute Sea Level Monitoring System. It will focus on a joint US NOAA and the DOE pilot programme to establish two regional absolute sea level networks, in Hawaii and along the Atlantic Coast of the United States and Bermuda, which take advantage of new geodetic techniques and improved measurement systems.

Radio interferometric surveying – VLBI and GPS

In geodetic VLBI (Carter and Robertson, 1986), radio telescopes (typically 10–30 m in diameter) separated by as much as several thousand kilometres, simultaneously track the same extragalactic radio source (quasar). Each observatory records the time of receipt of the microwave signals, with respect to independent local clocks, on magnetic tapes. These tapes are sent to a correlator centre, where they are replayed and the differences in the arrival times of the signals at the various stations (delays) and the time rate of change of the delays (delay rates) are determined. A typical observing session is

24 hours in length and a few hundred delays and delay rates distributed among 10–15 sources are amassed. Using least squares analysis, parameters such as the clock offsets and rates, source co-ordinates, and the components of the vectors between observatories can be estimated to the centimetre level.

In the phase-differenced GPS technique (Askenazi *et al.*, 1985) the signal transmitted by the NAVSTAR satellites (a constellation of satellites that will number 18 to 24 when the system is fully operational in 1991) are treated in much the same manner as the quasar signals in VLBI. However, because the satellite signals are many orders of magnitude stronger than those received from the quasars, relatively inexpensive, compact, highly transportable receivers are used. The GPS system is not as suitable as VLBI for connecting stations on global scales, i.e., over thousands of kilometres, because errors in the orbits limit the accuracy that can be achieved. But over distances of up to a few hundred kilometres, GPS is competitive in accuracy with VLBI, and the reduced costs make it possible to do many more stations. By combining the two techniques with classical levelling, a coherent global geodetic reference system can be developed at tolerable costs.

Geodetic positioning of tide gauges

Economic and technological considerations require that geodetic positioning of tide gauges be accomplished on three spatial scales: local, regional and global. All tide-gauges that are to be used to monitor global sea level must be connected to local level networks that are properly monumented and regularly re-surveyed – annually if possible. These level networks should have a minimum of 6 to 10 bench marks and extend over sufficient area to minimize the chances that they will be destroyed by local engineering projects or by natural causes. Care must be taken in designing the networks to take into account such factors as the geological setting (proximity of faults, depth of bedrock, type of soil), climatic factors (frost line, flooding), and historical engineering works (land fills, drainage systems). It may often be necessary to extend a network out to a few kilometres to tie the gauge to monuments that are less likely to be disturbed by coastal or harbour construction and public works projects and that better represent the inherent stability of the region. At least one of the bench marks in the local levelling network needs to be located at a site suitable for GPS observations, e.g., the horizon should be free of obstructions above 10 to 15 degrees elevation.

Tide gauges should be organized into regional networks, and GPS surveys need to be conducted to determine their relative positions with an accuracy of 1 cm or better. The extent of the regional networks depends on several factors, perhaps the most important being the proven range of accurate GPS surveys. This will change as the GPS system matures and improved satellite ephemerides and observational procedures are developed. Initially, the project worked with a small constellation of test satellites of the GPS and a limited set of tracking stations that restrict accurate surveys in most regions to a range of a few hundred kilometres. The impact of these constraints will be discussed later when the designs of the Hawaiian and Atlantic Coast–Bermuda surveys are presented.

Finally, the regional networks should be tied into the global VLBI network. Fig. 5.1 shows the locations of existing and planned permanent VLBI observatories which form the global network. When the distances are not too great, GPS can be used to tie

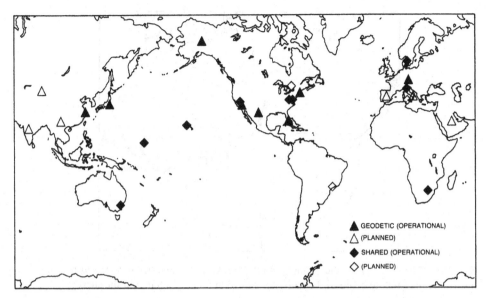

Fig. 5.1 The locations of existing and planned VLBI observatories. Those marked with triangles are dedicated geodetic VLBI facilities while those marked with diamonds are used for both geodetic and astrometric observing programmes.

directly to a fixed VLBI observatory. At greater distances, it may be necessary to use a mobile VLBI unit to establish a station nearer the tide gauge which, in turn, can be used for a local GPS survey, or the VLBI station may be tied directly to the tide gauge by conventional levelling.

The Hawaiian network

The Hawaiian Islands were chosen as one of the pilot regional networks because they present many of the challenges that will have to be met in positioning island tide-gauges over the large scales of the ocean basins and that are believed to be essential for monitoring global sea level. Also, Hawaii already possesses vital related installations that made it conducive to early installation of the pilot regional network, thereby yielding possible early payoffs. The National Aeronautics and Space Administration (NASA) operates a VLBI observatory at Kokee Park, Kauai, and a satellite–lunar laser ranging station at Haleakala, Maui. These stations are part of the NASA Crustal Dynamics Project, and are used to study the motion of the Pacific plate relative to other tectonic plates. The Navy Astronautics Group (NAG) operates a Doppler satellite tracking station at Wahiawa, Oahu, that regularly tracks oceanographic satellites equipped with microwave altimeters used to study variations in the sea surface. For decades, NOAA has operated tide gauges throughout the Hawaiian Islands, and has already begun to upgrade a number of the stations by installing the Next Generation Water Level Measurement System (NGWLMS) (Scherer, 1986).

The most serious difficulties encountered in implementing the Hawaiian regional network were caused by the fact that the constellation of GPS satellites is optimized for positioning within the continental United States, and that sufficiently accurate satellite

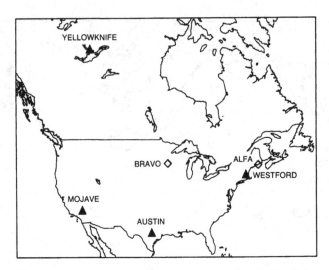

Fig. 5.2(*a*) The locations of the permanent GPS tracking stations (triangles) used to determine the satellite orbits and two NAG sites (diamonds) included in the survey.

orbits are not routinely available. Another concern was that the VLBI and laser ranging stations are at the summits of mountains (thousands of metres higher than the tide stations) and the differences in the atmospheric effects, particularly water vapour, could significantly degrade the GPS relative positioning accuracy. The solution to these problems, tested during the 1987 initial epoch survey, was to operate a set of 'fiducial' stations (i.e., GPS receivers placed at accurately known locations) throughout the surveys, enabling the project to improve the satellite ephemerides, to collect meteorological data (including water vapour radiometer observations) to correct the GPS observations, and to make several measurements at each tide gauge. The NOAA/DOE project has been able to implement this solution only because the US Geological Survey (USGS) and the NAVSTAR University Consortium (UNAVCO) have complementary observing programmes which have worked cooperatively with the project.

The first survey was conducted during the period March 30 to April 12, 1987. Fig. 5.2(*a*) shows the locations of the fiducial stations in the continental United States and Canada, as well as two additional NAG stations, that were observed during the period of the Hawaiian GPS survey. Fig. 5.2(*b*) shows the locations of the fiducial stations and tide-gauge stations in Hawaii. Analysis of the data indicates that the intra-island GPS connections, i.e., the connections between the fiducial station and tide gauges on any particular island, have an RMS scatter of about 1 cm. The inter-island connections, which are much longer, reaching lengths in excess of 500 km, scatter several times as badly. At least part of the problem with the inter-island connections is that the continental fiducial stations experienced technical problems, and on some days, there were not sufficient data to compute improved orbits by the methods currently being used. Subsequently, better methods were developed and a significant improvement in the results was seen when the data were reprocessed. The survey was repeated in 1988, again in cooperation with USGS and UNAVCO.

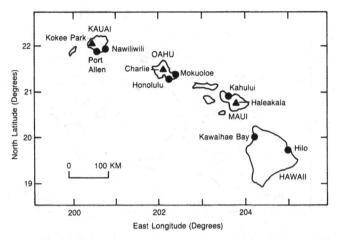

Fig. 5.2(*b*) The locations of the regional 'fiducial' stations (triangles) and the gauges (circles) of the Hawaiian network.

The Atlantic coast–Bermuda GPS survey

Fig. 5.3(*a*) shows the fiducial stations used and the tide-gauges visited during the Atlantic Coast-Bermuda GPS survey, conducted during June 1987. Fig. 5.3(*b*) is an enlargement of Bermuda, showing the locations of the three stations included in that part of the survey. Along the Atlantic Coast, there are three fixed VLBI observatories: two operated by NOAA as part of the Earth orientation monitoring programme (in Westford, Massachusetts and Richmond, Florida) and one, the Maryland Point Observatory, operated by the Naval Research Laboratories (NRL). The Westford Observatory also serves as a permanent GPS tracking station. A GPS receiver was located at the Richmond Observatory throughout the survey, and in November 1987, Richmond became a permanent GPS tracking station. Because there are no fixed VLBI observatories in Bermuda, a site that is suitable for the operation of a mobile VLBI unit was selected and included in the GPS survey. Mobile VLBI measurements were subsequently made and are discussed below.

The Bermuda GPS survey was conducted in four observing sessions, using three receivers simultaneously to get the best possible relative locations of the stations within Bermuda and of Bermuda itself relative to the fiducial stations in Massachusetts and Florida. Preliminary reductions within Bermuda indicate a repeatability over few-kilometre-long lines to be a fraction of a centimetre.

Bermuda mobile VLBI measurements

During August 1987, NOAA used a mobile VLBI unit to determine the location of a station in Bermuda. A total of four 24–hour observing sessions were conducted using the mobile unit and the Westford, Maryland Point, and the Richmond fixed observatories. Water vapour effects in the microwave region of the spectrum are the largest remaining source of error in VLBI measurements. Water vapour radiometers (WVR)

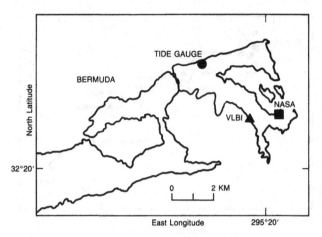

Fig. 5.3(a) The locations of the permanent GPS tracking and regional fiducial stations
(triangles) and the tide-gauge stations (squares) included in the Atlantic Coast–Bermuda
network.

were operated at Westford, Richmond, and Bermuda. Preliminary reductions of the
Bermuda station, without taking into account the WVR data, have a repeatability in
the latitude and longitude co-ordinates of 2.0 cm, while the RMS scatter in the height is
5.0 cm.

Absolute gravity measurements

Recent advances in the development of an absolute gravity meter at the Joint Institute
for Laboratory Astrophysics (JILA), University of Colorado, have produced a port-
able unit with a nominal accuracy of approximately five microgals and a repeatability
of one microgal. Working cooperatively with the Defense Mapping Agency (DMA),
NOAA is using the meter to establish a national absolute gravity reference system to
support, among other applications, geophysical and geodynamical research (Peter *et
al.*, 1986). Absolute gravity measurements are proving to be a relatively inexpensive
and independent method to check the vertical crustal motions measured at tide-gauge
stations with VLBI and GPS. Potentially, gravity measurements may be capable of
resolving vertical crustal motions of 0.3–1.0 cm, that is, if the effects of mass redistribution (particularly the subsurface mass changes caused by seasonal variations in the
local water-table and soil moisture content) are adequately accounted for. There are
two possible approaches to this problem: judicious selection of sites (e.g., locating the
stations on exposed bedrock), or monitoring the changes in the groundwater-table and
soil moisture content (this requires drilling wells at the station and using a neutron
probe to sample the moisture content in the upper levels of the soil). Variations in the
distribution of the atmosphere and the ocean masses must also be taken into account.
While accounting for these redistributions of mass complicates the interpretation of
absolute gravity measurements, it was concluded that the potential returns warranted
the investment of the required resources, and absolute gravity measurements were
included in the Hawaii and Atlantic Coast-Bermuda projects.

The following sites for absolute gravity stations along the Atlantic Coast were

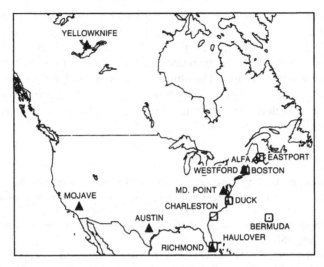

Fig. 5.3(*b*) An enlargement showing the locations of the mobile VLBI (triangle), the gauge (circle) and laser ranging facility (square) in Bermuda.

selected: Richmond, Florida; Atlanta, Georgia; Charlottesville, Blacksburg, Great Falls Park, and Herndon, Virginia; Boston, Massachusetts; and Bangor, Maine. A station was also selected in Bermuda and on the islands of Kauai, Oahu, and Maui, Hawaii. The Atlanta, Blacksburg, Great Falls Park, Herndon, and Bermuda stations were surveyed and the three sites in Hawaii were re-surveyed during October 1987; the remaining Atlantic Coast stations were re-surveyed in 1988. Reductions of repeat measurements at the Great Falls Park and Herndon stations agree with the intial epoch values within 0.2–1.7 microgals.

The next generation water level measuring system (NGWLMS)

The prospect of relating sea level measurements to a global geodetic reference system which is accurate to the centimetre level introduces the requirement for co-located measurements of oceanographic and meteorological variables that affect the sea level signal at a particular location. Consequently, there is a need to provide for the measurement of wind speed and direction, barometric pressure, salinity, and air and water temperature at the global sea level stations. New gauges of the acoustic type, as well as data collection platforms, ancillary meteorological and oceanographic sensors and telemetry, and newly designed protective tide-gauge wells have been developed as part of the NGWLMS and deployed. A single measurement, the mean of 181 samples collected over a 3 minute period, from the NGWLMS is accurate to approximately 1 cm, about an order of magnitude improvement over the existing measuring system.

Concluding remarks

In the future, the Atlantic Coast–Bermuda regional network will be expanded to include several more tide gauges, perhaps including a number in the Caribbean islands.

A new tide-gauge facility, in a more open location, has been planned for Bermuda. NGWLMS units will be installed as quickly as they become available. As the resources become available, the NOAA programme will be expanded to include additional regional networks, and efforts will be made to work with other countries to make the technology more widely available. The ultimate goal is to assist in the establishment of an internationally operated Global Absolute Sea Level Monitoring System that is sufficiently accurate to determine conclusively the rate of change in sea level, and if that rate of change is increasing with time.

References

Askenazi, V., Agrotis, L.G. & Yau, J.H. (1985). GPS interferometric phase algorithms. *Proceedings First International Symposium on Precise Positioning with the Global Positioning System, Rockville, MD, May, 1985*, US Department of Commerce, Vol. II, 299–314.

Barnett, T.P. (1984). The estimation of 'global' sea level change: a problem in uniqueness. *J. Geophys. Res.*, **89**, 7980–8.

Carter, W.E. & Robertson, D.S. (1986). Studying the earth by very-long baseline interferometry. *Sci. Am.*, **254**, No. 11, 46–54.

Carter, W.E., Robertson, D.S., Pyle, T.E. & Diamante, J.M. (1986). The application of geodetic radio interferometric surveying to the monitoring of sea level. *Geophys. J. Royal Astron. Soc.*, **87**, 3–13.

Douglas, B.C., McAdoo, D.C. & Cheney, R.E. (1987). Oceanographic and geophysical applications of satellite altimeters. *Rev. Geophys.*, **25**(5), 875–80.

Peltier, W.R. (1986). Deglaciation-induced vertical land motion of the North American Continent. *J. Geophys. Res.*, **91**, 9099–123.

Peter, G., Moose, R.E. & Beruff, R.B. (1986). New US absolute gravity programme, *EOS*, **67**, 1393.

Scherer, W.D. (1986). National Ocean Service's New Generation Water Level Measurement System. FIG, International Congress of Surveyors, June 1–11, Toronto, Canada. 4, 232–43.

6

Long term changes in eustatic sea level

M.J. Tooley

Abstract

The concept of eustasy is explored and its history is selectively reviewed. The implications of geoid topography on world-wide changes of sea level of the same magnitude are considered, and the range of variations of sea level records confirms the conclusion that the world's oceans can be divided up into regions with distinctive sea level histories. Sea level data have suffered from a lack of rigorous collection, analysis and screening, and stress is laid on the need for the application of a uniform methodology which will allow correlations within and between areas. A research methodology for sea level investigations is proposed. The evidence for and nature of sea level change is considered and evidence is advanced for periods of rapid sea level rise associated with the catastrophic melting of ice caps and discharge of meltwater: rates in excess of 20 mm/yr have been recorded compared with 1–2 mm/yr over the past 100 years. The linkages with climatic change are explored and a graph is presented showing the relationship between sea level, temperature and carbon dioxide during the past 160,000 years.

Introduction

The definition of eustasy as a uniform change of sea level throughout the world has been questioned. This has resulted from the theoretical and empirical discovery of deformations of the geoid relief and the effects of the distribution of density of earth materials on sea level altitude following the redistributions of water masses over the Earth during a glacial–interglacial cycle (Mörner, 1976, 1987).

Explanations of sea level changes comprise changes in ocean-basin volume, ocean-water volume, ocean-mass distribution and dynamic sea level (Mörner, 1987). These explanations are appropriate at different temporal and spatial scales, but in the history of sea level investigations a single explanation has been pre-eminent at one time or another. Diastrophism was a popular explanation for changes in sea level in the nineteenth century. Glacial eustasy became popular in the succeeding century, although its recognition was delayed (Meyer, 1986).

In recent decades, the complex factors affecting the recorded altitudes of the sea level surface have been recognized by many. Nonetheless, at different spatial and temporal scales the existence of a globally valid sea level curve and an overriding control of sea

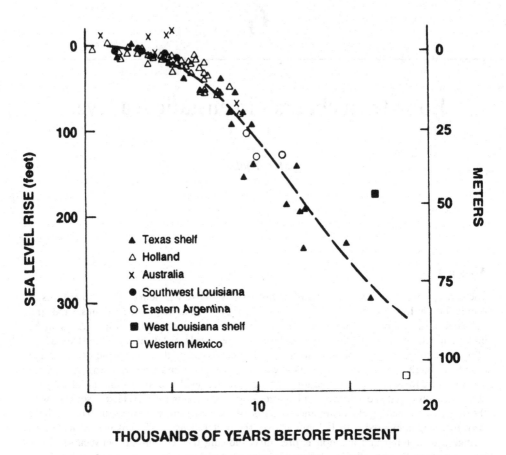

Fig. 6.1 A time/altitude graph, adapted from Shepard (1963), employing sea level index points from 'relatively stable areas' throughout the world (Reprinted from *Responding to Changes in Sea Level*, 1987, with permission from the National Academy Press, Washington, DC).

level altitude continues to be promoted. For example, Gornitz and Lebedeff (1987) refer to an average eustatic sea level rise for the past century, although they note large regional variations. Colquhoun and Brooks (1986) describe significant eustatic components in sea level changes recorded from the southeastern coast of the USA. The National Research Council (1987) has adapted Shepard's (1963) sea level graph to show a sea level curve which is assumed to be globally coherent (see Fig. 6.1). In China, eustatic sea level curves are *de rigueur* (e.g., Yang and Xie, 1984).

It is perhaps opportune therefore to re-iterate the statements on eustasy and the explanations of sea level changes that have been made during the last decade within a selective historical context. Fuller reviews have been given by Devoy (1987a), Fairbridge (1961), Kidson (1982), Mörner (1987) and Tooley (1987). Rates of sea level change and the coupling of sea level and climate will be considered, and the statement by Smith (1981) that 'many sea level changes (for instance in the Quaternary) take place during a much shorter time-scale than can be resolved by any known correlation technique' will be addressed. The time period for these considerations will be the Late Quaternary, although reference will be made to sea level changes in pre-Quaternary times.

Eustasy in the history of sea level investigation

Eustasy has been defined by Visser (1980) as 'a world-wide change in sea level'. This definition follows that of Gary *et al.* (1972) in the *Glossary of Geology*, who defined eustasy as 'the world-wide sea level regime and its fluctuations, caused by absolute changes in the quantity of sea water'.

Such definitions originate from Suess (1888) who described eustatic movements as changes affecting the level of the strand. The changes took place to an approximately equal height, whether in a positive or negative direction over the whole globe. Suess concluded that there were two kinds of eustatic movements: spasmodic negative movements caused by the formation of sea basins and a subsidence of the Earth's crust; and positive movements caused by the accumulation of marine deposits.

Daly (1934) elaborated on the causes of sea level change and supplemented Suess's explanation with the following: the addition of new or juvenile water through erosion and volcanic activity; the abstraction of water during sedimentation; the 'robbery and restitution' of water during a glacial/interglacial cycle; the displacement of the Earth's centre of gravity (first harmonic); and changes in the speed of the rotation of the Earth (second harmonic). Daly grouped these causes into two classes of change: primarily local and primarily eustatic. The former involved local uplift and subsidence as the dominant factor. The latter involved positive and negative changes affecting the World Ocean. However, Daly elaborated on this latter class by drawing attention to the fact that eustatic sea level is displaced rhythmically in hemispheres or alternating zones of the Earth as a result of the displacement of the centre of gravity and changes in the speed of rotation of the Earth. In allowing for unequal changes in eustatic sea level, Daly anticipated by about forty years conclusions that redirected sea level investigations at the opening of the last quarter of this century (Walcott, 1972, 1974; Mörner, 1976; Farrell and Clark, 1976).

In the intervening years, several investigators addressed the problem of eustatic sea level change and produced different solutions. Fairbridge (1961) employed a variety of sea level index points both from the coastline and continental shelves of the world and plotted them on a time/altitude graph from which an oscillating eustatic sea level curve was derived. This curve was modified in 1976 and some of the oscillations corroborated by dates from shell middens on the Brazilian coast and the dated Littorina transgressions in Blekinge, Southern Sweden (Berglund 1971). Further updating and refinement resulted in a third curve (Fig. 6.2) published in 1977 (Fairbridge and Hillaire-Marcel, 1977) incorporating data from North-West England (Tooley, 1974) and republished as the Richmond Gulf eustatic record by Hillaire-Marcel and Fairbridge (1978).

Shepard (1960) summarized the limitations and uncertainties of the sea level data base, and restricted sampling to bivalves from the Texas shelf (Fig. 6.3). The resulting eustatic curve on the time/altitude graph contrasted markedly with Fairbridge's curve, with its high-amplitude and high-frequency oscillations, although Shepard (1960, 1963) did allow for high-amplitude and low-frequency oscillations which he correlated with glacial interstadial events.

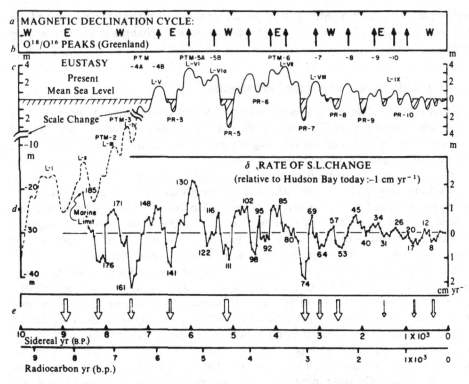

Fig. 6.2 Fairbridge's eustatic oscillations (*c*) enhanced with sea level data from Sweden (PTM-2 to PTM-10, and PR-3 to PR-10, Mörner, 1969) and Northwest England (L-I to L-IX, Tooley, 1974) and compared with the rate of sea level change relative to present day sea level in the Hudson Bay (*d*). These variations are compared with $^{18}O/^{16}O$ from the Camp Century ice core (*b*) and magnetic declination cycles (*a*). (Fairbridge & Hillaire-Marcel, 1977. Reprinted by permission of *Nature*, **268**, Copyright © 1977 Macmillan Magazines Ltd.).

Jelgersma (1961) applied an objective, unified methodological approach to derive a graph showing relative changes of sea level from a restricted geographical area, using data from the basal peats overlying Pleistocene sands beneath the coastal lowlands and dunes of The Netherlands. This graph has been only slightly modified, although it has been extended back into time by using data from the continental shelf (Jelgersma, 1979), thereby mixing data with different post-depositional histories, but sustaining a continuous sea level rise curve. Jelgersma (1966) argued that the lithostratigraphy of the world's coastal lowlands, rather than a derived sea level curve, would contribute to an understanding of climatic fluctuations. Wet periods are associated with transgressional facies and dry periods with regressional facies.

Subsequently, sea level investigations cleaved either to one solution or the other, as described by Jelgersma (1966) and addressed by Kidson (1982). Both solutions have utility. Fairbridge's compilations include climate change summaries, and the correlations between climate and sea level are explicit and persuasive. However, Fairbridge (1961) noted that although one of the causes of eustatic sea level change during the Quaternary was a climatically controlled glacial eustasy, geodetic change associated with either the shape of ocean basins or the shape of the geoid is a second major effect.

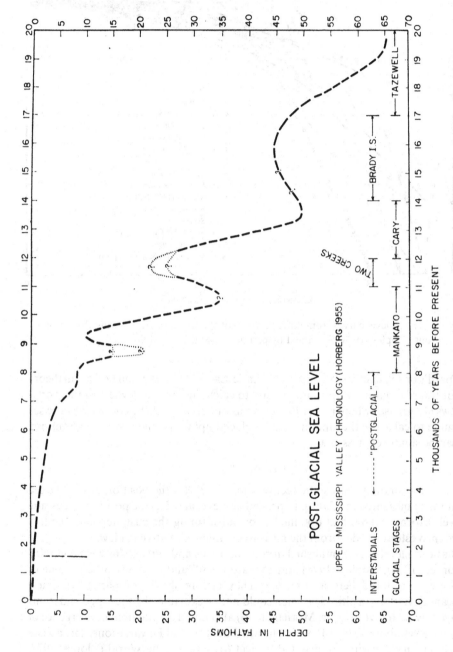

Fig. 6.3 A time/altitude graph showing apparent fluctuations of sea level based on evidence from the Texas and West Louisiana continental shelf (Shepard, 1960).

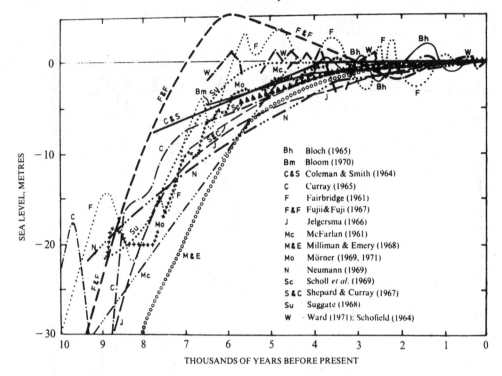

Fig. 6.4 Fifteen sea level curves from different parts of the world ocean showing the range of interpretations (Hopley, 1978. Reprinted by permission of the Royal Society).

Shepard's or Jelgersma's exponential eustatic sea level curves could be used in theory without transformation of the data directly to calculate uplift anywhere in the world. Andrews (1970) used Shepard's (1963) curve as an *estimate* of the correct eustatic sea level curve to calculate the amount of post-glacial uplift and the rate of sea level rise. This can be simply expressed as

$$U_t = R_t - E_t$$

where U is the amount of isostatic recovery in time t, R is the position of relative sea level at time t measured in relation to present sea level and E is the position of eustatic sea level. Mörner (1969) had obtained U by substituting the eustatic values he had derived in Western Sweden from the measured values of relative sea level.

Eustatic sea level curves published during the 1960s and early 1970s were compiled by Mörner (1976), Hopley (1978) and Everard (1980) and by many others, some of which are reproduced here as Fig. 6.4. They clearly show (for particular times) significant deviations in altitudes which could not be explained by the range of datums employed by each investigator. Mörner (1976) also noted an apparent convergence of sea level curves about 7,000 BP, which he attributed to geoidal variations. The atlases of sea level curves arising from IGCP Project 61 covering the World (Bloom, 1977, 1982), the scholarly World Atlas arising from IGCP Project 274 (Pirazzoli, 1991) and the regional atlases (e.g., for Japan, Ota *et al.*, 1981) reinforced the conclusion, first posited by Mörner (1976), that 'regional' eustasy, rather than 'global' eustasy, was a more appropriate description for the range of sea level curves of Holocene Age, and that regional eustasy was explained by diversity of interrelated phenomena, including changes in the geoid configuration, amongst other eustatic variables.

Eustatic variables

Donovan and Jones (1979) have described possible causes of global sea level changes, whereas Fairbridge (1961) identified four categories of processes that will affect the eustatic sea level:

a. Tectono-eustasy, a movement of world-wide sea level controlled by the deformation of the ocean basins. The downwarping of deep ocean basins could arise from sediment loading, an increase in water load following deglaciation and geodetic changes resulting from a redistribution of mass on the Earth and consequential change in gravity and the rate of rotation.

b. Sedimento-eustasy, a movement controlled by the addition of sediments of either pelagic or terrigenous origin (one of Suess' original explanations).

c. Glacial eustasy, a climatically controlled movement described by Daly as the 'robbery and restitution' of water during an interglacial/glacial cycle.

d. Steric change, an expansion/contraction of the mass of seawater and the consequential rise/fall of sea level, caused by changes in ocean density due to changes in temperature or salinity. Fairbridge also referred to the addition of juvenile water from volcanoes under this heading.

Mörner (1987) has drawn together all the eustatic variables in a useful table (Table 6.1) and redefined eustasy as absolute sea level changes regardless of causation and including both the main family of vertical and horizontal geoid changes and dynamic changes. He concluded that sea level changes could no longer be regarded as either world-wide, or simultaneous or parallel, and that in each region of the world ocean, it was necessary to define a regional eustatic curve, which could then be employed in comparison with others for resolving problems related to the strength and behaviour of the crust and mantle.

Mörner (1976) drew particular attention to the geoid, its variations in space and time and the implications for the registration of sea level at the same time in different parts of the world. Devoy (1987b) pointed out that in sea level literature, the use of the term palaeogeoid is misleading, in that it assumes that the geoid must be unstable spatially and temporally, whereas this has not been demonstrated empirically.

The distribution of geoid relief (Fig. 6.5) has been obtained using satellite orbit data from Goddard Earth Model 10³ (Marsh and Martin, 1982). This shows a range of altitudes, measured in relation to the best-fitting ellipsoid of the Earth, of -104 m south of India to $+74$ m over Irian Jaya and Papua-New Guinea. This range is similar to the measured differences between the low relative sea levels of the last glacial age and high relative sea levels of this interglacial age (Tooley, 1978a), although it has been argued rather surprisingly that the last glacial sea level geoid exhibits a relief of more than 360 m, which exceeds the highest estimated glacio-eustatic rise by a factor of two (Pirazzoli, 1988). This argument is based on an incorrect definition of the palaeogeoid. The range of altitudes for late glacial sea level in, for example, Northwest Europe (Newman and Bateman, 1987) has been caused by uneven isostatic uplift.

Table 6.1. *A summary of the eustatic variables according to the new definition of eustasy as ocean-level changes (Mörner, 1987. Reprinted by permission of Basil Blackwell)*

					EARTH-VOLUME CHANGES	
E U S T A S Y — OCEAN LEVEL CHANGES — VERTICAL AND HORIZONTAL GEOID CHANGES	OCEAN BASIN VOLUME	TECTONO-EUSTASY	TECTONICS	OROGENY	
				MID-OCEANIC RIDGE GROWTH	
				PLATE TECTONICS	
				SEA FLOOR SUBSIDENCE	
				OTHER EARTH MOVEMENTS	
			ISOSTASY	SEDIMENT IN-FILL	
				LOCAL ISOSTASY	
				HYDRO-ISOSTASY	
				INTERNAL LOADING ADJUSTMENT	
	OCEAN WATER VOLUME			GLACIAL EUSTASY	
				WATER IN SEDIMENT, LAKES AND CLOUDS, EVAPORATION, JUVENILE WATER	
	OCEAN MASS/LEVEL DISTRIBUTION	GEOIDAL EUSTASY		GRAVITATIONAL WAVES	
				TILTING OF THE EARTH	
				EARTH'S RATE OF ROTATION	
				DEFORMATION OF GEOID RELIEF (DIFFERENT HARMONICS)	
	DYNAMIC SEA LEVEL CHANGES (DYNAMIC CHANGES)			METEOROLOGICAL	
				HYDROLOGICAL	
				OCEANOGRAPHIC	

The distribution of geoidal lows and highs is meridional and appears to be associated with the global pattern of positive and negative gravity anomalies, shown in Fig. 6.6 (Fifield, 1984). Positive gravity anomalies, where there is a concentration of more dense Earth materials in the crust and mantle, are associated with geoidal highs, and negative anomalies with geoidal lows. The negative anomalies over Laurentide and Fennoscandia are associated with areas of former glaciation, and gravitational equilibrium will return when relaxation is complete and the crust returns to its shape and density prior to the last glaciation.

The abstraction and return of water from and to the world's ocean basins during a glacial/interglacial cycle will have an effect on the shape of the geoid, as well as on the eustatic sea levels, through hydro-isostatic, geodetic and steric processes. Walcott (1972) described changes in the global distribution of sea level and identified three regions within which the behaviour of sea level during the last 18,000 years varied. The first region was one in which rapid uplift occurred following deglaciation; the second region was one in which submergence occurred in the area immediately outside the glacial limits; and the third region, remote from ice loading, in which coastal tilting

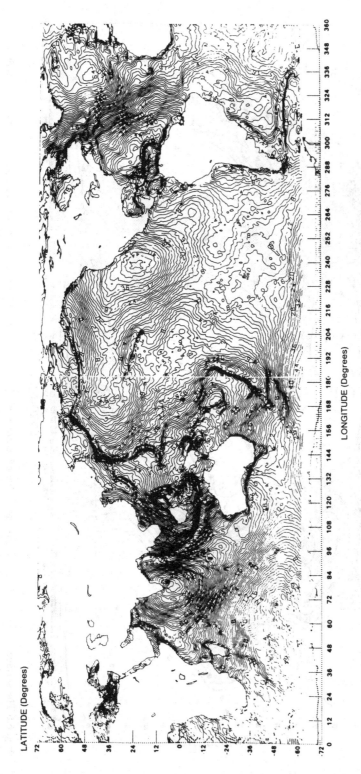

Fig. 6.5 A map to show the topography of the geoid surface derived from the altimeter data from the NASA SEASAT mission in 1978. The long-wavelength components of the geoid reveal the undulations with a maximum amplitude of ~ 180 m. (Marsh and Martin, 1982).

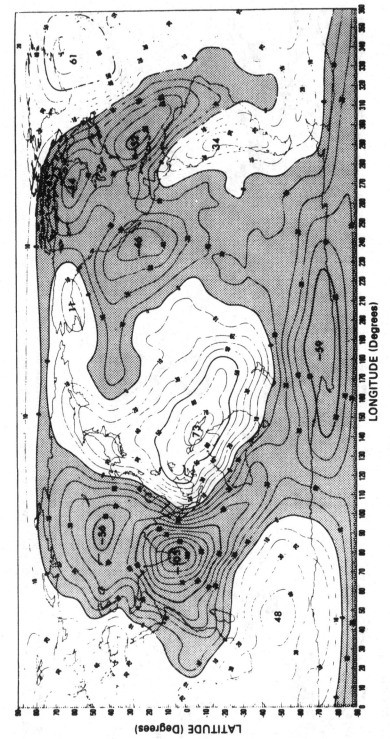

Fig. 6.6 The surface of the geoid, with depressions shaded, compared with the pattern of large-scale gravity anomalies. Negative anomalies are shown as shaded areas, which mirror the depressions in the geoid. Geoid surface computed from the GEM 10 model (height in metres above the mean ellipsoid, f = 1/298.257).

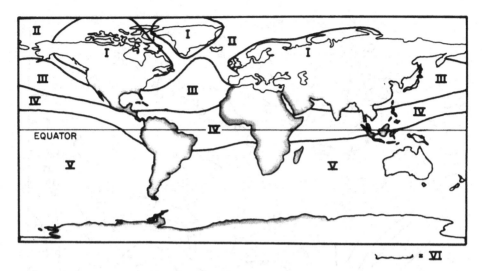

Fig. 6.7 The six sea level zones with differing characteristic sea level responses assuming a uniform 1000 m of ice melt instantaneously at 18,000 BP, an eustatic sea level rise of 77.5 m and a change in relative sea level during the following 10,000 years as the Earth relaxes. The sea level signatures are the same in each zone: emerged beaches are predicted in Zones I, III, V and VI; submergence is characteristically predicted in Zones II and IV (Clark *et al.*, 1978. Reprinted by permission of Academic Press Inc.).

results in emergence on continental shorelines because the rise of continents relative to sea level is greater than the fall of the ocean floor due to enhanced water load. Clark *et al.* (1978) indicate that Walcott's major contribution was to demonstrate how sea level was affected in the ocean basins far from the glaciated regions by loading of the ocean floor. Walcott (1972) calculated that a 100 m eustatic sea level rise would produce an observed sea level change in the centres of the world's oceans of 120–130 m but less than 100 m on the North African coast. Clark *et al.* (1978) support Walcott's conclusions but increased the number of sea level regions from three to six. These six regions and the predicted sea level curves for the last 6,000 years are shown in Fig. 6.7. Emerged beaches are predicted for Zones I, III, V and VI, and submergence is predicted in Zones II and IV. To a certain extent empirical data support these models (for example, Nunn, 1986; Martin *et al.*, 1985; but also see critical discussion in Ireland, 1987).

Clark and Primus (1987) applied this methodology to predict sea level changes consequent upon the partial melting of the Greenland ice sheet and to the retreat of the Antarctic ice sheet to AD 2100. They showed that observations of sea level rise would vary from − 200% to + 120% of the equivalent global mean value (Figs. 6.8(*a*), 6.8(*b*)). These results are in accord with earlier work on sea level and earth models which show that the attenuation of ice sheets does not result in a globally uniform rise in sea level, thus vindicating Mörner's (1976) argument for regional eustatic studies.

Of all the eustatic variables discussed here only two are directly related to climate: glacial eustasy and steric changes. These are regarded as the most important causes of a possible eustatic rise of sea level in the near future (National Research Council, 1987). The high-amplitude and high-frequency sea level curves as exemplified by Fairbridge (1961, 1976); Fairbridge and Hillaire-Marcel (1977) and Mörner (1969) lent themselves

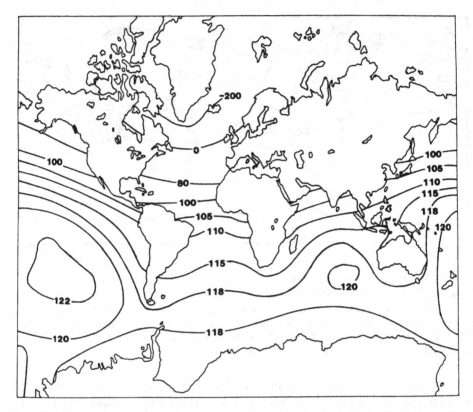

Fig. 6.8(*a*) A map showing the amount of sea level change caused by the melting of part of the Greenland ice sheet if sea level rose 1 m. Contours are in centimetres (Clark & Primus, 1987. Reprinted by permission of Basil Blackwell).

to correlation with climatic variables, with glacial eustasy being the overriding control. Oscillating sea level curves appeared to serve as a proxy record for climate change although Shennan (1982a) has drawn attention to the fact that no statistical tests have been employed to measure the strength of correlations between curves showing variations of different phenomena. However, the introduction of the concept of regional eustasy arising from considerations of the geoid topography and the migration of the geoid, and well-founded arguments about the altitudinal and temporal limitations of sea level index points, made both correlation of sea level data and the coupling of sea level curves and climatic data problematic. A more cautious approach prevailed. Kidson (1982) concluded that '... progress in resolving the considerable difficulties which remain in understanding the nature of ... sea level change must depend on greater rigour in acquiring and interpreting data than has frequently been shown in the past'.

Sea level data

More rigorous approaches have been adopted recently in the acquisition and interpretation of sea level data both from the Cainozoic and earlier eras.

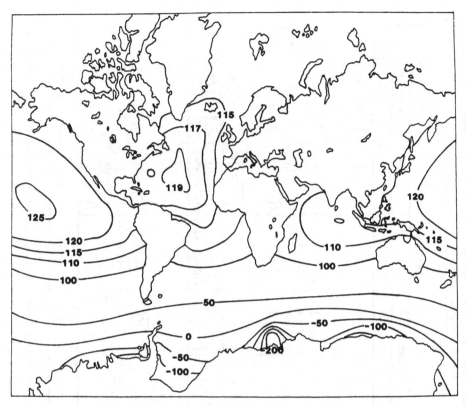

Fig. 6.8(*b*) A map showing the amount of sea level change caused by the melting of part of the Antarctic ice sheet if sea level rose by 1 m. Contours are in centimetres (Clark & Primus, 1987. Reprinted by permission of Basil Blackwell).

Considerable advances have been made in the past ten years in the subsurface exploration of sedimentary rock sequences using seismic stratigraphy, outcrop sections and well-log data to aid hydrocarbon exploration (Vail *et al.*, 1977; Hallam, 1981; Haq *et al.*, 1987). Arising from an analysis of primary seismic reflections from bedding surfaces and unconformities with velocity–density contrasts, chronostratigraphic boundaries can be identified which permit a global stratigraphic framework and correlations between different sedimentary basins. Vail *et al.* (1977) have used sediment packages or depositional sequences bounded by unconformities as primary units; unconformities are interpreted as the consequences of onlap and downlap. Patterns of onlap, downlap, truncation and basinward shifts of coastal onlap are used to infer relative sea level changes along continental margins. A relative rise of sea level is inferred from coastal onlap (transgressive overlap, as defined by Tooley, 1982 and Shennan, 1982b for the Holocene), and the magnitude of sea level rise is calculated by the amount of coastal encroachment and aggradation (Hallam, 1981).

The relative changes of coastal onlap for the past 65 million years (Palaeocene to Holocene), from which the long- and short-term eustatic curves are derived, are shown in Fig. 6.9. Corroboration for Vail's short-term sea level changes (Haq *et al.*, 1987), based on global data and hence comparable to the sea level data base employed by

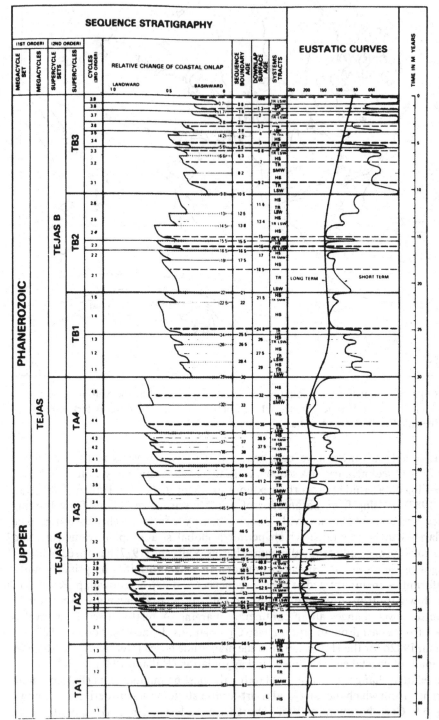

Fig. 6.9 Long-term and short-term eustatic sea level curves for the past 65 million years, showing also the scaled relative changes of coastal onlap (Haq *et al.*, 1987).

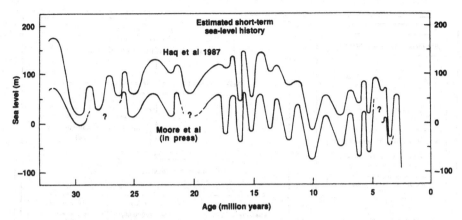

Fig. 6.10 Short-term sea level changes for the past 30 million years based on global data (Haq *et al.*, 1987) and a single site (Moore *et al.*, in Kerr, 1987). (Reprinted from *Science*, **235**, © 1987, American Association for the Advancement of Science).

Fairbridge (1961, 1976; Fairbridge and Hillaire-Marcel, 1977), has come from a single-site study (Moore *et al.*, in Kerr, 1987); both sea level curves are reproduced in Fig. 6.10. This has led Kauffman (in Kerr, 1987) to the conclusion that 'global eustatic control (overrides) local control in 90% of the cases'. These corroborative data, together with the release of additional data to the public domain, go some way towards addressing the criticisms of the curves summarized by Devoy (1987c). However, Burton *et al.* (1987) have argued that two of the three processes (sediment accumulation, tectonic subsidence and eustasy) must be specified in order to determine the third and, at present, this is not possible. They conclude that an accurate sea level variation chart cannot be made.

For Quaternary sea level changes, problems relating to sea level data acquisition, analysis and correlation have been addressed. In 1977, Tooley (1978b) drew attention to different interpretations of similar coastal lithostratigraphic sequences and argued that the correlations of marine events and of marine events with other global phenomena, such as climate, could not proceed without explicit criteria for data selection. These criteria, reiterated in 1985, were that sea level variates should originate from small homogeneous areas; that they should originate from similar palaeoenvironments and have the same indicative meaning in relation to a water level indicator (van de Plassche, 1977, 1986); and that any 'absolute' dating technique should be capable of independent corroboration, either by seriation or by biostratigraphic or lithostratigraphic analyses.

At the same time, the methodology of sea level investigations and the operational definitions were overhauled largely due to the work of Shennan (1980, 1982a,b, 1983a,b, 1987a,b); Shennan *et al.* (1983). Shennan (1983b) has elaborated on a model of sea level research methods (Fig. 6.11) which permits robust correlations within areas, between areas and between sea level variates and other variates at different spatial scales. He has quantified some of the errors that affect the age and altitude estimations of a sea level data base, and has utilized and developed the sea level tendency concept. Examples of within-area correlation are given from the Fenland and between-area correlations from the Fenland, Northwest England and the Tay estuary (Fig. 6.12). In

M.J. Tooley

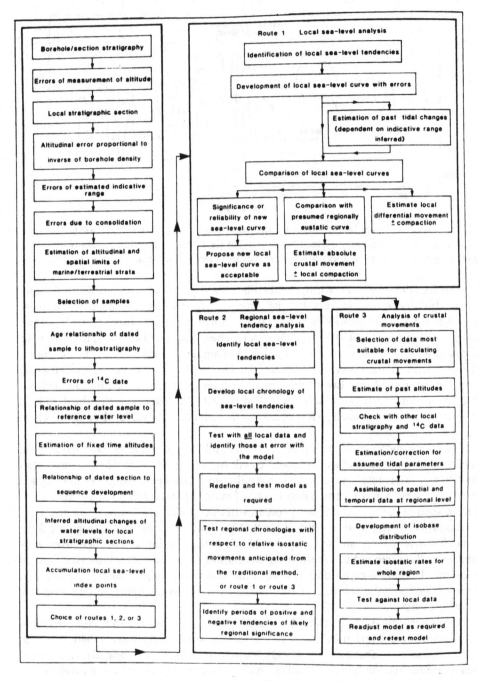

Fig. 6.11 A model of sea level research methods with routes to be followed as a necessary prerequisite for correlation (Shennan, 1983b).

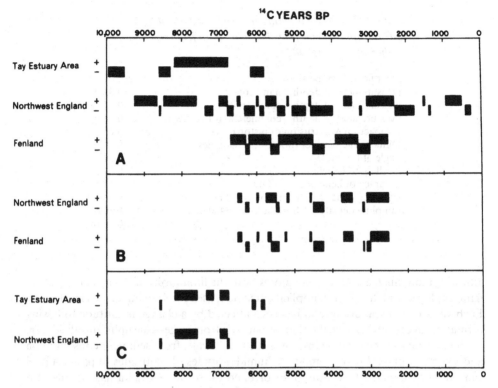

Fig. 6.12 Regional tendencies of sea level movement indicated by correlated data from the Tay estuary, Northwest England and the Fenland. The complete chronologies, to date, are shown in A, and the regionally significant tendencies are shown in B and C. (Shennan, 1986a,b; Shennan *et al.*, 1983. Reprinted by permission of *Nature*, **302**, copyright © 1983 Macmillan Magazines Ltd).

the latter case, Shennan *et al.* (1983) have shown that the number of matches is greater than that expected from two random sequences when the period during which either the Fenland chronology indicates negative tendencies or the Northwest England chronology indicates positive tendencies are compared. The next stage will be to separate the climatic signal (whether glacial eustatic, weather-induced storm surge conditions or steric) from the noise of palaeotidal changes, varying sedimentation rates, coastal processes, terrigenous sediment input and varying river discharges, that may be human-induced. The successful application of this methodology to low latitude coastal lowlands (e.g., Brazil; Ireland, 1987) demonstrates its wider application and utility. In the meantime, Smith's (1981) conclusion that many sea level changes take place during a shorter time-scale than can be resolved by any known correlation technique needs re-evaluation.

Rates of sea level change

Calculating rates of sea level change, either 'eustatic' or relative, is not simply resolved by taking the difference in altitude for a unit time from a sea level curve on an age/

Table 6.2. *Errors affecting the measured altitude of stratigraphic boundaries based on data from the Fenland (Shennan, 1980, 1982)*

identification of boundary	± 0.01 m.
measurement of depth – hand coring	± 0.01 m.
measurement of depth – commercial U4	± 0.05 m.
measurement of depth – commercial (disturbed)	± 0.25 m.
compaction & extrusion of piston cores	up to 0.06 m.
Duits gouge sampler (not for ^{14}C samples)	up to − 0.20 m.
angle of borehole	up to + 0.04 m.
levelling to benchmark	up to ± 0.02 m.
accuracy of benchmark to OD	± 0.15 m.
sampling density – 1 bore-hole per 2 m^2	*c.* ± 0.06 m.
sampling density – 1 bore-hole per 5400 m^2	*c.* ± 0.14 m. ($\hat{\sigma}$)
	∴95% limits = ± 0.30 m.

altitude graph. Such a calculation ignores both the limitations of the age and altitude data employed and the criteria, implicit or explicit, used for drawing the sea level curve. Each variate on a sea level graph has been affected by a change in altitude following formation: at different scales, the change can arise from large-scale uplift or subsidence, from variations in the tidal regime and, at the site scale, from sediment consolidation and sediment influx. In cool temperate humid latitudes, the altitudes of past sea level can only be estimated to within an error band of ∼1 m. Shennan has summarized (Table 6.2) all the altitudinal errors to which a sea level index point in the humid temperate latitudes, such as the Fenlands in Great Britain, is subject. In tropical and subtropical waters, vermetid gastropods are ideal sea level indicators with an error range as small as ±10 cm (Laborel, 1979–80). The accuracy of age estimations is determined by the degree of contamination and by the counting error if the decay rate of a radioactive isotope isolated from the sediment is used. In the Holocene, standard errors on radiocarbon dates from sea level variates as great as 385 years have been employed on age/altitude graphs (e.g., Tooley, 1978a). Significant age differences arise if the ages of boundaries on transgressive and regressive overlaps are determined by simple averaging compared with an average that is weighted by the standard errors of the dates (Shennan, 1987b). In sea level investigations, there are wider dating issues that have been treated by Sutherland (1987).

Employing a different methodology and time-scale, Haq *et al.* (1987) refer to different rates of sea level rise and fall during pre-Quaternary ages. Rates are inferred by the prominence of the unconformity. Long- and short- term eustatic sea level curves have been published, and from these rates of change can be calculated. For example, the significant short-term fall of sea level during the Upper Turonian Stage of the Upper Cretaceous was ∼0.23 mm/yr, and was preceded by a rise, the rate of which was of a similar order. Hancock and Kauffmann (1979) give possible measures of Late Cretaceous rates of sea level rise and fall, ranging from 90 m/My to 170 m/My.

Cronin (1983) has reviewed the evidence for, and nature of, sea level changes during the last 250,000 years and concluded that sea level has risen and fallen at rates of 10–30 mm/yr and probably at greater rates during periods of rapid climate change. For the

Holocene, Fairbridge and Hillaire-Marcel (1977) have calculated rates of sea level change based on an analysis of 185 Holocene strandlines in Richmond Gulf, Eastern Hudson Bay. Rates of sea level rise and fall are given (Fig. 6.2(*d*)) as in excess of 2 cm/yr 6,500, 5,200 and 3,000 radiocarbon years ago. There is some doubt about the basis and nature of the calculation, and this criticism (Shennan, 1982a) needs addressing. Earlier, Fairbridge (1961) argued that a smoothing of his oscillating sea level curve yielded a mean rise of sea level of 9 mm/yr for 10,000 until 6,000 BP, but on increasing the resolution to 500-year periods rates exceeding 24 mm/yr were noted.

Yang *et al.* (1987) have calculated rates of sea level fluctuations for Eastern China (Fig. 6.13(*a*)). Nine cycles of sea level fluctuation are recognized, and the calculated rates of fluctuation during semi-cycles H2 and H3 (correlated with the culminating stages of the last glaciation) of between 20 and 75 mm/yr are the highest rates recorded from any area. Earlier solutions (Yang and Xie, 1984) showed (Fig. 6.13(*b*)) lower rates for this period ranging from 0.8-18.2 mm/yr.

In Northwest Europe, values lying between 20 and 75 mm/yr have been recorded, and are associated with the catastrophic disintegration of the Laurentide ice sheet about 8,000 years ago. In Denmark, Petersen (1981) has recorded a rise of sea level of ~ 27 m between 8,260 and 7,680 radiocarbon years ago, and this yields a rate of sea level rise of 46 mm/yr. In Northwest England, rates of rise at this time range from 44 mm/yr in Morecambe Bay and 34 mm/yr in the Ribble estuary (Tooley, 1978a, 1991). In the Fenland of Great Britain, Shennan (1987a) has utilized a modelled version of Mörner's (1980) regional eustatic curve and the field evidence from the Fenland to calculate a linear subsidence rate of 0.9 m/1000 years for the area. From this a relative sea level curve has been computed and compared with the sea level tendency chronology for the Fenland (Fig. 6.14). Four periods of negative tendency based on empirical field data are associated with computed periods showing negative rates of sea level change. Shennan (1987a) notes that there appears to be a time lag between sea level change and the registration of a negative tendency in the sedimentary record. There are important implications for coastal change in the Fenland: the critical sea level rise rate is 5 mm/yr and, when sustained, net coastline retreat occurs (Shennan, 1987b). The relationship between rate of sea level change and coastline advance or retreat requires further exploration and application to other coastal lowlands.

Sea level and climate

The linkages between sea level changes and changing climate parameters, such as temperature, are through the processes of glacial eustasy and steric effects. At the global and regional scale, glacial eustasy has probably been the most important process affecting sea levels and coastlines during the last 2 My. At the local scale, where sea level index points are registered, a complex of factors will determine age and altitude, which are most satisfactorily portrayed within error boxes or bands (see reviews in Kidson, 1982; Tooley, 1986; Shennan and Tooley, 1987), with a quoted quantitative error estimate.

Large-amplitude and long-period sea level changes have been correlated, usually by visual inspection, with changes in volume of high and middle latitude glaciers and ice caps inferred from terminal ice stages in the Quaternary. In pre-Quaternary periods, the presence of glacial diamicts (tillites) in strata is the classical basis of inference for

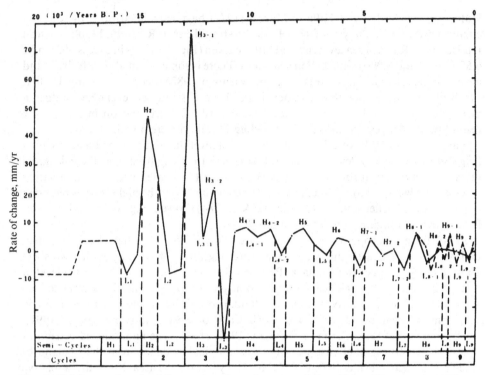

Fig. 6.13(a) Rates of sea level fluctuation along the coast of East China (Yang *et al.*, 1987).

ancient glaciations (Flint, 1975), and many glaciations with sea level changes have been inferred. Hallam (1981) explained the rise of sea level during the early Silurian as a consequence of the melting of a Late Ordovician Saharan ice sheet; the Late Carboniferous cyclothems may reflect changes in the mass budget of Gondwanan ice. Haq *et al.* (1987) suggest that since the mid-Oligocene (last 30 My), short-term, marked sea level falls, frequently associated with major global unconformities, may be due to the increasing influence of glaciation; during this period 22 sea level cycles have been identified. For the last 255 My, from the Triassic to the Quaternary, 119 sea level cycles have been identified, of which 19 began with major sequence boundaries, which may be associated with glacial eustasy. The balance require other explanations, such as oceanbasin volume changes and geoidal eustasy. Mörner (1981) explained changes in sea level during the Cretaceous – a non-glacial system – as the consequence of tectonoeustasy and geoidal eustasy.

The Quaternary era is distinguished by widespread sedimentary and biological evidence of climatic change, sometimes rapid. Mid- and high- latitude glaciation, lowlatitude aridity and lowered sea levels are characteristic. Many sea level curves published in the past thirty years (see Figs. 6.1 to 6.5) show a close correlation with glacial stages (often of short duration and limited extent) and sea level changes.

Short periods (such as the last 10 kyr) and low amplitude (less than 1 m) sea level changes are open to a variety of interpretations. Correlation by visual inspection is less acceptable now than it was hitherto. Over longer time periods covering several glacial/ interglacial cycles during the Late Quaternary, broad changes in sea level and other

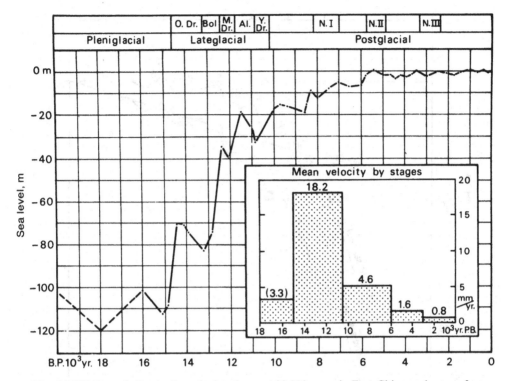

Fig. 6.13(*b*) Eustatic fluctuations during the past 20,000 years in East China and rates of sea level change during the past 18,000 years (Yang & Xie, 1984. Reprinted by permission of The Centre of Asian Studies, University of Hong Kong).

parameters such as temperature and CO_2 can be compared as a starting point for exploring causal relationships (Fig. 6.15). Chapell and Shackleton (1986) compared the ^{18}O record from Pacific core V19-30 with the sea level record from the Huon Peninsula and assumed that both records must be dominated by a common ice volume signal. The time-scale for the sea level record is based on assays of corals using ^{14}C and $^{230}Th/^{234}U$ methods: the time-scale for the ^{18}O record was developed iteratively, employing radiometric age determinations as a starting point. Although the time-scales are not directly comparable, the latter has been used to refine the ages of the Huon Peninsula corals and the resulting sea level curve is reproduced in Fig. 6.15(*a*). This curve is regarded as the best sea level record available for the New Guinea/Australian region, and peaks at 105, 83 and 60 kyr on the unrefined curve agree with estimates for elsewhere in the Pacific Ocean and in the Atlantic Ocean. This lends support to the claim that the sea level curve from the Huon Peninsula is a global curve (*sic*, Chappell and Shackleton, 1986, p. 137).

The refined sea level curve (Fig. 6.15(*a*)) is shown in relation to the deuterium profile from the Vostok ice core (Antarctica) which has been interpreted in terms of atmospheric temperature changes (Fig. 6.15(*b*)) (Jouzel *et al.*, 1987). Dating is not based on radiometric assay but on a glaciological model, and it is accepted that an underestimation of the accumulation rate and thinning could have resulted in an overestimation of duration. The carbon dioxide content of gas trapped in ice pores in the Vostok ice core

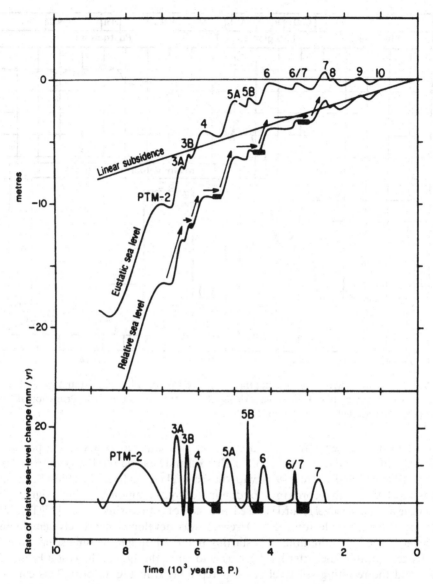

Fig. 6.14 Recalculated eustatic and relative sea level curves, linear subsidence and rates of sea level change in the Fenland, with periods of negative tendencies of sea level movement shown as black rectangles (Shennan, 1987a). PTM are the postglacial transgression maxima identified in West Sweden by Mörner. (Reprinted by permission of Basil Blackwell).

has also been measured (Barnola *et al.*, 1987) and high correlations noted between CO_2 concentrations and Antarctic climate with significant oscillations in CO_2 between interglacials (high CO_2 concentrations) and glacials (low CO_2 concentrations). Variations in CO_2 for the last 160 kyr are shown in Fig. 6.15(*c*) although uncertainties remain in the chronology. Visual inspection alone of Fig. 6.15 invites the conclusion that a relationship (explored by Geyh, 1971, 1980) exists between changes in the carbon

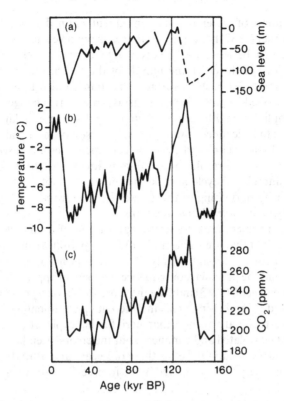

Fig. 6.15(*a*) Sea level changes during the past 160,000 years for the Huon Peninsula, recalculated after detailed correlation with the ^{18}O record from Pacific core V19–30 (Chappell & Shackleton, 1986. Reprinted by permission of *Nature*, **324**, copyright © 1986 Macmillan Magazines Ltd.). Fig. 6.15(*b*) Variation with time of the Vostock isotope temperature record as a difference from the modern surface temperature value. (Jouzel *et al.*, 1987. Reprinted by permission of *Nature*, **329**, copyright © 1987, Macmillan Magazines Ltd.). Fig. 6.15(*c*) CO_2 concentrations plotted against age in the Vostock ice core record (Barnola *et al.*, 1987. Reprinted by permission of *Nature*, **329**, copyright © 1987, Macmillan Magazines Ltd).

dioxide content of the atmosphere, local temperature variation and global sea level changes. Clearly, all three parameters require refinement and corroboration. A commonly measured chronology is essential and the problem of time lags must be addressed. It is noteworthy, though conjectural, that the peak value for CO_2 of 300 ppmv during the last 160 kyr occurred 135 kyr BP and is associated with a peak in the sea level curve. That the current level of CO_2 is about 345 ppmv implies a geologically imminent rise in sea level, if CO_2 variation is the driving force behind global temperature change and glacial eustasy.

Conclusions

The definition of eustasy as a world-wide, simultaneous, uniform change of sea level is obsolete and of historical interest only, since the discovery of deformation of the geoid relief and redistribution of water masses consequent upon density changes in the aesthenosphere. Mörner (1987) has argued that in each region eustatic curves should be defined; the differences between the curves would permit conclusions about the strength and behaviour of the crust and mantle of the Earth. These sea level curves would be an expression of the eustatic variables listed in Table 6.1 and would include the climatically controlled signals of glacial eustasy and steric changes. Isolating the signals for large-amplitude and long-period changes of sea level is largely accepted. Correlation amounts to little more than visual comparison of the apparent agreement between climatic and sea level records, even when time-scales are based on different criteria. At a higher level of resolution small-amplitude and short-period changes during a glacial or interglacial cycle may be the consequence of local effects and can readily be lost either by smoothing of the data or by the errors affecting the age and altitude data employed in sea level investigations.

A solution to this problem must lie in the application of a unified methodology, yielding comparable data that are sufficiently robust to permit strong and meaningful correlations that can be tested statistically. The exploratory approach to sea level research (Fig. 6.11) and the practical consequences of such an approach as exemplified by the work of Shennan *et al.* (1983) indicate that correlations are possible between sea level data from subsiding and rising areas in different epicontinental seas. Consistent signals point to supra-regional processes that merit further exploration. In this way, the climatic signal could be isolated and enhanced, but the errors attendant upon sea level data are such that time resolutions better than 100 years and altitudinal resolutions better than 1 m are unlikely to be improved in the foreseeable future for many parts of the world's coastlines.

References

Andrews, J.T. (1970). *A Geomorphological Study of Post-glacial Uplift, with Particular Reference to Arctic Canada*. Institute of British Geographers, Special Publication 2, London.

Barnola, J.M., Raynaud, D., Korotkevich, Y.S. & Lorius, C. (1987). Vostok ice core provides 160,000-yr record of atmospheric CO_2. *Nature*, **329**, 408–14.

Berglund, B.E. (1971). Littorina transgressions in Blekinge, South Sweden: a preliminary survey. *Geol. För. Stock. Förh.*, **93**, 625–52.

Bloom, A.L. (1982). *Atlas of Sea Level Curves: Supplement*. IGCP Project 61, Unpublished typescript.

Bloom, A.L. (1977). *Atlas of Sea Level Curves*. Cornell University, Ithaca, New York.

Burton, R., Kendall, C.G.St.C. & Lerche, I. (1987). Out of our depth: on the impossibility of fathoming eustasy from the stratigraphic record. *Earth Sci. Rev.*, **24**, 237–77.

Chappell, J. & Shackleton, N.J. (1986). Oxygen isotopes and sea level. *Nature*, **324**, 137–40.

Clark, J.A. & Primus, J.A. (1987). Sea level changes resulting from future retreat of ice sheets: an effect of CO_2 warming of the climate. In *Sea Level Changes*, M.J. Tooley & I. Shennan (eds.). Blackwell, Oxford, pp. 356–70.

Clark, J.A., Farrell, W.E. & Peltier, W.R. (1978). Global changes in post-glacial sea level: a numerical calculation. *Quat. Res.*, **9**, 265–87.

Colquhoun, D.J. & Brooks, M.J. (1986). New evidence from the Southeastern US for eustatic components in the late Holocene sea levels. *Geoarchaeology: an international journal*, **1**, 275–91.

Cronin, T.M. (1983). Rapid sea level and climatic change: evidence from continental and island margins. *Quaternary Science Reviews*, **1**, 177–214.

Daly, R.A. (1934). *The Changing World of the Ice Age.* Yale University Press, New Haven.

Devoy, R.J.N. (Ed.) (1987a). *Sea Surface Studies: a Global View.* Croom Helm, London.

Devoy, R.J.N. (1987b). Introduction: first principles and the scope of sea surface studies. pp. 1–30, in R.J.N. Devoy, (ed.), op.cit.

Devoy, R.J.N. (1987c). Hydrocarbon exploration and biostratigraphy: the application of sea level studies. pp. 531–68, in R.J.N. Devoy (ed.), op.cit.

Donovan, D.T. & Jones, E.J.W. (1979). Causes of world-wide changes in sea level. *J. Geol. Soc. Lond.,* **136**, 187–92.

Everard, C.E. (1980). On sea level changes. In *Archaeology and Coastal Change,* F.H. Thompson (ed.), London, The Society of Antiquaries, Occasional Paper (New Series) I, pp. 1–23.

Fairbridge, R.W. (1976). Shellfish-eating preceramic Indians in coastal Brazil. *Science,* **191**, 353–9.

Fairbridge, R.W. (1961). Eustatic changes in sea level. In *Physics and Chemistry of the Earth,* **4**, L.H. Ahrens, F. Press, K. Rankama & S.K. Runcorn (eds.). Pergamon Press, London, pp. 99–185.

Fairbridge, R.W. & Hillaire-Marcel, C. (1977). An 8,000-yr palaeoclimatic record of the 'Double Hale' 45-yr solar cycle. *Nature,* **268**, 413–16.

Farrell, W.E. & Clark, J.A. (1976). On postglacial sea level. *Geophys. J.R. Astron. Soc.,* **46**, 647–67.

Fifield, R. (1984). The shape of Earth from space. *New Sci.,* **1430**, (46)–50.

Flint, R.F. (1975). Features other than diamicts as evidence of ancient glaciations. In *Ice Ages: Ancient and Modern,* A.E. Wright & F. Moseley (eds.). Seel House Press, Liverpool, pp. 121–36.

Gary, M., McAfee, R., Jr. & Wolf, C.L. (1972). *Glossary of Geology.* American Geological Institute, Washington, DC.

Geyh, M.A. (1980). Holocene sea level history: case study of the statistical evaluation of ^{14}C dates. *Radiocarbon,* **22**, 695–704.

Geyh, M.A. (1971). Middle and young Holocene sea level changes as global contemporary events. *Geol. För. Stock. Förh.,* **93**, 679–92.

Gornitz, V. & Lebedeff, S. (1987). Global sea level changes during the past century. In *Sea Level Changes and Coastal Evolution,* D. Nummedal, O.H. Pilkey & J.D. Howard (eds.). The Society of Economic Palaeontologists and Mineralogists Special Publication No. 41, pp. 1–16.

Hallam, A. (1981). *Facies Interpretation and the Stratigraphic Record.* W.H. Freeman and Company, Oxford.

Hancock, J.M. & Kauffmann, E.G. (1979). The great transgressions of the Late Cretaceous. *J. Geol. Soc. Lond.,* **136**, 175–86.

Haq, B.U., Hardenbol, J. & Vail, P.R. (1987). Chronology of fluctuating sea levels since the Triassic. *Science,* **235**, 1156–67.

Hillaire-Marcel, C. & Fairbridge, R.W. (1978). Isostasy and eustasy of Hudson Bay. *Geol.,* **6**, 117–22.

Hopley, D. (1978). Sea level change on the Great Barrier Reef: an introduction. *Phil. Trans. R. Soc. Lond.,* **A291**, 159–66.

Ireland, S. (1987). The Holocene sedimentary history of the coastal lagoons of Rio de Janeiro State, Brazil. In *Sea Level Changes,* M.J. Tooley & I. Shennan (eds.). Blackwell, Oxford, pp. 25–66.

Jelgersma, S. (1979). Sea level changes in the North Sea basin. In *The Quaternary History of the North Sea,* E. Oele, R.T.E. Scuttenhelm & A.J. Wiggers (eds.). Uppsala, pp. 233–248.

Jelgersma, S. (1966). Sea level changes during the last 10,000 years. In *World Climate from 8,000 to 0 BC,* J.S. Sawyer (ed.). Royal Meteorological Society, London, pp. 54–69.

Jelgersma, S. (1961). Holocene sea level changes in the Netherlands. *Mededelingen van de Geologische Stichting Series C,* **VI** (7), 1–100.

Jouzel, J., Lorius, C., Petit, J.R., Genthon, C., Barkov, N.I., Kotlyakov, C.M. & Petrov, V.M. (1987). Vostok ice core: a continuous isotope temperature record over the last climatic cycle (160,000 yrs). *Nature*, **329**, 403–8.

Kerr, R.A. (1987). Refining and defending the Vail sea level curve. *Science*, **235**, 1141–2.

Kidson, C. (1982). Sea level changes in the Holocene. *Quat. Sci. Rev.*, **1**, 121–51.

Laborel, J. (1979–80). Les gasteropodes vermetides: leur utilisation comme marqueurs biologiques de rivages fossiles. *Oceans*, **5**, 221–39.

Marsh, J.G. & Martin, T.V. (1982). The SEASAT altimeter mean surface model. *J. Geophys. Res.*, **87**, 3269–80.

Martin, L., Flexor, J.M., Blitzkow, D. & Suguio, K. (1985). Geoid change indications along the Brazilian coast during the last 7,000 years. *Proc. Coral Reef Congress*, **3**, 85–90.

Mörner, N-A. (1987). Models of global sea level changes. In *Sea Level Changes*, M.J. Tooley & I. Shennan (eds.), Blackwell, Oxford, pp. 332–55.

Mörner, N-A. (1981). Revolution in Cretaceous sea level analysis. *Geol.*, **9**, 344–6.

Mörner, N-A. (1980). The Northwest European 'sea level laboratory' and regional Holocene eustasy. *Palaeog., Palaeoclim., Palaeoecol.*, **29**, 281–300.

Mörner, N-A. (1976). Eustasy and geoid changes. *J. Geol.*, **84**, 123–51.

Mörner, N-A. (1969). The Late Quaternary history of the Kattegat Sea and the Swedish West Coast. *Sveriges Geologiska Undersökning* C.640, 1–487.

Meyer, W.B. (1986). Delayed recognition of glacial eustasy in American Science. *J. Geol. Educ.*, **34**, 21–4.

National Research Council (US) (1987). *Responding to Changes in Sea Level: Engineering Implications*. National Academy Press, Washington, DC.

Newman, W.S. & Bateman, C. (1987). Holocene excursions of the Northwest European geoid. *Prog. Oceanogr.*, **18**, 287–322.

Nunn, P.D. (1986). Implications of migrating geoid anomalies for the interpretation of high-level fossil coral reefs. *Geol. Soc. Am. Bull.*, **97**, 946–52.

Ota, Y., Matsushima, Y. & Moriwaki, H. (1981). *Atlas of Holocene Sea Level Records in Japan*. Yokohama National University, Yokohama.

Petersen, K.S. (1981). The Holocene marine transgression and its molluscan fauna in the Skagerrak-Limfjord region, Denmark. *Spec. Publs. Int. Ass. Sediment.*, **5**, 497–503.

Pirazzoli, P.A. (1988). Sea level correlations: applying IGCP results, *Episodes*, **11**, 111–6.

Pirazzoli, P.A. (1991). *World Atlas of Holocene Sea Level Changes*. Elsevier, Amsterdam.

Shennan, I. (1987a). Holocene sea level changes in the North Sea region. In *Sea Level Changes*, M.J. Tooley & I. Shennan (eds.). Blackwell, Oxford, pp 109–151.

Shennan, I. (1987b). Impacts on the Wash of sea level rise. In *The Wash and its Environment*, P. Doody and B. Barnett (eds.). Nature Conservancy Council.

Shennan, I. (1986a). Flandrian sea level changes in the Fenland I. *J. Quaternary Science*, **1**, 119–53.

Shennan, I. (1986b). Flandrian sea level changes in the Fenland II. *J. Quaternary Science*, **1**, 155–79.

Shennan, I. (1983a). A problem of definition in sea level research methods. *Quaternary Newsletter*, **39**, 17–19.

Shennan, I. (1983b). Flandrian and Late Devensian sea level changes and crustal movements in England and Wales. In *Shorelines and Isostasy*, D.E. Smith & A.G. Dawson (eds.). Institute of British Geographers Special Publication, No. 16, Academic Press, London. pp. 255–83.

Shennan, I. (1982a). Problems of correlating Flandrian sea level changes and climate. In *Climatic Change in Later Prehistory*, A.F. Harding (ed.). Edinburgh University Press, Edinburgh, pp. 52–67.

Shennan, I. (1982b). Interpretation of Flandrian sea level data from the Fenland, England. In IGCP Project 61 Sea level movements during the last deglacial hemicycle (about 15,000 years), J.T. Greensmith & M.J. Tooley (eds.), *Proc. Geol. Ass.*, **93**, 53–63.

Shennan, I. (1980). *Flandrian Sea Level Changes in the Fenland.* Unpublished thesis, University of Durham.

Shennan, I. & Tooley, M.J. (1987). Conspectus of fundamental and strategic research on sea level changes. In *Sea Level Changes*, M.J. Tooley & I. Shennan (eds.). Blackwell, Oxford, pp. 371–90.

Shennan, I., Tooley, M.J., Davies, M.J. & Haggart, B.A. (1983). Analysis and interpretation of Holocene sea level data. *Nature*, **302**, 404–6.

Shepard, F.P. (1963). Thirty-five thousand years of sea level. In *Essays in Marine Geology in Honour of K.O. Emery*, T. Clements (ed.). University of South California Press, Los Angeles.

Shepard, F.P. (1960). Rise of sea level along Northwest Gulf of Mexico. In *Recent Sediments: Northwest Gulf of Mexico*, F.P. Shepard, F.B. Phleger & T.H. van Andel (eds.). American Association of Petroleum Geologists, Tulsa, Oklahoma.

Smith, D.G., (ed.) (1981). *The Cambridge Encyclopedia of Earth Sciences.* Cambridge University Press, Cambridge.

Suess, E. (1888). *Das Antlitz der Erde.* Translated by H.B.C. Solas 1906, *Volume II, Part III: The Sea.* Clarendon Press, Oxford.

Sutherland, D. (1987). Dating and associated methodological problems in the study of the Quaternary sea level changes. In R.J.N. Devoy (ed.), op.cit. 1987a, pp. 165–97.

Tooley, M.J. (1991). Recent sea level changes. In *Saltmarshes: Morphodynamics, Conservation and Engineering Significance*, J.R.L. Allen & K. Pye (eds.) Cambridge University Press, pp. 19–40.

Tooley, M.J. (1987). Sea level studies. In *Sea Level Changes*, M.J. Tooley & I. Shennan (eds.). Blackwell, Oxford, pp. 1–24.

Tooley, M.J. (1986). Sea levels. *Prog. Phys. Geogr.*, **10**, 120–9.

Tooley, M.J. (1982). Sea level changes in Northern England. *Proc. Geol. Ass.* **93**, 43–51.

Tooley, M.J. (1978a). *Sea Level Changes in Northwest England During the Flandrian Stage.* Oxford Research Studies in Geography, Clarendon Press, Oxford.

Tooley, M.J. (1978b). Interpretation of Holocene Sea level changes. *Geol. För. Stock. Förh.*, **100**, 203–12.

Tooley, M.J. (1974). Sea level changes during the last 9,000 years in Northwest England. *Geogr. J.*, **140**, 18–42.

Vail, P.R., Mitchum, R.M. & Thompson, S. (1977). Seismic stratigraphy and global changes of sea level, Part 4: Global cycles of relative changes in sea level. In Seismic stratigraphy – applications to hydrocarbon exploration, C.E. Payton (ed.), *Am. Assoc. Petrol. Geol.* Memoir 26, pp. 83–97.

van de Plassche, O., (ed.) (1986). *Sea Level Research: a Manual for the Collection and Evaluation of Data.* Geo Books, Norwich.

van de Plassche, O. (1977). *A Manual for Sample Collection and Evaluation of Sea Level Data.* Institute for Earth Sciences, Free University, Amsterdam, Unpublished typescript.

Visser, W.A., Ed. (1980). *Geological Nomenclature.* Bohn, Scheltema and Holkema, Utrecht.

Walcott, R.I. (1974). *Recent and Late Quaternary Changes in Water Level.* Keynote address presented at the 7th GEOP Research Conference on Coastal Problems related to water level. Ohio State University, Columbus, Ohio, unpublished typescript.

Walcott, R.I. (1972). Past sea levels, eustasy and deformation of the Earth. *Quat. Res.*, **2**, 1–14

Yang, Huai-jen & Xie, Zhiren (1984). Sea level changes in East China over the past 20,000 years. In *The Evolution of the East Asian Environment*, R.O. Whyte (ed.). Centre of Asian Studies, University of Hong Kong, Hong Kong, pp. 288–308.

Yang, Huairen, Chen Xiqing, Xie Zhiren (1987). Sea level changes since the last deglaciation and its impact on the East China Lowlands. In *Late Quaternary Sea Level Changes*, Qin Yunshan & Zhao Songling (eds.). China Ocean Press, pp. 199–212.

III

PROJECTIONS

III

7

Future changes in global mean temperature and sea level

T.M.L. Wigley and S.C.B. Raper

Abstract

Time-dependent global mean temperature changes in response to the increasing concentrations of greenhouse gases are estimated, along with past and future contributions to mean sea level rise due to thermal expansion of the oceans and the melting of land-based ice. Possible modelling strategies are reviewed first. Projections to the year 2100 are then made for the BAU, B and C scenarios for greenhouse gas forcing given by the Intergovernmental Panel on Climate Change (IPCC). In calculating future warming and sea level change, we ensure compatibility between modelled and observed changes in global mean temperature to date by making appropriate adjustments to the forcing. The sensitivity of the results to uncertainties in model parameters is assessed. Uncertainties in climate sensitivity and future forcing are the dominant controls on the range of possible future changes. The overall range of projected temperature change over 1990–2100 is 1–7 °C. For sea level, the 1990–2100 changes lie in the range 3–124 cm. Mid-value estimates based on IPCC forcing scenario B are 2.5 °C and 46 cm. These values correspond to a global mean warming rate some five times, and rate of sea level rise between two and four times the mean rates over the past 100 years. The inescapable warming commitment that arises because of the delaying effect of oceanic thermal inertia is considered and the concept is extended to the sea level rise problem. The sea level rise commitment is shown to be, in proportional terms, much larger than the warming commitment.

Introduction

Over the last 100 years, the world has warmed by about 0.5 °C and sea level has risen by 10–20 cm (Gornitz, this volume). These changes are intimately coupled: changes in global mean temperature result in changes in sea level through thermal expansion of the ocean and the melting of land ice. Due to increasing atmospheric concentrations of carbon dioxide and other greenhouse gases, the prospect for the future is an even warmer world and further rises in sea level. What are the projected rates of global warming due to greenhouse gas forcing and how fast is sea level likely to rise? These are the questions that we address in this paper, paying particular attention to the range of uncertainties attached to all estimates.

In the following sections we show model-based estimates of past and future temperature and sea level changes, together with analyses of the sensitivity of the results to assumptions regarding future forcing and climate/sea level model parameters. Some of the material presented in this paper has been given elsewhere (Raper *et al.*, 1991), but the emphasis here is different.

Greenhouse gas forcing

For calculating past forcing changes we have used observed greenhouse gas concentration histories and best-estimate relationships between concentration changes and radiative forcing for the various greenhouse gases, as given by IPCC (Shine *et al.*, 1990). The total forcing between pre-industrial times and 1990 is about 2.5 W/m^2 (i.e., equivalent to slightly more than a 1% change in solar irradiance!), of which about 1.5 W/m^2 is due to CO_2.

There are uncertainties in these figures of up to $\pm 10\%$. These are due mainly to uncertainties in the relationships between concentration changes and radiative forcing, but also to uncertainties in the radiative forcing contribution due to changes in ozone concentration which is not included in the above IPCC figures.

For future forcing we consider three scenarios from IPCC, their 'Business as Usual' (BAU) scenario and their scenarios 'B' and 'C' (Houghton *et al.*, 1990, Annex). The BAU scenario, more correctly referred to as an existing policies scenario, represents an upper limit to what might be expected in the future. Scenario B assumes that reasonably realistic policies are implemented to reduce future greenhouse gas emissions. Scenario C assumes somewhat stronger policies and probably represents a lower bound to what could be achieved, at least out to the middle of next century. These scenarios are illustrated in Fig. 7.1.

The temperature/thermal expansion model

General modelling considerations

An important contributor to past and future sea level rise is the steric effect associated with changes in the density of sea water. If a column of sea water of mass M experiences a mean density change, $\Delta\rho$, then its volume must change by $\Delta V = M\Delta\rho/\rho^2$, a change which will be manifest by an expansion or contraction of the column and a change in sea level. Since the density of sea water is dependent on its temperature and salinity, a change in either will produce a steric change in sea level (thermosteric or halosteric changes, respectively).

At its present salinity, round 35‰ sea water has a maximum density at close to 0 °C. Thus, as temperature increases, density decreases and sea water expands. The amount of expansion per unit rise in temperature is determined by the thermal expansion coefficient, which is a complicated, but well-known function of temperature, salinity and pressure (e.g., Bryan and Cox, 1972). Near the ocean's surface and at a salinity of 35‰, the thermal expansion coefficient varies considerably, from 114×10^{-6} per °C at 5 °C (typical of high latitudes) to 297×10^{-6} per °C at 25 °C (typical of tropical latitudes). In other words, at these temperatures, if a 500 m deep layer of ocean were warmed uniformly by 1 °C, the sea level would rise by 5.7 cm or 14.6 cm, respectively. (The rise would actually be slightly more than these values because the expansion coefficient increases slightly with depth and because the warming would, itself, increase the expansion coefficient.)

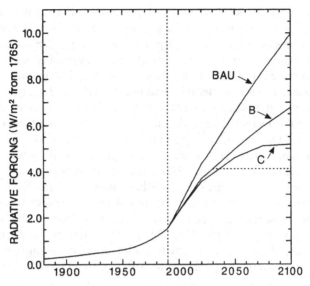

Fig. 7.1 Past and future forcing. Past forcing from Shine *et al.* (1990). Future scenarios are IPCC 'BAU', 'B' and 'C' (Houghton *et al.*, 1990, Annex). The horizontal dashed line is the scenario used in Fig. 7.7 – following scenario B to 2030, then constant.

Global-scale temperature changes will also produce a small indirect halosteric effect because of the non-linear dependence of density on salinity. Any addition of fresh meltwater from land-based ice due to global warming must lower the mean salinity and hence the density of the oceans, leading to an increase in the volume of the initial mass of ocean water (i.e., a sea level rise) independent of the added volume of the meltwater. In global mean terms this is a negligible effect, much less than the thermal expansion contribution. It is largely offset by the slight decrease in the volume of the added water due to its increase in salinity when mixed into the ocean.

If one wished to model sea level rise due to past and future thermal expansion effects comprehensively, it would be necessary to determine the time-dependent changes in the three-dimensional ocean temperature field. In principle, this could be done using an oceanic general circulation model (OGCM) coupled to an atmospheric general circulation model (AGCM), forced with appropriate forcing. Coupled O/AGCMs are still in their early states of development (e.g., Stouffer *et al.*, 1989) and no realistic sea level change simulations have been published. In any event, such simulations would be model-specific, and they would not immediately allow one to assess the magnitude of uncertainties due to, for instance, uncertainties in the climate sensitivity. For these reasons, the recent IPCC assessment relied on much simpler models to make their projections (Warrick and Oerlemans, 1990). The same models as used by IPCC will be used here.

Specifically, we will use a one-dimensional energy-balance climate model that parameterizes ocean mixing processes as a pure diffusion or upwelling-diffusion process. Models of this type have been widely used to study the time-dependent response of global mean temperature to external forcing, and to estimate the associated uncertainties. As a by-product, some of these models give oceanic thermal expansion information.

These models do, of course, have major shortcomings. Even if they were able to model near-surface or mixed-layer temperature changes correctly, they may fail in their predictions of changes in temperature at depth within the ocean simply because of the crude parameterizations of vertical heat transport that they employ. Such errors could lead to significant inaccuracies in thermal expansion results. Furthermore, because the expansion coefficient varies considerably, and non-linearly, with temperature, global mean expansion is not uniquely determined by global mean temperature profile changes. These problems introduce additional uncertainties – but these can, to some extent, be quantified by carrying out appropriate sensitivity studies.

The first analysis of thermal expansion effects with such a model was that of Gornitz *et al.* (1982) who used a pure diffusion (PD) model. PD models can produce unrealistic results because they have an unrealistic, isothermal steady state. Even if a realistic initial vertical temperature profile is used, because the model temperature profile tends to relax towards the isothermal steady-state profile, this must eventually lead to an unrealistic vertical profile of expansion. This problem may be largely overcome by using an upwelling-diffusion (UD) parameterization of ocean mixing (e.g., Wigley and Raper, 1987). (For a comprehensive review of PD and UD models, see Hoffert and Flannery, 1985.)

Model parameters

The model we use to make global mean temperature and thermal expansion projections is an energy-balance model in which land and ocean regions in both hemispheres are distinguished by separate, but interacting 'boxes' (Wigley and Raper, 1987, 1990a). The oceans each have an isothermal mixed layer of depth h, below which lies an upwelling-diffusion deeper ocean (total ocean depth is $(h + 3,900)$m). Land-ocean and inter-hemispheric heat transport coefficients have been tuned to match observed annual temperature cycles in each box. The main model parameters that determine the response to external forcing changes are the feedback parameter, λ, and the ocean diffusivity, K. The feedback parameter is specified by the equilibrium global mean temperature change for a CO_2 doubling, $\Delta T_{2\times}$, which is generally thought to lie in the range 1.5–4.5 °C (Mitchell *et al.*, 1990). λ and $\Delta T_{2\times}$ are related by $\Delta Q_{2\times} = \lambda \Delta T_{2\times}$ where $\Delta Q_{2\times}$ is the forcing change for a CO_2 doubling, taken here as 4.37 W/m^2 (from IPCC). From tracer measurements, K is believed to be about 1 cm^2/s (see, e.g., Hoffert and Flannery, 1985). Results are also sensitive to changes in the deep ocean temperature (which is controlled by the deep water formation temperature, and specified by the ratio of high-latitude to global mean temperature change, π), and to changes in upwelling rate, w.

For thermal expansion calculations, the results may be sensitive to the non-linear effects of the dependence of the expansion coefficient, α, on temperature. We have attempted to minimize this problem in the following way. First, we divided the ocean into six regions, 0–20°N and S, 20–40°N and S and > 40°N and S. For each hemisphere, the modelled hemispheric mean ocean temperature change profile is used to estimate a temperature change profile in each of the three zones, using proportionality values from Manabe and Stouffer's (1980) equilibrium $2 \times CO_2$ results. Warming is therefore greater in the high-latitude zones. In each zone, an initial temperature profile is assumed, matched to the observations of Levitus (1982). Time-dependent thermal expansion coefficients are then calculated from the time-dependent temperature profiles using the equation of state given in Gill (1982), assuming a salinity of 35‰. This leads to a different thermal expansion in each zone, because the different temperature change profiles combined with the corresponding initial temperature profiles lead to expansion coefficient profiles which differ noticeably from zone to zone. Finally, a global mean expansion-related sea level rise is calculated by area weighting of the zonal values.

In carrying out our simulations, we use a range of model parameter values. The 'best-guess' values used are similar to those used by IPCC, but there are some important differences. For climate sensitivity we use the same three $\Delta T_{2\times}$ values which were adopted as key reference values, viz. 1.5, 2.5 and 4.5 °C. The other parameters which affect the magnitude of the model's response to external forcing, together with the ranges of uncertainty adopted here, are: the vertical diffusivity K (0.5–2.0 cm²/s), the mixed-layer depth h (70–110 m) and the polar-sinking water to global mean temperature change ratio π (0.0–0.4). The upwelling rate, w, is linked to the diffusivity through the length scale of the steady-state vertical temperature profile in the ocean with w/K kept constant at 4 m/yr per cm²/s. IPCC adopted a different and smaller range of model parameters for their simulations: they varied $\Delta T_{2\times}$ between 1.5 °C and 4.5 °C, but kept other model parameter values fixed at $K = 0.634$ cm²/s, $h = 70$ m, $\pi = 1.0$ and $w = 4$ m/yr. The choice of $\pi = 1.0$ is very difficult to justify, as explained by Wigley and Raper (1991). Fortunately, most results out to the year 2100 are relatively insensitive to π.

Tuning

For most values of $\Delta T_{2\times}$ there is a validation problem; past global mean temperature changes simulated by the model are inconsistent with observations. For example, if the climate sensitivity were high (say, $\Delta T_{2\times} = 4.5$ °C) then the modelled global mean warming between 1880 and today is considerably higher than the observed change of 0.5 °C; specifically, 1.14 °C for $K = 1$ cm²/s, $h = 90$ m, $\pi = 0.2$ and $w = 4$ m/yr. One might infer from this that the climate sensitivity must be considerably less than 4.5 °C. However, when all uncertainties are properly accounted for, it can be shown that even a high climate sensitivity may be consistent with the observed warming (Wigley and Raper, 1991). In other words, although there are differences between model predictions and observed data, the uncertainties are such that the full range of $\Delta T_{2\times}$, 1.5–4.5 °C, is still possible. Still, the inconsistency remains and should be accounted for in some way.

Since the analysis of Wigley and Raper (1991) shows that the observations cannot be used to limit the range of possible model parameter values, consistency must be achieved by other means. We therefore assume that the greenhouse warming has been offset by other external forcings and/or natural internal variability of the climate system. To produce model projections that are consistent with past observations, we supplement the past greenhouse forcing, $\Delta Q(t)$, with some additional (usually negative) forcing, $\Delta N(t)$, so that the 1880–1990 global mean warming is 0.5 °C. $\Delta N(t)$ therefore represents the combined effect of other external forcings and natural internal variability.

For any given set of model parameters, and given observed data uncertainties, there is no unique solution for $\Delta N(t)$. Because of the effect of ocean thermal inertia, long time-scale temperature changes over 1880–1990 could be noticeably affected by forcing changes that occurred before 1880. For example, the 0.64 °C discrepancy noted above for $\Delta T_{2\times} = 4.5$ °C could either be due to the climate system being in an anomalously warm disequilibrium state around 1880 (which would result in a natural cooling trend after 1880, offsetting the greenhouse warming), or to the existence of strong negative forcing after 1880 partially offsetting the greenhouse forcing. Most probably, $\Delta N(t)$ has shown variations of similar magnitude both prior to and after 1880. Although pre-1880 variations could have an influence, those occurring more recently are assumed to be more important in determining the 1880–1990 temperature change.

There are a number of ways we could choose to specify $\Delta N(t)$. For example, we could choose $\Delta N(t)$ to match the observed temperature fluctuations on a 10-year time-scale, or a 100-year time-scale – or anything in between. What we have done is to concentrate on longer time-scales by assuming simply that $\Delta N(t)$ is proportional to $\Delta Q(t)$. This is equivalent to scaling $\Delta Q(t)$ by a factor chosen to give the correct 1880–1990 warming. Other approaches give very similar results.

If we invoke natural forcing changes up to 1990, then some assumption must be made regarding their future changes. For large $\Delta T_{2\times}$, $\Delta N(t)$ in 1990 is an appreciable negative quantity (just how much depends on the chosen model parameters, particularly $\Delta T_{2\times}$). Yet, on average, $\Delta N(t)$ should be near zero. The most likely future change in $\Delta N(t)$, therefore, must be a relaxation back towards zero. We have assumed that this additional forcing is removed linearly between 1990 and 2020. This means that, for most values of $\Delta T_{2\times}$, the change in the forcing between now and 2020 exceeds the change in the greenhouse gas forcing shown in Fig. 7.1, albeit by a relatively small amount. Because $\Delta N(t)$ begins at zero in 1765 and returns to zero in 2020, the total forcing up to 2020 is precisely the same as the total greenhouse forcing shown in Fig. 7.1. The overall effect on future temperature changes is, for most values of $\Delta T_{2\times}$, to produce a small additional future warming over and above that due to the greenhouse effect, representing the return of the system towards its untuned state.

How important are these assumptions? To answer this question, we compare the tuned and untuned cases. Figure 7.2 shows these alternative forcing scenarios, superimposed on scenario B greenhouse forcing from Fig. 7.1, together with the implied changes in global mean temperature and thermal expansion for a specific set of model parameters ($\Delta T_{2\times} = 2.5$ °C, $K = 1$ cm^2/s, $h = 90$ m, $\pi = 0.2$, $w = 4$ m/yr). The top panel shows the reduced forcing to 1990 in the tuned case and the additional forcing required to return the system to the untuned state in 2020. As shown in the middle panel, the net effect of tuning is a smaller warming to 1990 (with a 0.5 °C change from 1880) and more rapid warming over 1990–2030. Tuned and untuned cases are virtually identical after 2030, although the former gives slightly warmer temperatures due to the 'history effect' (Wigley and Raper, 1990b). The effect of tuning on thermal expansion (lower panel of Fig. 7.2) is less noticeable than for surface temperature because of the additional lag due to the effect of integrating over the whole ocean column.

Modelling the ice melt contributions

Global warming changes sea level not only through thermal expansion of the oceans but also by changing the net mass balance of small glaciers and large ice sheets. The ice melt effects are quantified in this section. In addition, the stability of the West Antarctic ice sheet and uncertainties due to the present state of the mass balance of Antarctica are considered.

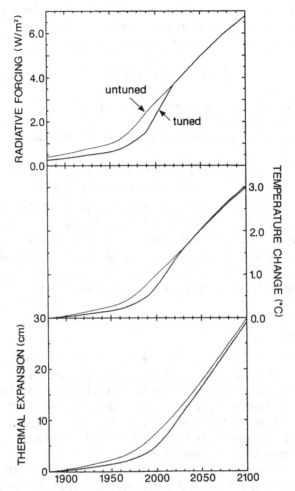

Fig. 7.2 Raw and tuned forcing (top panel) and consequent global mean temperature (middle panel) and thermal expansion (bottom panel) effects. Raw forcing results, thin lines; tuned results, thick lines. Simulations use scenario B forcing and best-guess model parameters ($\Delta T_{2\times} = 2.5°C$, $K = 1$ cm²/s, $h = 90$ m, $w = 4$ m/yr, $\pi = 0.2$). Greenhouse gas forcing prior to 1990 is multiplied by a constant factor (here, 0.63) to give a 1880–1990 warming of 0.5°C, consistent with observations. The implied negative, non-greenhouse forcing in 1990 is then relaxed to zero by 2020.

Small glacier contribution

Over the last 100 years, there has been a general worldwide retreat in small 'alpine' glaciers (Grove, 1988). This has almost certainly contributed significantly to past sea level change (Meier, 1984; for review see Warrick and Oerlemans, 1990). For the calculation of the small glacier contribution to sea level rise we use a simple model which relates small glacier mass change to global mean temperature changes. A similar type of model was used by Oerlemans (1989). While such a model is an extreme simplification, it is preferred to other methods of estimating the ice melt contribution (see Kuhn, this volume) because it allows for the lag between forcing and glacier response, and keeps track of the total ice mass available for melting.

The model can be written in sea level equivalent as follows:

$$\frac{\mathrm{d}z}{\mathrm{d}t} = [-z + \{Z_0 - z\}\beta\overline{\Delta T}]/\tau$$

where z is the sea level change (initially zero) (m)

τ is a relaxation time (years)

β is a constant representing glacier melt sensitivity to temperature change (/°C)

Z_0 is the initial ice mass in sea level equivalent (m)

and $\overline{\Delta T}$ is the annual mean global air temperature change (°C).

The model is first calibrated using observed temperature data and Meier's (1984) small glacier melt estimate of 0.46 ± 0.26 mm/yr sea level equivalent between 1900 and 1961. This calibration exercise gives ranges of possible values for τ and β. Combinations of τ and β consistent with the uncertainty in Meier's estimates are then used to give low, medium and high sea level rise estimates, viz. $(\tau,\beta) = (30,0.10)$, $(20,0.25)$ and $(10,0.45)$. We assume a single value for the initial pre-industrial ice volume of 50 cm sea level equivalent.

Ice sheet contributions

For the Greenland ice sheet we follow IPCC (Warrick and Oerlemans, 1990) by using a sensitivity value for the surface mass balance expressed as a rate of change of sea level per degree warming, viz. $\beta_g = 0.3 \pm 0.2$ mm/yr/°C (cf. Oerlemans, 1989; Polar Research Board, 1985). The ice sheet was assumed to be in equilibrium in 1880. Temperature changes prior to 1980 are taken from observations (summer data), while temperature changes after 1980 are assumed to be 1.5 times the modelled global mean change (this scaling is based on results from Stouffer *et al.*, 1989).

In Greenland, the loss of ice is due to both calving and melting (runoff) in approximately equal proportions. In Antarctica, however, the loss of ice is almost exclusively due to calving, with negligible melting and runoff (Oerlemans, this volume). For the contribution to sea level change from the Antarctic ice sheet we consider three sources of mass balance change separately.

Mass balance is the difference between accumulation and ice loss. Thus, changes in mass balance, ΔB, may result from changes in accumulation (which we denote by ΔB_1) or changes in the rate of ice loss. The latter we divide into changes associated with the possible instability of the West Antarctic ice sheet, denoted by ΔB_2, and all other changes associated with large-scale ice-flow dynamics (ΔB_3). We assume ΔB_1 and ΔB_2 to be linearly dependent on temperature change (as in Warrick and Oerlemans, 1990). On the century time-scale, ΔB_3 can be assumed constant (i.e., independent of temperature change) because of the long response time of the bulk of the Antarctic ice sheet. Sea level change is therefore determined by

$$\frac{\mathrm{d}z}{\mathrm{d}t} = \Delta B = \Delta B_3 + (\beta_1 + \beta_2)\Delta T$$

where ΔT is the Antarctic temperature change.

We now need to quantify ΔB_3, β_1 and β_2. In the IPCC analysis, Warrick and Oerlemans (1990) assumed that the present total mass balance (ΔB) is 0.0 ± 0.5 mm/yr (in sea level equivalent) and that the Antarctic contribution to sea level rise over the last 100 years was 0.0 ± 5 cm. If ΔT over the last 100 years were near zero, a reasonable assumption in the absence of information to the contrary, then, as a best-estimate, all of the ΔB components, including ΔB_3, would be zero over this period. However, more recent work (Jones, 1990) provides strong evidence of warming in Antarctica over the past 100 years, so it is likely that the present values of ΔB_1 and ΔB_2 are *not* zero.

In this case, we need to decide whether to retain the assumption that the best-guess, present-day value of ΔB is zero or whether to take the assumption that $\Delta B_3 = 0$ as being more reasonable. For $\Delta T > 0$, $\Delta B = 0$ requires $\Delta B_3 > 0$ (since $\beta_1 + \beta_2$ is probably negative, see below), necessitating a long-term positive contribution to sea level rise of around 2 cm per century over the centuries prior to the 19th century. Alternatively, if ΔB_3 were zero, then the present value of ΔB must be negative. We choose to assume $\Delta B_3 = 0$. For the uncertainty in ΔB_3, we use the range assumed by IPCC for their estimate of the current ΔB value, i.e., we take $\Delta B_3 = 0.0 \pm 0.5$ mm/yr.

This assumption requires that the Antarctic ice sheet is not currently in balance, but *was* in balance at some time before the recent warming trend began (which we take as 1880). This differs from IPCC, but is consistent with our assumption for Greenland. The incorporation of past Antarctic warming also results in a 1990 ΔB value which differs from IPCC, but which is still consistent with observations.

For ΔB_1, the contribution from accumulation changes, there is both empirical and theoretical evidence to suggest that increased temperatures should lead to increased accumulation and therefore a negative contribution to sea level rise (Muszynski, 1985; Fortuin and Oerlemans, 1990; Oerlemans, this volume; see also the review by Warrick and Oerlemans, 1990). On the basis of these analyses, we assume $\beta_1 = -0.3 \pm 0.1$ mm/yr/°C. We note, however, that one GCM result gives a decrease in accumulation with increasing temperatures (Schlesinger, this volume), so the uncertainty in β_1 may be wider than indicated here. Finally, based on Budd *et al.* (1987), we assume $\beta_2 = 0.1 \pm 0.1$ mm/yr/°C to quantify the contribution arising from possible instability of the West Antarctic ice sheet (ΔB_2).

The above formula is used to predict both past and future contributions to sea level change from the Antarctic ice sheet. There remains the question of how Antarctic temperatures will change in the future, relative to the global mean. As noted earlier, there is evidence that Antarctic temperatures may have increased at around twice the rate of the global mean over the past century (Jones, 1990). However, there is no way of knowing how much of the past increases in either Antarctic or global mean temperature was related to the increase in greenhouse gas concentrations. Thus, the factor two cannot be applied to the future. Equilibrium results from General Circulation Models (GCMs) when CO_2 concentrations are doubled suggest a value of 1.5 for the ratio of Antarctic to global mean temperature change. In order to be consistent with the treatment of Greenland, however, we base our choice on the transient GCM results of Stouffer *et al.* (1989) and assume that, for the future, temperature changes over Antarctica will be the same as the global mean. Clearly, future Antarctic effects are sensitive to this assumption. It should also be noted that estimates of both past and future sea level change contributions from Antarctica are quite sensitive to the assumed 1880–1990 temperature change. We assume a single value of 1 °C.

Table 7.1. *Modelled 1880–1990 changes before and after the use of the forcing tuning factor which constrains the temperature change to 0.5 °C. The results are shown for the pre-tuning parameter combinations which maximize the range of the temperature response, and for the parameter combinations which maximize the range of the thermal expansion response. A middle value is also given. ΔT is the temperature change and Δz is the thermal expansion*

ΔT response	–	Min		Max	–
Δz response	Min	–		–	Max
$\Delta T_{2\times}$ (°C)	1.5	1.5	2.5	4.5	4.5
h (m)	70	110	90	70	110
K (cm^2/s)	0.5	2.0	1.0	0.5	2.0
w (m/yr)	2	8	4	2	8
π	0.0	0.4	0.2	0.0	0.4
ΔT (before tuning)	0.58	0.49	0.79	1.32	0.94
Tuning Factor	0.86	1.02	0.63	0.38	0.53
ΔT (after tuning)	0.50	0.50	0.50	0.50	0.50
Δz (before tuning)	3.17	5.37	6.04	7.43	10.54
Δz (after tuning)	2.73	5.48	3.80	2.77	5.56

Temperature and sea level change results

Past changes

Past changes in temperature and sea level were simulated using observed forcing changes from IPCC (Shine *et al.*, 1990) and the full range of possible model parameter values. For temperature, tuning ensures that the 1880–1990 change is 0.5 °C for all model parameter-set values. There is, however, a range of results for the various sea level change components.

In general, different parameter values are required to give extreme values of *untuned* temperature change and thermal expansion. This is because, for any given forcing and $\Delta T_{2\times}$, larger values of K, h or π result in lower mixed-layer (surface) temperature changes but higher column-mean ocean temperature changes and hence larger thermal expansion. For example, in the high climate sensitivity case ($\Delta T_{2\times} = 4.5$ °C) the parameter set $h = 70$ m, $K = 0.5$ cm^2/s and $\pi = 0.0$ gives the maximum 1880–1990 warming (viz. 1.3 °C), but does not maximize thermal expansion (it gives $\Delta z = 7.4$ cm). To maximize the thermal expansion over this interval, the model parameter values must be $h = 110$ m, $K = 2.0$ cm^2/s and $\pi = 0.4$. This gives a warming of 0.9 °C and a thermal expansion of 10.5 cm.

The full set of results over 1880–1990 is summarized in Table 7.1. Model tuning has the effect of considerably reducing the range of possible thermal expansion values (from 3–11 cm before tuning, to 3–6 cm after tuning), and of making the range almost independent of the climate sensitivity. Both tuned and untuned expansion values are maximized by large values of h, K and π.

The model-generated temperature changes were used to obtain estimates of the small glacier contribution to sea level rise over the last century (Table 7.2). These results are independent of the choice of climate model parameters, since all climate model output

Table 7.2. *Model-based contributions to*
sea level rise, 1880–1990 (cm)

Contributors	Low	Mid	High
Thermal expansion	2.7	3.8	5.6
Small glaciers	1.5	3.9	7.4
Greenland*	0.7	2.3	3.9
Antarctica	−7.7	−1.1	5.5
Total	−2.8	8.9	22.4

Notes:
*1880–1980 only

was tuned to give the same global mean warming. The range of values obtained, 1.5, 3.9 and 7.4 cm, reflects only the uncertainty in Meier's estimates of the observed small glacier contribution, which is effectively scaled by the modelled temperature changes over 1880–1990 relative to 1900–61.

For the Greenland ice sheet, model-based estimates were obtained (to 1980 only) using observed summer temperatures (P.D. Jones, personal communication). These results, shown in Table 7.2, cover a range which reflects uncertainty in the sensitivity parameter β_g only.

To estimate the past contribution to sea level change from the Antarctic we need to know past temperature changes. We assume the 1880–1990 change to be a linear increase of 1.0 °C (based on Jones, 1990) and that temperature changes prior to 1880 have no long-term trends so that the 1880 level can be taken as a primary reference level. Results are shown in Table 7.2.

Table 7.2 summarizes the 1880–1990 sea level change estimates for all of the components considered above. Thermal expansion and small glacier melt contributions are constrained to be consistent with a global mean warming of 0.5 °C over the period 1880–1990, as described earlier. The estimates for Greenland and Antarctica, on the other hand, are based on the observed average summer temperatures in Greenland and on the estimated warming trend in Antarctica, respectively. The overall modelled range of sea level rise (−3 to +22 cm) encompasses the estimated range from the tide gauge data (10 to 20 cm; see Gornitz, this volume, and Woodworth, this volume). The model-based values are largely independent estimates, and their consistency with observations provides support for the use of the same modelling approach to project future sea level change.

Future changes

Here we consider the sensitivity of the results to assumptions regarding future forcing, climate sensitivity, diffusivity and deep water temperature changes. We do not show any results for h sensitivity, since this is minimal. π sensitivity is illustrated using a wider range of values than we consider realistic in order to cover the IPCC 'best-guess' value of $\pi = 1$.

Results for the three forcing scenarios are shown in Fig. 7.3 for the best-guess set of

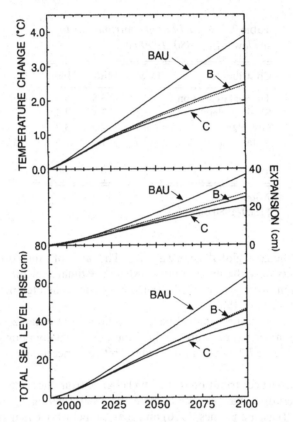

Fig. 7.3 Future global mean temperature, thermal expansion and total sea level changes, 1990–2100: effect of forcing uncertainties. The results shown are for best-guess model parameters only, $\Delta T_{2\times} = 2.5$ °C, $K = 1$ cm²/s, $h = 90$ m, $w = 4$ m/yr, $\pi = 0.2$. Forcing values are tuned to ensure a 0.5 °C global mean warming between 1880 and 1990. Also shown are results for the scenario B forcing case for $w = 0$ (dashed lines), in order to illustrate the different results obtained by a pure diffusion model compared with a more realistic upwelling-diffusion model.

model parameters, $\Delta T_{2\times} = 2.5$ °C, $K = 1$ cm²/s, $h = 90$ m and $\pi = 0.2$. As can be seen from Fig. 7.1, scenarios B and C are very similar initially, and this is reflected in the temperature and sea level simulations. For the overall uncertainty (scenario C versus scenario BAU), there are large differences: a factor of two in temperature change by 2100 (1.9–3.8 °C), somewhat smaller for the sea level (38–63 cm). The importance of using an upwelling diffusion model is also illustrated in this Figure, for the B scenario forcing case only. The PD results shown here (dashed lines) use tuned forcing to ensure direct comparability with the UD results. (The tuning required for the PD case differs from that for the UD case.) Relative to the UD results, the PD parameterization of ocean mixing underestimates the surface warming and overestimates the sea level rise. The differences are relatively small on this time-scale, but these are reduced by the tuning.

In Figs. 7.4 and 7.5, we show the effects of climate sensitivity and diffusivity on the

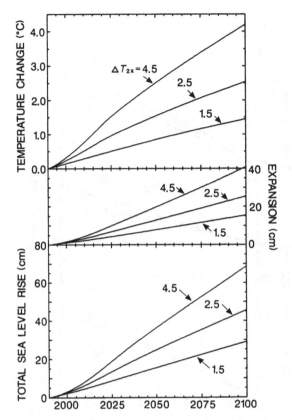

Fig. 7.4 Future global mean temperature, thermal expansion and total sea level changes, 1990–2100: effect of climate sensitivity (ΔT_{2x}) uncertainties. The results are for scenario B forcing only, with model parameter values $\Delta T_{2x} = 1.5$, 2.5 or 4.5 °C, $K = 1$ cm²/s, $h = 90$ m, $w = 4$ m/yr and $\pi = 0.2$.

results, for the scenario B forcing case only and using $h = 90$ m and $\pi = 0.2$. The influence of climate sensitivity is clear-cut and of primary importance. Greater sensitivity leads to greater future warming. The range of temperature changes varies by a factor of three, more or less the same as the climate sensitivity range in proportional terms (1.4–4.2 °C in 2100). Sea level change uncertainty is slightly less than a factor of three (29–68 cm in 2100). Climate sensitivity uncertainty is even more important than the uncertainties arising from future forcing.

The influence of the assumed diffusivity value is smaller and more complex. For temperature, greater K gives smaller surface warming since there is greater heat penetration into the deeper ocean. This leads to an increase in the total amount of heat absorbed by the ocean. This is because the total heat absorbed is the time-integral of $\Delta Q - \lambda \Delta T$, where $\Delta Q(t)$ is the forcing, λ is the feedback parameter, and $\Delta T(t)$ is the mixed-layer temperature change. The increase in heat penetration produces a net increase in warming and expansion below the mixed layer that more than compensates for the reduced expansion due to lower mixed-layer and nearby upper level temperature changes. The net effect is greater overall expansion for larger K. For total sea level

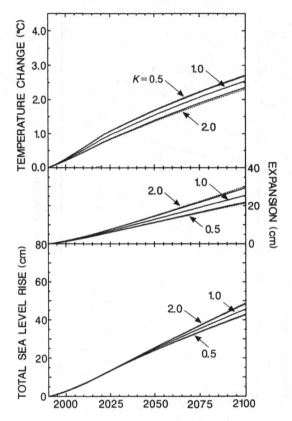

Fig. 7.5 Future global mean temperature, thermal expansion and total sea level changes, 1990–2100: effect of uncertainties in ocean mixing as parameterized by the diffusion coefficient (K, in cm^2/s). The results are for scenario B forcing only, with model parameter values $\Delta T_{2\times} = 2.5$ °C, $h = 70$ m and $\pi = 0.2$. The full lines show the results that obtain when w is varied so as to maintain a constant w/K ratio – i.e., all cases have the same steady-state temperature profile. The dashed lines are for the upwelling rate fixed at 4 m/yr. Note the opposite direction of the effects on temperature and expansion, leading to reduced K-sensitivity for total sea level rise.

rise, this expansion effect is partly offset by the smaller ice melt contribution with larger K, so that overall sensitivity of sea level change to K is small. Diffusivity effects are quite sensitive to the climate sensitivity (results not shown). For temperature changes, the results are almost independent of K at low $\Delta T_{2\times}$, while for high $\Delta T_{2\times}$, K-sensitivity is higher. For sea level rise, the K-dependence in relative terms is greater for low $\Delta T_{2\times}$.

Figure 7.6 shows the effect of π on the results for forcing scenario B. We use an expanded range of parameter values in order to cover the value $\pi = 1.0$ assumed by IPCC. Within the range of π values we consider likely, π has little effect: ± 0.1 °C and ± 1 cm in 2100. Even allowing $\pi = 1.0$, the effects are negligible out to 2050. Sensitivity to π is somewhat greater for forcing scenarios that level off or begin to decline with time. To illustrate this we consider a case more restricted than scenario C, following B to 2030 and with constant forcing thereafter. The effect of a difference in π for this case, 0.2 versus 1.0, is shown in Fig. 7.7.

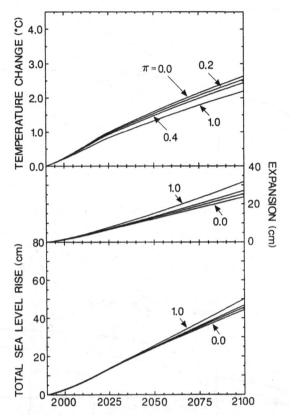

Fig. 7.6 Future global mean temperature, thermal expansion and total sea level changes, 1990–2100: effect of uncertainties in the polar-sinking water temperature change (characterized by π). As for diffusivity, the effects on temperature and expansion are in the opposite direction, leading to reduced π-sensitivity for total sea level rise.

The expansion commitment

One of the most important consequences of oceanic thermal inertia is that, at any given time, there must be a noticeable warming 'commitment'. The realized or transient warming lags behind the instantaneous equilibrium warming by an amount which varies with time and which depends on the forcing history and on model parameters, particularly the climate sensitivity. Thus, if the forcing increase could be suddenly curtailed, warming would continue for many decades as the climate system relaxed towards a new equilibrium state. Just how rapidly the unrealized warming manifests itself depends critically on the value assumed for π, since this determines how much heat is sequestered in the deep ocean. This effect is illustrated in Fig. 7.7 in which, since there is no forcing change after 2030, all we are seeing after that date is the realization of the commitment accrued up to 2030. Fig. 7.7 shows results for $\pi = 0.2$ and $\pi = 1.0$ to demonstrate that, by choosing a high value of π, one could obtain a false idea of the warming commitment.

The warming commitment issue is rather complicated and open to a number of different interpretations. Consider a specific example, namely the 1990 commitment

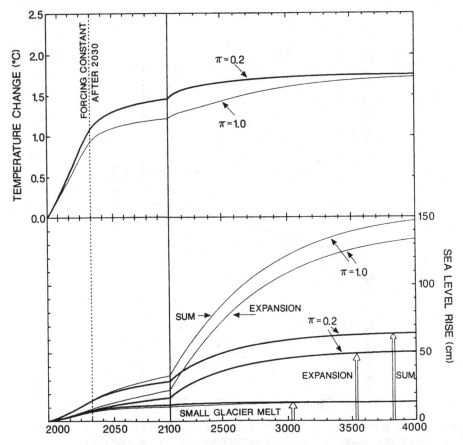

Fig. 7.7 Temperature and sea level rise commitments. In this illustration, radiative forcing changes stop in 2030. Warming and sea level rise continue long after 2030, as the system tends towards a new steady state. For sea level, the commitment is relatively much larger and continues to be realized for many centuries. Its magnitude depends critically on the π value – large π requires substantial warming of the whole ocean column and hence a large eventual expansion. (Results shown are for $\Delta T_{2\times} = 2.5\,°C$, $K = 1\,cm^2/s$, $h = 90\,m$, $w = 4\,m/yr$ and $\pi = 0.2$ and 1.0.)

under the assumption of a climate sensitivity of $\Delta T_{2\times} = 4.5\,°C$ with model diffusivity $K = 1\,cm^2/s$, mixed layer depth 90 m, upwelling rate 4 m/yr and $\pi = 0.2$. The equilibrium warming between our nominal pre-industrial date of 1765 and 1990 is 2.5 °C. The transient warming over this period is around 1.4 °C, implying a further commitment of 1.1 °C even if greenhouse gas concentrations remained constant at their 1990 levels. However, since the observed warming to 1990 is only about 0.7 °C (assuming a change between 1765 and 1880 of 0.2 °C, which the above model parameters would imply), the potential commitment may be as large as 1.8 °C (2.5 °C minus 0.7 °C). For the 4.5 °C sensitivity to be correct, there must have been other processes that ought eventually to disappear leading to an additional warming; it is this that accounts for the difference between 1.8 °C and 1.1 °C. In addition, there is a further commitment to warming over and above these effects – simply because we are committed to further greenhouse gas concentration increases no matter what we do. Because of these factors, there are many ways in which the warming commitment could be defined.

Rather than use a particular definition, we illustrate the effect with a specific example. We also show that, alongside the warming commitment, there is a parallel, and relatively much larger, sea level rise commitment.

For illustrative purposes we will assume the following model parameters: $\Delta T_{2x} = 2.5$ °C, $K = 1 \text{ cm}^2/\text{s}$, $h = 90 \text{ m}$, $w = 4 \text{ m/yr}$, $\pi = 0.2$. We also restrict ourselves to the combined contribution of thermal expansion and small glacier melting, because of the great difficulties in making valid projections for the large ice sheets beyond the next century or so. For the forcing we use scenario B up to 2030, with no forcing change after this date. The warming and sea level rise after 2030 are expressions of the commitment existing in the year 2030. It should be noted that the thermal expansion commitment in particular is very sensitive to the chosen value of π. This is because at equilibrium, if π is small, the deep ocean warms very little, whereas if π is large the whole ocean column warms (to the same extent if $\pi = 1$).

The results are shown in Fig. 7.7. For changes from 1990, the 2030 transient temperature increase is 62% of the final equilibrium warming for $\pi = 0.2$. More than one-third of the warming occurs after the forcing changes stop. For sea level rise, the small glacier component reaches 54% of its equilibrium value in 2030. For thermal expansion, however, only 16% of the final value is seen by 2030. Future sea level increases would be substantial and unavoidable for many centuries. To exacerbate this problem, because of the very large response times of the major ice sheets, the sea level rise commitment from this source on time-scales of centuries or longer may also be very large.

Projections and conclusions

Estimates of future changes in global mean temperature and sea level rise have wide ranges of uncertainty, which arise for the following reasons:

1. Uncertainties in future greenhouse gas concentrations and the resulting radiative forcing changes.
2. Uncertainties in both past and future natural climate changes. We have attempted to account for one aspect of natural variability by tuning past forcing changes to ensure compatibility with observations, but the details of future variability are unpredictable.
3. Uncertainties in the climate sensitivity (i.e., ΔT_{2x}).
4. Uncertainties in the damping of, or delay in the response to external forcing changes due to ocean mixing, as parameterized in our model simulations by the diffusion coefficient.
5. Uncertainties in the high-latitude downwelling flux of heat into the deep ocean due to uncertainties in changes in the temperature of the sinking water (i.e., uncertainties in π).
6. Uncertainties arising from possible ocean circulation changes, which we have not attempted to consider. In particular, changes in the intensity of the thermohaline circulation could affect global mean temperature and sea level. Other circulation

Table 7.3. *Model parameter values used to obtain extreme and mid-value estimates of future global mean temperature (* $\triangle T$ *) and sea level change (* $\triangle z$ *). Units:* $\triangle T_{2\times}$ *, °C; K, cm²/s.; w, m/yr; h, m; π, dimensionless; τ, yr;* $\triangle B_3$ *, mm/yr; β, βg, β₁, β₂, mm/yr/°C*

	Temp./expansion model					Ice-melt models					
	$\triangle T_{2\times}$	K	w	h	π	τ	β	β_g	$\triangle B_3$	β_1	β_2
Low $\triangle T$	1.5	2.0	8.0	110	0.4						
Mid $\triangle T$	2.5	1.0	4.0	90	0.2						
High $\triangle T$	4.5	0.5	2.0	70	0.0						
Low expansion	1.5	0.5	2.0	70	0.0	30	0.10	0.1	−0.5	−0.4	0.0
Mid expansion	2.5	1.0	4.0	90	0.2	20	0.25	0.3	−0.0	−0.3	0.1
High expansion	4.5	2.0	8.0	110	0.4	10	0.45	0.5	0.5	−0.2	0.2

changes could cause marked regional effects on sea level (Mikolajewicz et al., 1990).

7. Large uncertainties in the various ice-melt model parameters.

We have quantified at least some of these uncertainties individually. By considering extreme values of each parameter, we can thereby estimate the extreme range of likely future changes. The parameter values used are shown in Table 7.3. Note that it is most unlikely that all of the extreme model parameter values will be realized together, so the calculated extremes for temperature and sea level change represent very low probability combinations. Furthermore, as noted earlier, parameter values that maximize or minimize the temperature changes (and probably, thereby, the ice-melt contribution to sea level rise) do not necessarily maximize or minimize the thermal expansion contribution to sea level rise.

The results are shown in Figs. 7.8 (temperature) and 7.9 (sea level), for each of the three forcing scenarios. In Fig. 7.8, the temperature change extremes are shown as full lines, while the temperature changes that arise from model parameter values chosen to maximize the thermal expansion contribution to sea level rise are shown dashed. In Fig. 7.9, the latter cases are shown full.

The temperature results differ noticeably from those given by IPCC. For example, our upper extreme (i.e., temperature-change-maximized) temperature increase in 2100 of 7 °C is considerably larger than the 4.9 °C given by IPCC using the same forcing (BAU) and climate sensitivity ($\Delta T_{2\times} = 4.5$ °C). For scenario B forcing (middle panel of Fig. 7.8), the temperature increase of 2.5 °C in 2100 can be compared with 2.0 °C from IPCC with both cases using $\Delta T_{2\times} = 2.5$ °C. Our lowest temperature change prediction (warming of about 1.0 °C by 2100) is similar to the value given by IPCC using scenario C forcing with $\Delta T_{2\times} = 1.5$ °C. The differences here arise partly from the different model parameters used in this study (IPCC considers uncertainties in $\Delta T_{2\times}$ only, while we account for uncertainties in all model parameters) and also from the tuning we perform in order to account for model inconsistencies prior to 1990. Note, however, that the probability that one might assign to our high temperature change estimate must be less than would be assigned to the highest IPCC estimate (although both must have low probabilities attached to them).

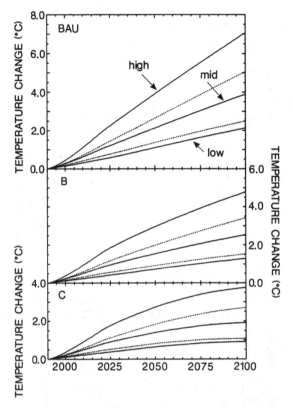

Fig. 7.8 Full range of projections of global mean temperature change under IPCC forcing scenarios BAU (top panel), B (centre) and C (bottom). The three full lines in each panel show low, mid and high estimates, based on the following model parameter values:

low, $\Delta T_{2\times} = 1.5$ °C, $K = 2.0$ cm^2/s, $h = 110$ m, $w = 8$ m/yr, $\pi = 0.4$;
mid, $\Delta T_{2\times} = 2.5$ °C, $K = 1.0$ cm^2/s, $h = 90$ m, $w = 4$ m/yr, $\pi = 0.2$;
high, $\Delta T_{2\times} = 4.5$ °C, $K = 0.5$ cm^2/s, $h = 70$ m, $w = 2$ m/yr, $\pi = 0.0$.

Parameter values for the low and high estimates are chosen to give extremes of temperature change. The dashed lines show results with model parameters chosen to give extreme values of thermal expansion (parameter values given in Fig. 7.9 caption.)

Total sea level rise projections are given in Fig. 7.9. Compared to temperature, there is reduced dependence of the results on model parameter values other than $\Delta T_{2\times}$, particularly for high $\Delta T_{2\times}$. Prior to around 2060, maximizing thermal expansion does not maximize total sea level rise, because of the lesser relative importance of the expansion component in the early decades. Over 1990 to 2030 the contributions from thermal expansion and small glacier melt are of comparable size (see, e.g., Fig. 7.7), while the contributions from the large ice sheets are relatively small. By 2100, however, thermal expansion becomes the dominant contributor. This is primarily due to the increase in the rate of heat flux into the ocean in response to the increase in the degree of disequilibrium of the surface temperature. Also, the relative contribution from the

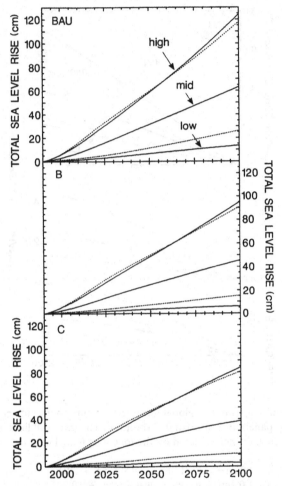

Fig. 7.9 Full range of projections of total global mean sea level rise under IPCC forcing scenarios BAU (top panel), B (centre) and C (bottom). The three full lines in each panel show low, mid and high estimates, with the two extremes based on parameter sets chosen to give extreme values of thermal expansion. The parameter values used are:

low, $\Delta T_{2\times} = 1.5\,°C$, $K = 0.5\ cm^2/s$, $h = 70\ m$, $w = 2\ m/yr$, $\pi = 0.0$;
mid, $\Delta T_{2\times} = 2.5\,°C$, $K = 1.0\ cm^2/s$, $h = 90\ m$, $w = 4\ m/yr$, $\pi = 0.2$;
high, $\Delta T_{2\times} = 4.5\,°C$, $K = 2.0\ cm^2/s$, $h = 110\ m$, $w = 8\ m/yr$, $\pi = 0.4$.

Temperature changes corresponding to the low and high cases in this Figure are shown dashed in Fig. 7.8. The dashed lines in Fig. 7.9 are results for the parameter sets giving the high and low temperature limits shown in Fig. 7.8. For $\Delta T_{2\times} = 4.5\,°C$ prior to about 2060, maximizing temperature levels leads to a higher total sea level rise than maximizing thermal expansion because the relative contribution from ice melt is greater in earlier decades.

small glaciers decreases as the small glaciers themselves are depleted. Since the contributions from the large ice sheets are proportional to the time-integral of the temperature change curve they become increasingly more important with time. For further details, see Raper *et al.* (1991).

Our high, middle and low estimates of total sea level rise from 1990 to 2100 (124, 46 and 3 cm, respectively) can be compared to the 'equivalent' IPCC estimates for BAU high (109 cm), scenario B mid (47 cm) and scenario C low (18 cm). The larger range found here is due to inclusion of more of the uncertainties. While for each forcing scenario the range is larger, the middle estimates are nearly identical to those given by IPCC, despite model differences. For the extremes, the probabilities associated with our values are almost certainly lower than those associated with the IPCC extremes. The low estimate is relatively much more sensitive to the extra uncertainties than the high estimate. Raper *et al.* (1991) provide a more extensive discussion.

Although our results for any given forcing scenario have wide ranges of uncertainty, the full range may well be even larger because of uncertainties which have not been included. In tuning the climate model we did not account for the uncertainty in the past temperature rise but took the single value of 0.5 °C. In the small glacier model we accounted for the uncertainties in Meier's estimates, but we did not include the uncertainty in the initial ice volume. For Greenland, we did not include any uncertainty for the initial 1880 mass balance which was assumed to be zero. Future contributions from both the Greenland and Antarctic ice sheets are sensitive to the assumed past temperature increase over the ice sheets. We did not include any uncertainty for the past temperature rise over Greenland or for the even more uncertain estimate of warming over Antarctica. Also, we did not include a range of uncertainty for future Greenland or Antarctic temperatures relative to the global mean change. Finally, we have not considered possible melt contributions from the Antarctic Peninsula separately (see Paren, this volume). Clearly, if all the uncertainties were combined, the range of possibilities would be even larger than shown here, but the upper and lower extremes would have extremely low probabilities attached to them.

While we have stressed the major uncertainties in these projections, we must also note that most of the projected values are extremely large compared with past variations. For our mid-range projection (forcing scenario B and best-guess model parameters), the rate of global mean warming over 1990–2100 is more than 0.2 °C/decade, some five times the mean rate of warming over the past 100 years. For total sea level change, the corresponding rate of rise is around 4 cm/decade, which is two to four times the estimated rate of rise over the past 100 years. Furthermore, accompanying these changes are inescapable and increasingly large warming and sea level rise commitments. For sea level, even if we could stabilize the increase in forcing by 2030, substantial rises in sea level are likely to continue for centuries into the future.

Acknowledgements

This work has been supported by grants from the Atmospheric and Climate Research Division of the US Department of Energy, and the Commission of the European Communities' Climatology and Natural Hazards Research Programme.

References

Bryan, K. & Cox, M.D. (1972). An approximate equation of state for numerical models of ocean circulation. *J. of Physical Oceanography*, **2**, 510–17.

Budd, W.F., McInnes, B.J., Jenssen, D. & Smith, I.N. (1987). Modelling the response of the West Antarctic ice sheet to a climatic warming. In *Dynamics of the West Antarctic Ice Sheet*, C.J. van der Veen and J. Oerlemans (eds.). Reidel, pp. 321–58.

Fortuin, J.P.F. & Oerlemans, J. (1990). Parameterization of the annual surface temperature and mass balance of Antarctica. *Annals of Glaciology*, **14**, 78–84.

Gill, A.E. (1982). *Atmosphere–Ocean Dynamics*. Academic Press, New York, 662 pp.

Gornitz, V., Lebedeff, S. & Hansen, J. (1982). Global sea level trends in the past century. *Science*, **215**, 1611–14.

Grove, J.M. (1988). *The Little Ice Age*. Methuen, London.

Hoffert, M.I. & Flannery, B.P. (1985). Model projections of the time-dependent response to increasing carbon dioxide. In *Projecting the Climatic Effects of Increasing Carbon Dioxide*, M.C. MacCracken and F.M. Luther (eds.). US Department of Energy, Carbon Dioxide Research Division, pp. 149–90.

Houghton, J.T., Jenkins, G.J. & Ephraums, J.J. (eds.) (1990) *Climate Change: The IPCC Scientific Assessment*. Cambridge University Press, 364 pp.

Jones, P.D. (1990). Antarctic temperatures over the present century – A study of the early expedition record. *Journal of Climate*, **3**, 1193–203.

Levitus, S. (1982). *Climatological Atlas of the World Oceans*. NOAA Professional Paper, 13, US Government Printing Office, Washington, DC.

Manabe, S. & Stouffer, R.J. (1980). Sensitivity of a global climate model to an increase of CO_2 concentration in the atmosphere. *J. Geophys. Res.*, **85**, 5529–42.

Meier, M.F. (1984). Contributions of small glaciers to global sea level. *Science*, **226**, 1418–21.

Mikolajewicz, U., Santer, B.D. & Maier-Reimer, E. (1990). Ocean response to greenhouse warming. *Nature*, **345**, 589–93.

Mitchell, J.F.B., Manabe, S., Meleshko, V. & Tokioka, T. (1990). Equilibrium climate change – and its implications for the future. In *Climate Change: The IPCC Scientific Assessment*, J.T. Houghton, G.J. Jenkins and J.J. Ephraums (eds.). Cambridge University Press, pp. 131–72.

Muszynski, I. (1985). The dependence of Antarctic accumulation rates on surface temperature and elevation. *Tellus*, **37A**, 204–8.

Oerlemans, J. (1989). A projection of future sea level. *Climatic Change*, **15**, 151–74.

Polar Research Board (PRB) (1985). *Glaciers, Ice Sheets and Sea Level: Effect of a CO_2-induced Climatic Change*. Report of a Workshop held in Seattle, Washington, September 13–15, 1984. US DOE/ER.60235–1.

Raper, S.C.B. Wigley, T.M.L. & Warrick, R.A. (Global sea level rise: past and future. In *Proceedings of the SCOPE Workshop on Rising Sea Level and Subsiding Coastal Areas*, J.D. Milliman (ed.), John Wiley and Sons, Chichester. (in press).

Shine, K.P., Derwent, R.G., Wuebbles, D.J. & Morcrette, J-J. (1990). Radiative forcing of climate. In *Climate Change: The IPCC Scientific Assessment*, J.T. Houghton, G.J. Jenkins and J.J. Ephraums (eds.). Cambridge University Press, pp. 41–68.

Stouffer, R.J., Manabe, S. & Bryan, K. (1989). Interhemispheric asymmetry in climate response to a gradual increase of atmospheric CO_2. *Nature*, **342**, 660–2.

Warrick, R.A. & Oerlemans, J. (1990) Sea level rise. In *Climate Change: The IPCC Scientific Assessment*, J.T. Houghton, G.J. Jenkins and J.J. Ephraums (eds.). Cambridge University Press, pp. 257–82.

Wigley, T.M.L. & Raper, S.C.B. (1987). Thermal expansion of sea water associated with global warming. *Nature*, **330**, 127–31.

Wigley, T.M.L. & Raper, S.C.B. (1990a). Natural variability of the climate system and detection of the greenhouse effect. *Nature*, **344**, 324–7.

Wigley, T.M.L. & Raper, S.C.B. (1990b). Climatic change due to solar irradiance changes. *Geophysical Research Letters*, **17**, 2169–72.

Wigley, T.M.L. & Raper, S.C.B. (1991). Detection of the enhanced greenhouse effect on climate. In *Climate Change: Science, Impacts and Policy*, J. Jäger and H. Ferguson (eds.). Cambridge University Press pp. 231–42.

8

Possible future contributions to sea level change from small glaciers

M. Kuhn

Abstract

Past contributions of small glaciers to sea level rise are reviewed, based largely on the work of Meier. This work suggests that between 20% and 50% of global mean sea level rise over the past 100 years may have come from small glacier melting. Future contributions are estimated in a variety of ways using a climate scenario of a 4 °C warming over the next 100 years. Of the available small glacier ice mass (33–61 cm msl equivalent), empirical methods suggest that a large fraction could disappear. More realistic calculations based on energy balance considerations and possible equilibrium line altitude changes indicate that under the assumed scenario the small glacier contribution to changing mean sea level is likely to be around 20 cm over the next 100 years. All estimates, however, are subject to considerable uncertainty. The contribution could be as small as 9 cm or higher than 30 cm.

Introduction

The strong, world-wide retreat of mountain glaciers in the past 100 years suggests that they react significantly to even minor global-scale climatic fluctuations, thereby adding to or subtracting from the water reservoir of the world's oceans. It is now generally accepted that global mean sea level has risen by 10–15 cm in the past 100 years (see, e.g., Gornitz *et al.*, 1982, and Gornitz, this volume) and that glaciers outside of Antarctica and Greenland, small though they are, may have contributed 20 to 50% of this rise (e.g., Meier, 1984).

Since global mean temperature is expected to rise significantly over the next 100 years, further changes in the mass balance of mountain glaciers should occur. In order to estimate future sea level changes, it is therefore crucial to investigate the reaction of small glaciers to the regional climatic changes associated with future global warming. Only crude methods can be applied for this purpose, for a number of reasons. First, there are still uncertainties regarding the transfer function between regional climatic change and glacier response. Second, the regional details of future climatic change are highly uncertain. Third, the basic data sets on relevant glacier properties are inadequate – specifically, a detailed knowledge of the area/mass/altitude distribution and of surface albedo and roughness is still lacking. Fourth, while the total area of glaciers outside Antarctica and Greenland is fairly well established at about 540,000 km², estimates of their mass (and hence their potential contribution to sea level rise) differ by almost a factor of two, from 120,000 km³ to 220,000 km³ water equivalent (see section below concerning the effects of equilibrium line changes and total mass of small glaciers). These uncertainties place a severe limitation on the estimates of the future contribution from glaciers to sea level change. The uncertainty is increased by the wide range of climatic situations in which ice exists on land.

In classifying land-based ice, the usual distinction is between Antarctica and Greenland on one side and 'all other' or 'small ice caps and mountain glaciers' on the other. This is unsatisfactory because it masks a wide range of conditions within the 'all other' category, although it is a viable compromise until a better global ice inventory is established. For the sake of brevity, small ice caps and glaciers outside of Greenland and Antarctica will be referred to as 'small glaciers' in the rest of this paper. There are small, partly temperate glaciers in Greenland and on the Antarctic Peninsula as well. Their area may add 10 to 20% (depending on the definition) to the total small glacier area (Weidick, 1985). These are, however, not considered here as their fate is so closely linked to the climatic conditions of the large ice sheets that it seems logical to treat them as marginal features of the latter.

Contributions in the past 100 years

Gornitz, Barnett and others (see Gornitz, this volume) have concluded that global sea level rose 10–15 cm during the past 100 years. Meier (1984) ventured to estimate the contribution of small glaciers to this observed rise. He concluded that a total of 28 mm, or 0.46 ± 0.26 mm/yr were added to global sea level by the melting of small glaciers over the period 1900–61.

Meier identified 31 glacier regions, of which 13 had useful data spanning 1900–61 (from a total of 25 glaciers). His primary aim was to estimate the total glacier volume change over 1900–61, in terms of the equivalent mean sea level change. Because of the seminal nature of Meier's work, his method is described here in detail.

For any particular glacier, the mass change (ΔM) over interval Δt is related to the average mass balance ($\bar{b}$) by

$$\Delta M = \rho \Delta V = \rho \bar{b} \bar{A} \Delta t \tag{1}$$

where ρ is the density of water, ΔV is the water-equivalent ice volume change, $\bar{A}$ is the mean glacier area over the interval Δt, and $\bar{b}$ is the time average of the areal mean specific mass balance, usually specified in water-equivalent metres of ice per year.

In much of the glaciological literature values are referred to one balance year, i.e., b [kg/m²] is the annual specific balance. Hydrologists prefer to express the mass balance in equivalent water depth b[mm] or [m]. As changes in sea level are considered in this report, the dimension of equivalent water depth [m] is preferable. Also, as the results are to be applied to a large number of years (e.g., 100 years), it is helpful to express the balance of an individual year in terms of equivalent water depth per year [m/yr]. These units will be used for b throughout the paper.

If equation (1) is summed over all glaciers of the world, then the resulting sea level change will be Δh given by

$$A_0 \Delta h = -\sum_j b_j \bar{A}_j \Delta t \tag{2}$$

where A_0 is the area of the world's oceans (approx. 360×10^6 km^2).

Since neither ΔM nor b data are generally available, Meier estimated b values for *regions* using information on the time-average amplitude (or 'intensity') of the seasonal mass balance cycle, viz. '$\bar{a}$' given by

$$\bar{a} = 0.5(b_w - b_s) \tag{3}$$

where subscripts w and s indicate winter and summer. Using estimates of b and $\bar{a}$ from 13 regions, he calculated the mean ratio

$$<b/\bar{a}> = \tfrac{1}{13}\sum_{j=1}^{13} (b_j/\bar{a}_j) \tag{4}$$

obtaining a value of -0.23 with standard deviation of 0.12. He then assumed that, in general,

$$b_j = <b/\bar{a}> \bar{a}_j \tag{5}$$

Substituting equation (5) into equation (2) and summing over all 31 regions gives a sea level change, Δh, of

$$\Delta h = -[<b/\bar{a}>\sum_{j=1}^{31} (\bar{A}_j \bar{a}_j)]\Delta t / A_0 \tag{6}$$

The sum, $\sum \bar{A}_j \bar{a}_j$ (~ 661 km^3/yr) may be interpreted as the annual turnover in global small glacier ice volume. In order to estimate $\bar{A}_j$ and account for changes in A_j over the period 1900–61, Meier inflated present-day values (for which $\sum A_j = 540,000$ km^2) by 10%. The final result is an estimate of Δh of 28 mm, i.e., 0.46 mm/yr over 1900–61, with a subjectively estimated accuracy of 0.26 mm/yr.

This procedure rests heavily on the validity of relating b to $\bar{a}$, on the accuracy of the $\bar{a}$ estimates in regions with no b data, and on the accuracy of b estimates. Wet regions like the mountains surrounding the Gulf of Alaska gain considerable weight (one-third of global annual turnover takes place in this region), whereas the extensive, but dry areas in the Canadian Arctic become less important for sea level changes (approximately 7% of the total). The b:$\bar{a}$ link has a correlation of only 0.55, which shows that the 'relation is only crudely supported by the data' (Meier, 1984, p. 1419). It is not clear whether the uncertainty implied by this low correlation is included in the stated standard deviation for Δh over 1900–61. Regardless, the uncertainty is large and there must be a small probability that Δh is zero or negative.

Meier next estimated year-to-year variations in Δh and extrapolated the 1900–61 record to span 1885–1974 using data from three well-documented glaciers in temperate, partly maritime climate regions (see Table 8.1). It should be stressed that these reference glaciers themselves have been studied for less than 30 years and their time series have been extended to the past century only by using so-called hydrometeorological models, i.e., by using empirical relationships between mass balance and present climate and subsequent extrapolation based on past climatic records. Based on the first three glaciers in Table 8.1, Meier estimated that changes in sea level outside the 1900–61 period were at a somewhat smaller rate than 0.46 mm/yr.

Table 8.1 *Mass balance (water-equivalent metres of ice per year) of reference glaciers reconstructed from past climate records.*

	1900–1961	1884–1975
South Cascade glacier (North Cascades, USA)	− 1.13	− 0.82
Storbreen (Norway)	− 0.22	− 0.17
Glacier de Sarennes (France)	− 0.57	− 0.41
Average	− 0.64	− 0.47
Hintereisferner (Austria)	− 0.55	
(1900–1961, Finterwalder and Rentsch, 1980).		

Possible future contributions

Apart from uncertainties regarding future climatic change, extrapolation of past observations to predict future glacier behaviour is restricted by several factors:

a. The mass/altitude distribution is known only for a small number of glaciers. All extrapolations must be based on surface areas only, and even the area/altitude distribution is not well-documented.
b. Because of these inventory deficiencies, the dynamics of ice flow and adjustment to new mass balance conditions can only be modelled for a handful of glaciers, and no global mean figure can be derived.
c. Future contributions to sea level can only be inferred directly from well studied reference glaciers. These, however, are small mid-latitude glaciers so the results may not be representative of large glaciers or for polar or semi-arid regions.

For these reasons, the application of anything but simple models is not justified. Indeed, to use complex models could be deceiving to the users of such information if they are not familiar with glaciology. Examples of such simple methods will be given here, based, first, on the use of empirical relationships between mass balance and summer temperature only, and, second, on hydrometeorological and energy-balance models using a scenario of various climatic variables.

Prediction based on temperature alone

The correlation of annual balance with summer temperature is high in some regions. In the case of Hintereisferner, for instance, 55% of the mass balance variance is explained by the variance of temperature (of positive degree days, to be precise; Hoinkes and Steinacker, 1975). It is therefore tempting to use empirical relationships between observed (or estimated) glacier melt and summer temperature with predicted temperatures in order to predict future glacier melt.

The contribution to sea level rise over Δt from a single glacier (of mean area $\bar{A}$) with mean mass balance $\bar{b}$ is (cf. equation (2))

$$\Delta h = -(\bar{b}\bar{A}/A_0)\Delta t \qquad (7)$$

If $\bar{b}$ is controlled by temperature then, if the temperature increases by ΔT over a time Δt, the mean value of $\bar{b}$ over this period will be

$$\bar{b} = b_o + \tfrac{1}{2}\Delta T(\overline{\partial b/\partial T}) \tag{8}$$

where b_o is the initial value. Hence, the total sea level change due to the warming is

$$\Delta h = -\Delta t[\textstyle\sum_j \bar{b}_j \bar{A}_j]/A_o \tag{9}$$

where $\bar{b}_j$ is given by equation (8). As a rough approximation this may be written as

$$\Delta h = -\Delta t \textstyle\sum_j \bar{A}_j[<\bar{b}_o> + \tfrac{1}{2}\Delta \bar{T}<\partial b/\partial T>]/A_o \tag{10}$$

where $\Delta \bar{T}$ is the global temperature change and $<\bar{b}_o>$ and $<\partial b/\partial T>$ are representative values.

Estimates of $<\partial b/\partial T>$ may be obtained from hydrometeorological models based on interannual variability: Glacier de Sarennes, -0.6 m/yr/°C (Martin, 1978); South Cascade glacier, -1.0 m/yr/°C (Tangborn, 1980) and Hintereisferner, -0.5 m/yr/°C when referred to the mean temperature of the ablation period (Kuhn et al., 1985). These values imply that $<\partial b/\partial T>$ lies in the range -0.5 to -1.0 m/yr/°C.

Before applying these results, it must be stressed that they are only crude approximations which deserve criticism for a number of reasons:

a. The mass balance sensitivities derived from hydrometeorological models are based on annual values. This method may not be correct, since interannual data need not be applicable to longer time-scales.

b. Climate parameters other than temperature (e.g., precipitation), which may, in the analyses used here, explain only insignificant amounts of the $\bar{b}$ variance under present climatic conditions, may undergo important changes in the future and may be involved in non-linear feedbacks. Methods based on temperature alone, therefore, are more reliable for short range predictions than for the time-scale of greenhouse gas-induced climate change.

c. In equations (7) to (10), the problem was treated as if all glaciers had reached equilibrium size at the end of each year. In reality, the response time or e-folding time, τ, of glacier mass (M) is of the order of years for small glaciers and of decades-to-centuries for larger mountain glaciers.

$$M(t) = M_o + \Delta M(\Delta T)(1 - e^{-t/\tau})$$

What was used as 'initial condition' b_o in equations (7) to (10), strictly is a transient value ($b = 0$ for a steady-state glacier), and Δh, as formulated in equation (10), would be true if ΔT were the linear change from one steady-state to another and Δh were measured after a period several times (or even orders of magnitude) longer than Δt.

Only after mentioning these shortcomings is it fair to attempt predictions for future changes. In this case, a change of $\Delta T = +4$ °C in 100 years will be used as a climate scenario. From hydrometeorological models, $<\partial b/\partial T> = -0.5$ to -1.0 m/yr/°C. For the b_o contribution we use -0.20 m/yr (see below). Hence, from equation (10) we have

$$\Delta h = 100(540 \times 10^3)[0.20 + \tfrac{1}{2} \times 4(0.5 \text{ to } 1.0)]/360 \times 10^6$$

i.e., $\Delta h = 0.17$–0.33 m. Slightly higher values may be obtained if one were to use some form of weighted glacier area.

The above values are probably overestimates since they do not account for changes in glacier area or transient response effects. The upper estimate (0.33 m) is close to the total mass of water thought to be stored in small glaciers (see below). Only a qualitative conclusion can be drawn, therefore, namely that, if a 4 °C warming did occur over the next 100 years, then a large fraction of the currently existing small glacier mass would disappear over this period.

A second estimate for Δh may be obtained directly by scaling Meier's result. By dividing Meier's $\Delta h = 28$ mm for the period 1900–61 by the simultaneous ΔT, viz. $+0.3$–0.4 °C based on Jones *et al.* (1986a,b,c), we obtain

$$\Delta h/\Delta T = \frac{28 \pm 16}{0.35 \pm 0.05} = (30 - 147)\text{mm/}°\text{C}$$

Hence, for a 4 °C warming, $\Delta h = 0.12$–0.59 m. This result is quantitatively consistent with the earlier estimates.

Predictions using energy-balance models

A more realistic approach to long-range estimation of glacier melt is based on changes in mean mass balance and equilibrium line altitude as determined by an energy-balance model. This method is described in detail by Kuhn (1981, 1989). It makes use of the fairly well established relationships between changes in the various energy and mass balance components, q_i, with altitude, z, in order to determine the shift of equilibrium line altitude, δELA, following changes in the energy or mass balance, δq_i:

$$\delta ELA = -\sum \delta q_i / \sum (\partial q_i / \partial z) \tag{11}$$

In this equation, the δq_i are the independent variables while the denominator is supposed to be known. Values of q_i (see Table 8.2) can be expressed in terms of climate variables and their changes can then be calculated for any assumed scenario of climatic change. Values of δq_i can also be converted into mass balance terms using the latent heat of fusion (L_f)

$$b = \sum \delta b_i = \sum \delta q_i / L_f \tag{12}$$

The following climate change scenario was used (Table 8.2):

- a doubling of CO_2
- a temperature increase of 4 °C
- an increase of water vapour pressure (e) by 1 hPa
- an increase of cloudiness (w) by one-tenth
- an increase of accumulation (c) by 100 kg/m²/yr
- five additional summer snowfalls per year, increasing albedo from 0.4 to 0.8 on each of these days.

These changes lead to changes in sensible heat fluxes, δH, in downward long-wave radiation, δL, global short-wave radiation, δG, and latent heat fluxes, δLE.

The mass balance change predicted for this scenario is equivalent to -2.7 m/yr at the end of the 100–year period, i.e., a mean of -1.35 m/yr if all processes are assumed to proceed linearly with time from an initial condition in equilibrium. A more realistic initial condition is $b_0 = -0.20$ m/yr, so the mean value over 1986–2085 becomes $b = -1.55$ m/yr.

Table 8.2. *Changes of energy fluxes (δq_i), associated annual mass balance changes (δb_i), and changes of equilibrium line altitude (δELA) following a doubling of CO_2 (from Kuhn, 1985)*

Energy flux change* with causal climatic disturbance in brackets**	δq_i (MJ/m²/day)	δb_i (kg/m²/yr)	δELA (m)
$\delta H (T)$	+ 6.00	− 1790	+ 180
$\delta L (T)$	+ 1.40	− 420	+ 40
$\delta L (CO_2)$	+ 0.35	− 100	+ 10
$\delta L (w)$	+ 0.82	− 240	+ 25
$\delta G(w)$	− 0.63	+ 190	− 20
$\delta G (r)$	− 0.30	+ 90	− 10
$\delta LE (e)$	+ 1.70	− 510	+ 50
δc	− 0.33	+ 100	− 10
Total effect	9.01	− 2680	+ 265

Notes:
* Energy fluxes concerned:
      global radiation G
      atmospheric downward radiation L
      sensible heat H
      latent heat (condensation) LE
** temperature T
  cloudiness w
  albedo r
  vapour pressure e

What does this mean in terms of global sea level rise? From Meier's analysis (his Table 1), a 'global mean' value of b over 1900–61 can be taken as either 0.34 m/yr (unweighted average of his 13 regions) or 0.37 m/yr (unweighted average of 25 glaciers). These values correspond to a sea level rise of 0.46 ± 0.26 mm/yr so that $\Delta h/b = -1.35 \pm 0.76$ or -1.24 ± 0.70 mm/m. Thus, in rounded figures, Δh over the period 1986–2085 for a 4 °C warming is $1.55(100)[(1.3 \pm 0.7)/1000] = 0.20 \pm 0.11$ m.

This value is still likely to be an overestimate. If an average loss of $b = -1.5$ m/yr continued for 100 years, many glaciers will have disappeared before the end of that period, while others will undergo dynamic readjustment. This means that a transient maximum of $\partial h/\partial t$ is likely to occur (Kuhn, 1985) so that the 100-year Δh is likely to be less than the estimate given above. Equilibrium line arguments provide another reason for expecting the above Δh value to be an overestimate (see following section).

The effects of equilibrium line changes and total mass of small glaciers

The multiplication of uncertainties in the section concerning prediction based on temperature alone resulted in some extremely high estimates of Δh which need an independent check. Such a check is possible from two lines of reasoning: from the predicted change of equilibrium line altitude (ELA) and from a consideration of the total mass of small glaciers.

Table 8.3. *Total water equivalent of glaciers and small ice caps outside of Antarctica and Greenland and corresponding sea level rise. Note that not all are independent estimates*

Authors	Water equivalent (km³)	Maximum possible sea level rise (cm)
Shumskiy *et al.* (1964)	130,000	36
Hoinkes (1968)	220,000	61
Flint (1971)	180,000	50
UNESCO (1974)	124,000	34
Barry (1984)	120,000	33

Table 8.2 predicts an upward shift of *ELA* by 265 m. Although a quantitative statement is not possible today, it is certain that a large number of glaciers extend upward for more than 265 m above their present equilibrium line. These glaciers cannot melt completely under the conditions of the scenario in Table 8.2. With this *ELA* constraint, the total *Δh* predicted must remain substantially smaller than the *Δh* resulting if all small glaciers were to melt completely.

Unfortunately, even the upper limit for *Δh* corresponding to the melting of *all* small glaciers is not satisfactorily known. Recent publications of total water equivalent of small glaciers range from 120,000 to 220,000 km³. Shumskiy *et al.* (1964) state that 0.54% of the total land ice mass is stored in small ice sheets and mountain glaciers. If this percentage refers to a total of 24×10^6 km³, then the small glacier mass is about 130,000 km³. UNESCO (1974) and Barry (1984) seem to follow these authors by stating volumes of 124,000 and 120,000 km³, respectively. Flint (1971) states that all ice outside of Antarctica and Greenland makes up 180,000 km³. Hoinkes (1968) gives a higher value, 220,000 km³, which is again quoted by Untersteiner (1975). These values and the sea level changes corresponding to total melt are summarized in Table 8.3. No judgement of the quality of these data is attempted here.

Conclusion

Over the past 100 years 'small glaciers' may have contributed about 5 cm to the observed global sea level rise of 10–15 cm. While it is believed that this figure is accurate to ± 50%, all attempts to predict future contributions will be far less certain. A future climate scenario can be translated into cryospheric sea level changes only after conversion into the height-dependent mass distribution of glacier ice. Since the latter is not sufficiently well-known today, only crude approximations, with no temporal resolution of transient stages, can be applied to the problem. In view of the lack of basic information on glacier distribution and of uncertainities inherent in climatic predictions, values of possible future sea level rise due to ice melt cannot be given to an accuracy better than decimetres per century or millimetres per year.

Of the two methods explored in the section concerning prediction based on temperature alone, the more reliable is likely to be that based on the temperature dependence of mass balance, i.e., $< \partial b / \partial T >$, derived from calibrated hydrometeorological models. For $\Delta T = +4\,°C$, a value of Δh in the range 17–33 cm is predicted by this method. Even if the assumed temperature change is correct, these values may well be too high since they are calibrated on annual rather than centennial glacier response and since they do not account for reductions in glacier area.

Calibration of observed sea level changes with observed global temperature changes results in even higher estimates of Δh, the reason presumably being a comparatively small temperature change in the reference period and high regional variability. In general, predictions based on temperature alone are oversimplified even under the data constraints mentioned and do not do justice to the present state of the art.

In the section concerning predictions using energy-balance models, a more physically based energy-balance approach was used. A 100–year scenario of future climate was converted to future mass balance changes which in turn were empirically converted to a sea level change of around 20 cm. This may be the most reasonable approach presently feasible. It can be further improved by consideration of equilibrium line changes relative to existing glacier extent (final section), and by the limitation of total small glacier ice mass.

References

Barry, R.G. (1984). Possible CO_2-induced warming effects on the cryosphere. In *Climatic Changes on a Yearly to Millennial Basis*, N.-A. Mörner and W. Karlen (eds.). D. Reidel Publ. Co., Dordrecht, pp. 571–604.

Finsterwalder, F. & Rentsch, H. (1980). Zur Höhenänderung von Ostalpengletschern im Zeitraum 1969–1979. *Zeitschrift für Gletscherkunde und Glazialgeologie*, **16**, 111–15.

Flint, R.F. (1971). *Glacial and Quaternary Geology*. J. Wiley & Sons, New York.

Gornitz, V., Lebedeff, S. & Hansen, J. (1982). Global sea level trend in the past century. *Science*, **215**, 1611–15.

Hoinkes, H. (1968). Das Eis der Erde. *Umschau*, **10**, 301–6.

Hoinkes, H. & Steinacker, R. (1975). Zur Parametrisierung der Beziehung Klima-Gletscher. *Rivista Italiana di Geofisica*, **1**, 97–104.

Jones, P.D., Raper, S.C.B., Bradley, R.S., Diaz, H.F., Kelly, P.M. & Wigley, T.M.L. (1986a). Northern Hemisphere surface air temperature variations: 1851–1984. *J. Clim. Applied Met.*, **25**, 161–79.

Jones, P.D., Raper, S.C.B. & Wigley, T.M.L. (1986b). Southern Hemisphere surface air temperature variations: 1851–1984. *J. Clim. Applied Met.*, **25**, 1213–30.

Jones, P.D., Wigley, T.M.L. & Wright, P.B. (1986c). Global temperature variations between 1861 and 1984. *Nature*, **322**, 430–4.

Kuhn, M. (1989). The response of the equilibrium line altitude and mass balance profiles to climate fluctuations. In *Glacier Fluctuations and Climate Change*, J. Oerlemans (ed.). Kluwer Academic Publishers, pp. 407–17.

Kuhn, M. (1985). Reaction of mid-latitude glacier mass balance to predicted climatic changes. In *Glaciers, Ice Sheets, and Sea Level: Effects of a CO_2-induced Climatic Change*. Report of a Workshop held in Seattle, WA, Sept. 1984. National Academy Press, Washington, DC, pp. 248–54.

Kuhn, M. (1981). Climate and glaciers. In *Proceedings of the Canberra Symposium on Sea Level, Ice and Climatic Change*. Int. Association of Hydrological Science, Publ. No. 131, pp. 3–20.

Kuhn, M., Kaser, G., Markl, G., Nickus, U. & Pellet, F. (1985). *Hydrologische und Glaziologische Untersuchungen im Ötztal, 1952–1982*. Internal report, available from the Institute of Meteorology and Geophysics, University of Innsbruck, 104 pp.

Martin, S. (1978). Analyse et reconstitution de la serie des bilans annuels du Glacier de Sarennes, sa relation avec les fluctuations du niveau de trois glaciers du massif du Mont-Blanc (Bossons, Argentiere, Mer de Glace). *Zeitschrift für Gletscherkunde und Glazialgeologie*, **13**, 127–53.

Meier, M.F. (1984). Contribution of small glaciers to global sea level. *Science*, **226**, 1418–21.

Shumskiy, P.A., Krenke, A. & Zotikov, J. (1964). Ice and its changes. In *Research in Geophysics II*, H. Odishaw (ed.). The M.I.T. Press, Cambridge, Mass., pp. 425–60.

Tangborn, W.V. (1980). Two models for estimating climate-glacier relationships in the North Cascades, Washington, USA. *J. Glaciology*, **25**, 3–21.

UNESCO (1974). *Atlas of World Water Balance*. Hydrometeorological Publishing House, The UNESCO Press, Paris.

Untersteiner, N. (1975). Dynamics of sea ice and glaciers and their role in climatic variations. GARP Publ. Series No. 16, *The Physical Basis of Climate and Climate Modelling*. ICSU, Paris and WMO, Geneva, pp. 206–24.

Weidick, A. (1985). *The Ice Cover of Greenland*. Gletscherhydrologiske Meddelelser, nr. 85/4. The Geological Survey of Greenland, Copenhagen.

9

Possible changes in the mass balance of the Greenland and Antarctic ice sheets and their effects on sea level

J. Oerlemans

Abstract

A large part of the Earth's freshwater is stored in the ice sheets of Greenland and Antarctica: melting all the ice and spreading it over the world ocean would raise sea level by about 70 m. The ice sheets thus must be considered with great care in assessing the problem of greenhouse warming and its implications for sea level change.

In this contribution the basic physical principles governing the growth and decay of ice sheets are summarized. This involves both a schematic description of the plastic flow of ice, as well as the nature of the surface mass balance. Large differences between the Greenland and Antarctic ice sheets are evident. Antarctica experiences a much colder climate, implying that the mass balance is between accumulation and iceberg calving; ablation is negligible. In contrast, on the Greenland ice sheet the loss is roughly equally partitioned between ablation (and subsequent runoff) and iceberg calving. The implication of this difference with respect to greenhouse warming is discussed.

In broad view, one expects Antarctic accumulation to increase with temperature, while a decrease in mass balance is most likely for the Greenland ice sheet. However, the uncertainties in existing estimates of sensitivity are very large. The possibility of ice flow instabilities that can be triggered in the event of a large warming, in particular in West Antarctica, add further to this uncertainty.

Introduction: physical characteristics of ice sheets

Of all the land ice on Earth, over 99% is stored in the ice sheets of Greenland and Antarctica. Part of this ice is very 'old', i.e., was accumulated at the surface 10,000 or more years ago, and is now found in the deeper central parts of the ice caps. Some physical characteristics are summarized in Table 9.1.

The Antarctic ice sheet is about ten times as large as the Greenland ice sheet, but this does not necessarily imply that the latter is less important with regard to greenhouse warming and sea level changes. The Greenland ice sheet is in a much warmer climate, indicated by the role ablation plays in the mass budget; it contributes just as much to ice loss as calving (see Table 9.1).

Many figures in Table 9.1 are considerably uncertain; an attempt has been made to quantify this (asterisks). In particular, calving rates for the major ice sheets are very

Table 9.1. *Some data on the land ice masses of the earth. Data from Wilhelm (1975), Flint (1971), Ambach (1980), Radok (1982), Drewry (1983), Orheim (1985). Uncertainty generally within 10%, but * indicates an uncertainty of the order of 15%, ** 30%! Ablation refers to surface processes. Melting at the bottom of ice shelves is more important, but a figure is hard to give*

	Antarctica	Greenland	Other
Area (10^6 km²)	11.97	1.80	0.60*
Volume (10^6 km³)	29.33	3.0	0.18**
Mean thickness (m)	2 488	1 667	300**
Largest thickness (m)	4 700	3 400	—
Mean elevation (m)	2 000	2 080	—
Eq. sea level (m)	82.5	8.3	0.50**
Act. sea level (m)	65*	7.5*	0.50**
Accumulation (km³/yr)	2 200*	500*	—
Ablation (km³)	very small	290*	—
Calving	2 000**	210**	—
Equilibrium-line altitude (m)	− 200*	1 500	—
Turnover time (yr)	13 331	6 000	100–1 000

poorly known, making it impossible to arrive at a reliable estimate of their present-day mass budget.

The turnover time (bottom line of Table 9.1) is defined as the ice volume divided by accumulation. It gives an initially crude estimate of the time required by the ice sheets to equilibrate after a change in mass input. Apparently, the time-scale is of the order of thousands of years.

Two figures are given that refer to sea level. Equivalent sea level means total water spread out uniformly over the present ocean surface ignoring the fact that some portion of ocean volume is displaced by ice mass (i.e., the ice mass below sea level). Actual sea level represents an estimate that takes into account that displacement (geodynamics further complicate the picture!). However, on the time-scale of primary interest here, the distinction is not so relevant.

Globally speaking, the mass budget of an ice sheet can be formulated as

$$\frac{dV}{dt} = C + \int_s M ds \qquad (1)$$

Here t is time, V total ice volume, C loss of ice by calving at the margin and M the mass balance at the surface (resultant of snow deposition [precipitation and drift], ablation, evaporation, etc.); s denotes the surface of the ice sheet. When calving rates and the spatial distribution of the mass balance are known, then the volume change can, in principle, be calculated for a given geometry. However, at present the uncertainties in estimates of M and particularly C are very large, as indicated in Table 9.1. The question of the present balance is taken up again later in the text.

Although a complete description of the flow of an ice sheet is a complicated matter, some general statements can be made. For the background of the following, the reader

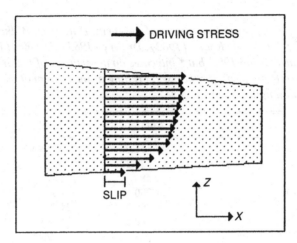

Fig. 9.1 Typical velocity profile in a glacier or ice sheet. Both slip and internal deformation contribute to the discharge.

is referred to Paterson (1981), Hutter (1983) and Oerlemans and van der Veen (1984). First of all, the changes in ice sheets are very slow. Ice velocities are of the order of metres to, in exceptional cases, kilometres per year, and accelerations do not play a role in the balance of forces. This then reduces to a balance between body forces due to gravity, and stress gradients, resulting from the 'strength' of the ice.

In most parts of the polar ice sheets, ice flow approaches what is called simple shearing flow. Horizontal pressure gradients due to variations in surface slope drive the ice outwards to the margins in an almost planar fashion.

A typical velocity profile is shown in Fig. 9.1. Ice flows down the gradient of surface elevation (along 'flowlines'), and the motion consists of two parts: basal sliding (slip) and shear within the ice mass (strain or deformation). Both parts can be found from the driving stress τ (e.g., Oerlemans and van der Veen, 1984)

$$U = U_d + U_s = k_1 H \tau^3 + (k_2 \tau^3)/N \qquad (2)$$

In this equation U is the vertical mean horizontal ice velocity, d refers to deformation and s to sliding. H is the ice thickness and N the overburden pressure on the bedrock (weight of overlying ice minus pressure of subglacial water when present). The parameters k_1 and k_2 are flow parameters. Note that, since N is proportional to H, the deformation part increases more rapidly with ice thickness than the sliding part. The break-even point generally occurs for an ice thickness of a few hundred metres, implying that for polar ice sheets deformation is far more important (in contrast to valley glaciers, where U_d and U_s are of comparable magnitude). However, when thickness increases because of deeper bedrock, pressurized water may reduce N and fast sliding may result. This occurs in many ice streams and outlet glaciers, which are responsible for a large part of the Antarctic ice outflow.

The magnitude of the driving stress is given by

$$\tau = \rho g H \mu \qquad (3)$$

Ice density is denoted by ρ, gravitational acceleration by g, surface slope by μ.

So, given the distribution of ice thickness and surface elevation (or bed topography), the velocity field can be calculated from equations (2) and (3). It is noteworthy that the ice velocity is determined by *local* properties. This is the consequence of the approximation (made implicitly here) that the total horizontal force on an ice column is exactly balanced by the basal drag. So longitudinal stress gradients do not play a role.

The evolution of an ice sheet can be obtained from the velocity field by using mass continuity. Assuming incompressibility, vertical integration of the continuity equation yields (written down here for the one-dimensional case again)

$$\frac{\partial H}{\partial t} = \frac{-\partial}{\partial x}(UH) + M \tag{4}$$

Several exercises can be done with these equations. Direct application to diagnose existing ice masses is difficult, however. First of all, considerable uncertainty exists concerning the flow parameters k_1 and k_2. The hardness parameter k_1 depends on ice temperature, crystal fabric and impurity content. These quantities are known to vary widely in ice sheets. The sliding parameter k_2 is not better known. Even if a perfect theory were available to calculate the sliding velocity, the lack of input data would spoil the case: we only have a few spot measurements of the physical appearance of the bedrock under the moving ice bodies.

A useful application of the foregoing model is the calculation of a steady-state ice sheet profile, because it shows how strongly ice thickness actually depends on mass balance and the hardness parameter. Suppose that sliding is negligible, and that the mass balance M is constant. For an ice sheet of size L, and symmetric with respect to $x = L/2$, the steady-state mass balance can be formulated as

$$(x - L/2)M = HU \tag{5}$$

For a flat bed, the surface slope equals dH/dx. Inserting the resulting expression of the driving stress in equation (3) yields the ice velocity, and upon substitution in equation (5) we obtain

$$(x - L/2)M = BH[-\rho g H\frac{dH}{dx}]^3 \tag{6}$$

This equation can be integrated with respect to x to give the ice sheet profile

$$[\frac{H}{H_o}]^{7/3} + |\frac{2x}{L} - 1|^{4/3} = 1 \tag{7}$$

H_o is the ice thickness at the centre

$$H_o = [\frac{7}{4}\frac{1}{\rho g}]^{3/7}[\frac{M}{k_1}]^{1/7}[L/2]^{4/7} \tag{8}$$

Equation (7) shows that the ice sheet profile is similar to the upper half of an ellipse. It appears that this model gives quite a reasonable description of an ice sheet profile (e.g., Paterson, 1981). Fig. 9.2 shows how ice thickness depends on the mass balance M, for various values of L. Apparently, the dependence is rather weak: according to equation (8) it goes with the 1/7th power of M. This is related to the non-linear character of the flow law, of course. Since the driving stress increases so rapidly with slope and ice thickness, only a minor increase in H is needed to adjust the ice mass discharge to the

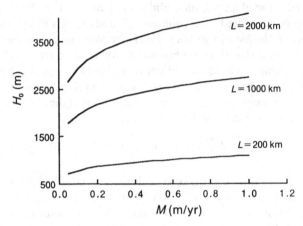

Fig. 9.2 Thickness in the centre of an ice cap (H_o), as a function of the mass balance (M). L is the size of the ice cap.

new mass balance. Also, the effect of changes in the flow parameter k_1 (or hardness $1/k_1$) is not very large. One should see this in perspective, however. When the interest is in variations of ice volume of a few percent, changes in ice hardness (associated for instance with changing temperatures or recrystallization) can be of significance.

As can also be seen in Fig. 9.2, the dependence of ice thickness on L is strong, and this presents a problem, since the coastal geometry is the decisive factor. A steep edge of the continental shelf will keep the ice margin in place even under substantially varying mass balance conditions. However, in most cases the situation is complex, i.e., a change in ice thickness will lead to a change in L. The foregoing analysis is the simplest model able to reproduce the shape of an ice sheet. It can be extended to account for axisymmetry and a spatially varying mass balance. However, real quantitative studies cannot be done without the use of numerical models, allowing full treatment of the geometry, varying mass balance, bedrock adjustment, etc.

Time-scales

In the literature there is confusion on this point, so first some definitions are outlined. The *turnover time*, as given in Table 9.1, is defined as the total volume of an ice sheet divided by the total accumulation. This is a natural first estimate of the time an ice sheet needs to react to any environmental change. The *relaxation time* is the time an ice sheet needs to come in balance again after being perturbed without any change in environmental conditions. However, the time-scale most relevant in the present discussion is the *response time*, which indicates how long it takes for the ice sheet to reach a new equilibrium after the boundary conditions have been changed. Finally, sometimes the word *growth time* is used for the time an ice sheet needs to grow from scratch (relevant in studying the Pleistocene glacial cycles, for instance).

It is to be expected that turnover time, relaxation time, and response time have the same order of magnitude. This order of magnitude is decades-to-centuries for mountain glaciers, and thousands of years for the Greenland and Antarctic ice sheets. This suggests that the problem of changes in ice volume associated with the greenhouse warming can be treated as a static problem, i.e.

$$\delta V(t) = \int_0^t \int_s \delta M(t') \mathrm{d}s \mathrm{d}t' \tag{9}$$

Here δV and δM are the changes in total ice volume and surface mass balance measured relative to the reference state at $t = 0$. So ice sheet dynamics do not come into play. This procedure is only correct for periods not longer than a few decades. As will be illustrated by an explicit model calculation later, changing mass balance *gradients* may lead to a significant dynamic response of an ice sheet margin on a time-scale of a century.

This concludes a necessarily brief introduction in the physical characteristics of ice sheets. No attention has been paid to marine ice sheets and ice shelves. This will be done later in the text.

Are the ice sheets in equilibrium now?

The large response time of an ice sheet implies that it is rarely in a dynamically steady state. Model calculations show that the deeper parts of the Antarctic and Greenland ice sheets must still be reacting to the 15,000 BP glacial–interglacial transition (e.g., Whillans, 1981).

Dahl-Jensen and Johnson (1986) have made a detailed study of a temperature profile from a deep borehole in central Greenland. The heat equation was solved for a vertical column, extending into the bedrock, for various time-dependent forcings. Comparison of measurements with the model-predicted temperatures showed that ice temperatures in the lower layers are still about 5 K below present-day equilibrium values. Consequently, on account of the temperature dependence of k_1 only, a few percent thinning of the ice sheet is yet to come. On the other hand, the softer (larger k_1) Wisconsinan ice found in the very deep layers will thin and be replaced by harder ice, thus leading to ice sheet thickening. However, these processes act on long time-scales, and will probably contribute to changes in sea level which are negligible in the present context.

Several attempts have been made to establish the overall mass balance of the Greenland ice sheet. In NAP (1985), pp. 25–8 and 155–71 (contribution by N. Reeh), it is stated that the overall tendency is a slight thickening of the ice cap in the interior parts and a slight thinning of the ice margin (ablation zone). This conclusion is mainly based on measurements along the EGIG line (central Greenland) and observations on outlet glaciers (trimlines marking maximum extents in the 19th century). In a recent study (Kostecka and Whillans, 1988) data from the EGIG-line, including ice-velocity measurements, were carefully re-analysed for comparison of surface accumulation with ice-mass discharge. Both the data from the EGIG-line and from the OSU-line (Oregon State University-line, in South Greenland from Dye 3) show that at least the inner part of the ice cap is close to equilibrium at present. Kostecka and Whillans do not agree with Grigoryan *et al.* (1985), who state that the inner part is thickening at a rate of 0.2 m/yr (based solely on a model calculation).

The final point to make with regard to Greenland concerns changes in the ablation zone. All outlet glaciers for which observations are available (this is mainly in central

West Greenland) have shown a marked retreat over the last 100 years (Weidick, 1967). This must be due to increased melting rates, and it seems likely that the vast ablation zones of the main ice cap have been subject to the same increase in melting rates. However, it is impossible to give a quantitative estimate of the amount of ice involved.

Altogether, there seem to be no indications that the Greenland ice sheet is grossly out of balance. However, the possibility of current wide-spread increased melting in the ablation zone cannot be excluded.

The Antarctic ice sheet will now be considered (see Fig. 9.3). It is generally accepted that the extent of grounded ice was larger during the last glacial, particularly in West Antarctica, and that grounding-line retreat in the Ross and Weddell seas has been initiated by sea level rise resulting from the waning of the Northern Hemisphere ice sheets (Thomas and Bentley, 1978; Denton *et al.*, 1986). An independent way of checking the Holocene deglaciation history of West Antarctica is through the interpretation of world-wide sea level data. A detailed study of this has recently been carried out by Peltier (1988). In this work a carefully designed Earth model was used to calculate changes in relative sea level due to Antarctic ice-sheet retreat. Comparison with proxy sea level data from almost 400 sites confirmed that deglaciation in Antarctica commenced significantly later than was previously assumed, probably around 13,000 BP. However, this does not give information on whether West Antarctic land ice is still retreating. The suggestion has been made that present isostatic rebound in the Ross Sea area is currently causing grounding-line advance (Thomas, 1976; Greischar and Bentley, 1980). The presented evidence is, however, not very convincing.

Interpretation of data from the Byrd core (Raynaud and Whillans, 1982) suggests that during the last glacial period the thickness of the West Antarctic ice sheet was not very much different from today. This is in accordance with geomorphological studies carried out in the Transantarctic Mountains: the evidence indicates grounded ice in the Ross Sea which was not very thick (Denton and Hughes, 1981). If this is correct, excursions from the present state are limited, and the present state could be in equilibrium with the prevailing climatic conditions. However, as will become clear later on, the dynamics of a marine ice sheet are a very complicated matter. Many feedback loops exist, and the possibility of intrinsic pulsating behaviour of the ice sheet–ice shelf system cannot be ruled out.

The East Antarctic ice sheet has the longest response time, but probably has also undergone only minor changes in boundary conditions over the last 100,000 years. It is generally accepted that accumulation (which is very low: about 0.1 m of water-equivalent/yr when averaged over the ice sheet) is mainly determined by air temperature, or better by the saturation vapour pressure in the lower atmosphere. The glacial–interglacial transition should thus have led to increasing accumulation and slight thickening of the ice. This has been confirmed by analysis of the Vostok core (Barnola *et al.*, 1987). A detailed calculation for a flow line passing through the Vostok station has been carried out (Huybrechts and Oerlemans, 1988). In this model the full thermo-mechanical coupling is taken into account. In one experiment the forcing function consisted of a 6 K warming in the period 16,000–10,000 BP, in accordance with the oxygen isotope signal from Vostok. The accumulation rate was changed accordingly. This led to a + 78 m change in ice thickness for the period 16,000–0 BP. In the case of an unaltered climate, another + 12 m increase will occur in the next 10,000 years, and then

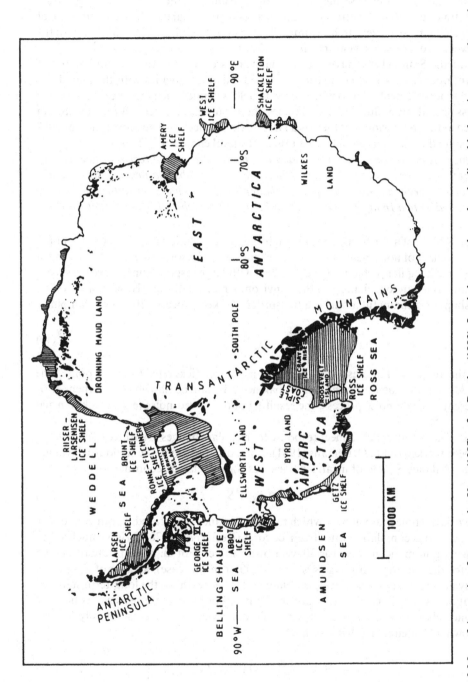

Fig. 9.3 Map of Antarctica, from van der Veen (1987), after Bentley (1983). Ice shelves are shown by shading, black regions are only partly glaciated (mountainous regions mainly). (Reprinted by permission of Kluwer Academic Publishers).

a 39 m drop will occur before the ice sheet settles to a steady state (this will take another 100,000 years!). In this case the effect of the warming of the deeper ice layers has a longer time-scale than the direct effect of increasing mass balance. The contribution of this to changing sea level can be estimated to be about -0.1 mm/yr. Compared to the sea level rise over the last century (about 1.5 mm/yr), this is negligibly small.

Budd and Smith (1985) have tried to make an assessment of the state of balance of the Antarctic ice mass by comparing so-called balance velocities with the actual ice-velocity measurements. The balance velocity field is defined such that the divergence of the associated mass flux exactly matches the accumulation field. Budd and Smith arrived at the conclusion that there is a large-scale balance, with an uncertainty of 20% (they state that this corresponds to a rate of sea level change of 1.2 mm/yr).

Hence, as in the case of the Greenland ice sheet, there is no direct evidence that the Antarctic ice sheet is out of balance. Unfortunately, the observations are scarce and a 20% error margin has to be accepted. This implies that a change in Antarctic ice mass corresponding to a 1mm/yr sea level rise or fall is undetectable with the present data sets.

The possibility of a stochastic component in ice sheet evolution also deserves attention. Due to a chain of non-linear interactions, climate exhibits sometimes regular, but most of the time irregular, behaviour. Subsystems with large response times, like ice sheets, may integrate random changes in their environmental conditions (Hasselmann, 1976). The simplest meaningful model is a first-order Markov process with linear damping

$$\frac{dV'}{dt} = \frac{V'}{T_R} + M' \tag{10}$$

V' is the quantity of interest (ice volume, or mean ice thickness), T_R the response time and M' the stochastic forcing (mass balance). It can be shown that, when $M'(t)$ represents a white-noise process, the resulting output is red-noise (e.g., Blackman and Tukey, 1958).

The idea of Antarctic ice sheet response to irregular changes in surface mass balance has been worked out in Oerlemans (1981). For a discrete series, the standard deviation of mass balance S_M and of mean ice thickness S_H are related according to

$$S_H = [t^* T_R]^{1/2} S_M \tag{11}$$

Here t^* is the time interval over which the mass balance is averaged. From data on the variability in accumulation rates from ice cores (Reeh *et al.*, 1978) S_M is as much as 4% of the long-term mean value for 30–year mean values. For an appropriate relaxation time for the Antarctic drainage systems (2,100 yr in this case), a value of $S_H = 1.4$ m (equivalent to 4.6 cm of sea level) is obtained. However, this is the long-term standard deviation. A question more relevant to the projection of future sea level is how the variance changes over time from a specific date when sea level is actually known. Following Hasselmann (1976) we have

$$S(t)^2 = S_\infty^2 [1 - \exp(-t/T_R)] \tag{12}$$

A simple calculation now shows that for the standard deviation of sea level change after 100 yr, for example, one should take 2.1 mm, which is really small.

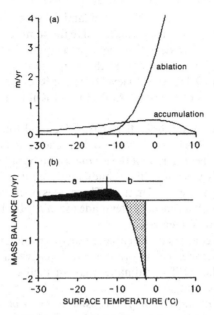

Fig. 9.4 (*a*) and (*b*) Generalized curves of ablation and accumulation in dependence of (surface) temperature. Depending on which part of the *M*-curve applies, a climatic warming may lead to increasing (Antarctic) or decreasing (Greenland) mass balance.

Changes in the surface mass balance of the Greenland and Antarctic ice sheets

Greenland and Antarctica will react differently to a climatic warming, because they are situated in entirely different climates. The complexity of the problem stems from the fact that the mass balance–temperature curve is multi-valued. This is illustrated in Fig. 9.4. The generalized ablation curve is monotonically increasing with annual temperature, but the accumulation curve shows a pronounced maximum. The ablation curve is based on measurements on the Greenland ice sheet (Ambach, 1979) and also incorporates the assumption that the seasonal cycle is constant under climate change. The accumulation curve should be considered as schematic, because it is a quantity that varies enormously from place to place (in particular associated with continentality and orographic effects). Nevertheless, for very low temperatures, the strongly reduced moisture content of the atmosphere ultimately suppresses accumulation. This is presently the case over the interior parts of the Antarctic continent, where accumulation is as little as a few centimetres of water equivalent per year.

With regard to the generalized mass balance curve, a difference between the Greenland and Antarctic ice sheets is apparent. The Antarctic ice sheet is in such a cold climate, that most of its surface is found to the left of the maximum in Fig. 9.4(*a*) and (*b*). Hence, for a moderate warming, an increasing total mass balance would be expected. On Greenland, however, a large part of the ice sheet surface is to the right of the maximum, and it is very likely that the total surface balance will decrease in the event of a climate warming. The question now arises whether these statements can be quantified. Very few attempts have been undertaken, in fact, and they are briefly reviewed below.

Ambach (1980) made an estimate for the Greenland ice sheet, essentially based on the measurements along the EGIG profile (obtained in the summers of 1959 and 1967). These measurements were used to derive the change in equilibrium-line altitude with (summer) temperature. Applying the results to the entire ice sheet, Ambach yielded a sensitivity corresponding to a 0.4 mm/yr sea level rise per 1 K temperature increase. In a recent, more detailed analysis of the data, Ambach and Kuhn (1989) end up with a figure of 0.34 mm/yr/K.

Another study was made by Bindschadler (1985). He used the so-called perfectly plastic ice sheet model (e.g., Weertman, 1961), to fit a schematic Greenland ice sheet. This allows the possible effects of retreat of the ice margin to be taken into account in a very crude way. For a 3.25 K warming, Bindschadler gives a rate of sea level rise of 1.3 mm/yr; for a 6.5 K warming a rate of 3.5 mm/yr. These rates correspond to 0.4 and 0.54 mm/yr/K. It should be noted that Bindschadler made use of Ambach's data on ablation and temperature lapse rate on Greenland.

Another aspect concerns changes in the calving rate of the outlet glaciers. Based on an empirical relationship between calving rate, water discharge and thickness at the grounding line, Bindschadler (1985) estimates that an increase in ablation would increase the calving rate by about 30%.

From the work mentioned above, 0.5 mm/yr is considered as the most likely value of sea level rise for a 1 K warming. Four points should be kept in mind, however: (1) the estimates refer to changes in summer temperature; (2) ablation data for the Greenland ice sheet are limited to a very small part of the central western area and so extrapolation may introduce a large error; (3) possible changes in accumulation rates are not considered; (4) it is not clear how universal the relation between calving rates and water discharge actually is. In view of this, the uncertainty in the estimate should be set to 50%!

A study for the Antarctic ice sheet was carried out some time ago by the author (Oerlemans, 1982). Here a numerical ice-flow model was used to simulate the evolution of the ice sheet for various scenarios of climatic change. This time-dependent ice sheet model [essentially based on the two-dimensional version of equations (2)–(4)] takes into account bed geometry and includes a fairly detailed parameterization of the mass balance [parameters: surface slope, elevation, sea level temperature, continentality]. The performance of this parameterization is shown in Fig. 9.5. The large-scale structure of the mass balance field shows up well, but some small-scale variations are certainly missing. However, this is not considered to be very serious, since the experiments discussed here are 'perturbation experiments' anyway. Although another approach could have been to impose the present-day mass balance as observed and perturb it, this was not done as it leaves out some important feedbacks (in particular the coupling of surface slope and accumulation).

Several runs were done with specific scenarios of sea level temperature and accumulation rates, and these were compared with a control run. This is necessary, since the model ice sheet drifts to a larger ice volume even without changing climate. The changes in sea level temperature and accumulation rates increase linearly in time during 100 yr, and are constant afterwards. The results are summarized in Fig. 9.6. The [+3 K, +12%] run corresponds to a doubled-CO_2 equilibrium climate as calculated by Manabe and Stouffer (1980).

The most striking result of these experiments is probably the irregular behaviour of

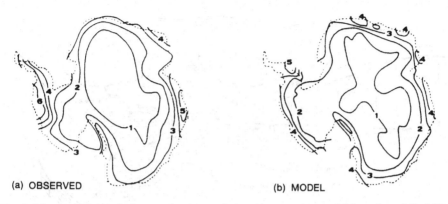

Fig. 9.5 Surface accumulation on the Antarctic ice sheet, as observed (*a*), and modelled (*b*). Units: 0.1 m ice depth/year. (Oerlemans, 1982. Reprinted by permission of the Royal Meteorological Society).

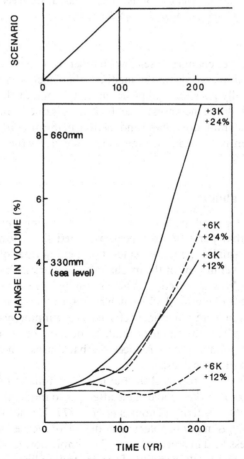

Fig. 9.6 Changes in ice volume (sea level) of Antarctica as calculated with a numerical ice sheet model. The imposed scenario is shown at the top. Four cases have been run; the curves are labelled accordingly.

J. Oerlemans

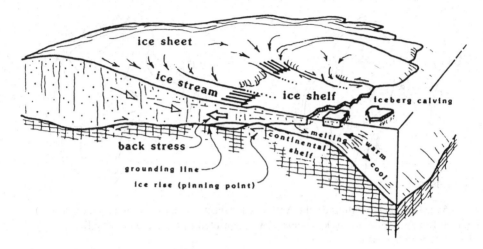

Fig. 9.7 Factors controlling the dynamics of the West Antarctic ice sheet. See text for further explanation. From NAP (1985), modified by van der Veen.

total ice volume for larger changes in sea level temperature. Ice volume first increases (increase in accumulation exceeds that in melting), then decreases slightly (increase in melting wins) and finally goes up again (the marginal zone of the ice sheet starts to thicken and the mass balance increases). The thickening is due to an increase in ice flux, forced by the larger accumulation inland and melting in the margin, i.e., by an enlarged surface slope. Apparently, even on the time scale of a century the dynamic response is not always negligible!

Possible ice-flow instabilities

It has been argued by several authors (Mercer, 1978; Thomas *et al.*, 1979; Hughes, 1981; Lingle, 1985) that parts of ice sheets grounded far below sea level may be extremely sensitive to small changes in sea level or melting rates of the buttressing ice shelves. A schematic overall view of the mechanisms at work is shown in Fig. 9.7. The situation is typical for West Antarctica. Where grounded ice starts to float an ice shelf forms (several hundreds of metres thick) which runs aground again in several places to form ice rises. Such points are sometimes referred to as pinning points and they act to produce a 'back stress' on the main grounded body of ice. It is conceivable that disappearance of the pinning points reduces the back stress and leads to increased melting at the bottom of ice shelves.

In qualitative terms the picture is clear, but it is very difficult to make quantitative statements. A number of workers have made attempts to model the ice sheet–ice shelf system and to study its sensitivity (Thomas *et al.*, 1979; Lingle, 1985; van der Veen, 1987; Budd *et al.*, 1987). A major problem in modelling of the ice sheet–ice shelf system is to deal with the stress field in a proper way. The simple flow model discussed earlier cannot be applied. In fact, in the vicinity of the grounding line all components of the stress field can become of comparable magnitude, making it impossible to use rigorous simplifications (see Herterich (1987) for an interesting discussion). Even a flow-line

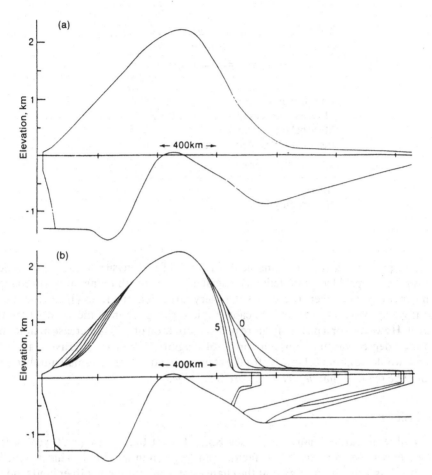

Fig. 9.8 Example of the retreat of the West Antarctic ice sheet as calculated by Budd *et al.* (1987). The flow line to which this calculation applies is shown in Fig. 9.4. Ice-shelf thinning rates were instantaneously set to three (Pine Island side, left) and seven (Ross Ice Shelf, right) times the present-day values at $t = 0$ year. (*a*) The initial profile, (*b*) profiles for every 100 years of integration, up to 500 years. (Reprinted by permission of Kluwer Academic Publishers).

model (i.e., a one-dimensional model) is already quite complicated. Here an extensive study of van der Veen (1986) is mentioned. He concludes that earlier estimates of the sensitivity of West Antarctica were too large. His model experiments suggest that a number of feedback processes, omitted in earlier work, tend to stabilize the position of the grounding line. Budd *et al.* (1987) give a fairly detailed discussion on the response of the West Antarctic ice sheet to a climatic warming. Their considerations are based on a large number of numerical experiments with flow-band models. An example is given in Fig. 9.8. The modelled flow band, shown in Fig. 9.4, is assumed to go through the most vulnerable parts of the West Antarctic ice sheet.

Starting from an initial profile, Budd *et al.* forced the model by prescribing thinning rates for the ice shelves. The larger the thinning rate, the quicker the retreat. Although

Table 9.2. *Estimates of rates of sea level change per degree warming, in mm/yr*

	Rate of change
Internal variability*	0.0 ± 0.02
Mass balance Greenland	+ 0.5 ± 0.25
Mass balance Antarctica	− 0.7 ± 0.35
Instability West Antarctica	+ 0.2 ± 0.5
Total from ice sheets	+ 0.0 ± 0.65

Notes:
*uncertainty 15%

the model probably gives a reasonable description of the physics of the ice sheet–ice shelf system, the problem is shifted: it is crucial now how the thinning rates will change when climate gets warmer. It appears that very large thinning rates (10 to 100 times present values) would be required to cause rapid disintegration of the West Antarctic ice sheet. However, for a probably more realistic situation of a 50% increase in thinning rate for a 1 degree warming (this is an order-of-magnitude estimate), the associated sea level rise would be about 0.1 mm/yr for the coming decades. This is a small number, but it should be realized that *uncertainties are very large*.

Conclusions

For small temperature changes (< 2 deg K) and short time-scales (< 30 years), the changes in sea level due to various factors can be given in mm/yr/K. This is done in Table 9.2. It is a curious finding that the changes on Antarctica and Greenland tend to cancel, thus leaving a negligible ice sheet contribution to a sea level rise. However, the uncertainties (standard deviations) do not! Roughly speaking, there is still a 30% chance of being outside the − 0.7 to + 0.7 mm/yr/K range.

For larger temperature changes and longer time-scales, it is not appropriate to work with constant figures.

The present figures are certainly within the range generally found in the literature. Precise comparisons are difficult to make, because of different approaches to the problem. To make a projection of future sea level, Robin (1986), for instance, actually used a linear relationship between global mean surface temperature and sea level, because it is 'the best one can do'. However, there are two problems with such an approach: (i) a small error in the input data for the regression analysis may lead to very large errors in predicted sea level, and (ii) it is doubtful whether a relation derived from data in a 0.5 K temperature range can be applied to temperature perturbations of up to 3 or 4 K. Still, there is no basic disagreement with the findings of this paper.

Table 9.2 shows that all listed processes have comparable uncertainties. One could conclude that work on anything is useful. A few points can be mentioned, however, that currently do not get enough attention:

i. Mass balance measurements on ice sheets and glaciers are very important and should be continued. Long series are needed to derive relations between mass balance and meteorological parameters.

ii. The margin of the Greenland ice sheet deserves special attention. Only a few ablation studies have been done, too few to assess the sensitivity of the entire ice mass.

iii. Monitoring of ice thickness (repeated levelling of selected transects) can be very helpful in early detection of significant changes. Here small ice caps, glaciers and marginal zones of the major ice sheets (where fixed points are available) offer many possibilities. Thickness changes of inland ice are a more complicated matter, but here satellites may help in the future.

References

Ambach, W. (1980). Anstieg der CO_2-Konzentration in der Atmosphäre und Klimaänderung: Mögliche Auswirkungen auf den Grönlandischer Eisschild. *Wetter und Leben*, **32**, 135–42.

Ambach, W. (1979). Zur Nettoeisablation in einem Höhenprofil am Grönlandischen Inlandeis. *Polarforschung*, **49**, 55–62.

Ambach, W. & Kuhn, M. (1989). Altitudinal shift of the equilibrium line in Greenland calculated from heat balance characteristics. In *Glacier Fluctuations and Climatic Change*, J. Oerlemans (ed.). Reidel, pp. 281–8.

Barnola, J.M., Raynaud, D., Korotkevitch, Y.S. & Lorius, C. (1987). Vostock ice core provides 160,000-year record of atmospheric CO_2. *Nature*, **329**, 408–14.

Bentley, C.R. (1983). West Antarctic ice sheet: diagnosis and prognosis. In *Proceedings: Carbon Dioxide Research Conference: Carbon Dioxide, Science and Consensus*. Conference 820970. Department of Energy, Washington DC, pp. IV.3–IV.50.

Bindschadler, R.A. (1985). Contribution of the Greenland ice cap to changing sea level: present and future. In *Glaciers, Ice sheets and Sea Level: Effects of a CO_2-induced Climatic Change*. National Academy Press, Washington, pp. 258–66.

Blackman, R.B. & Tukey, J.W. (1958). *The Measurement of Power Spectra*. Dover Publ., New York, 190 pp.

Budd, W.F., McInnes, B.J., Jenssen, D. & Smith, I.N. (1987). Modelling the response of the West Antarctic ice sheet to a climatic warming. In *Dynamics of the West Antarctic Ice Sheet*, C.J. van der Veen & J. Oerlemans (eds.). Reidel, pp. 321–58.

Budd, W.F. & Smith, I.N. (1985). The state of balance of the Antarctic ice sheet, an updated assessment 1984. In *Glaciers, Ice sheets and Sea Level: Effects of a CO_2-induced Climatic Change*. National Academy Press, Washington, pp. 172–7.

Dahl-Jensen, D. & Johnson, S.J. (1986). Palaeotemperatures still exist in the Greenland ice sheet. *Nature*, **320**, 250–2.

Denton, G.H. & Hughes, T.J. (1981). *The Last Great Ice Sheets*. Wiley–Interscience, New York.

Denton, G.H., Hughes, T.J. & Karlen, W. (1986). Global ice-sheet system interlocked by sea level. *Quat. Res.*, **26**, 3–26.

Drewry, D. (1983). *Antarctica, Glaciological and Geophysical Folio*. Scott Polar Research Institute, Cambridge.

Flint, R.F. (1971). *Glacial and Quaternary Geology*. John Wiley, New York, 892 pp.

Greischar, L.L. & Bentley, C.R. (1980). Isostatic equilibrium grounding line between the West Antarctic inland ice sheet and the Ross ice shelf. *Nature*, **283**, 651–4.

Grigoryan, S.S., Buyanov, S.A., Krass, M.S. & Shumskiy, P.A. (1985). The mathematical model of ice sheets and the calculation of the evolution of the Greenland ice sheet. *J. Glaciol.*, **31**, 281–92.

Hasselmann, K. (1976). Stochastic climate models. *Tellus*, **17**, 321–33.

Herterich, K. (1987). On the flow within the transition zone between ice sheet and ice shelf. In *Dynamics of the West Antarctic Ice Sheet*, C.J. van der Veen & J. Oerlemans (eds.). Reidel, pp. 185–202.

Hughes, T. (1981). The weak underbelly of the West Antarctic ice sheet. *J. Glaciol.*, **27**, 518–25.

Hutter, K. (1983). *Theoretical Glaciology*. Reidel, Dordrecht, 510 pp.

Huybrechts, Ph. & Oerlemans, J. (1988). Thermal regime of the East Antarctic ice sheet: a numerical study of the role of the dissipation-strain rate feedback with changing climate. *Ann. Glaciol.*, **11**, 52–9.

Kostecka, J.M. & Whillans, I.M. (1988). Mass balance along two transects of the West side of the Greenland ice sheet. *J. Glaciol.*, **34**, 31–9.

Lingle, C.S. (1985). A model of a polar ice stream and future sea level rise due to possible drastic retreat of the West Antarctic ice sheet. In *Glaciers, Ice sheets and Sea Level: Effects of a CO_2-induced Climatic Change*. National Academy Press, Washington, pp. 317–30.

Manabe, S. & Stouffer, R.J. (1980). Sensitivity of a global climate model to an increase of CO_2 concentration in the atmosphere. *J. Geophys. Res.*, **85**, 5529–42.

Mercer, J.H. (1978). West Antarctic ice sheet and CO_2 greenhouse effect: a threat of disaster. *Nature*, **271**, 321–5.

NAP (1985). *Glaciers, Ice sheets and Sea Level: Effect of a CO_2-induced Climatic Change*. National Academy Press, Washington, 330 pp.

Oerlemans, J. (1982). Response of the Antarctic ice sheet to a climatic warming: a model study. *J. Clim.*, **2**, 1–11.

Oerlemans J. (1981). Effect of irregular fluctuations in Antarctic precipitation on global sea level. *Nature*, **290**, 770–2.

Oerlemans, J. & van der Veen, C.J. (1984). *Ice Sheets and Climate*. Reidel, 217 pp.

Orheim, O. (1985). Iceberg discharge and the mass balance of Antarctica. In *Glaciers, Ice sheets and Sea Level: Effects of a CO_2-induced Climatic Change*. National Academy Press, Washington, pp. 210–15.

Paterson, W.S.B. (1981). *The Physics of Glaciers, 2nd edition*. Pergamon Press.

Peltier, W.R. (1988). Lithospheric thickness, Antarctic deglaciation history and ocean basin discretization effects in a global model of postglacial sea level change: a summary of some sources of non-uniqueness. *Quat. Res.*, **29**, 93–112.

Radok, U. (1982). *Physical Characteristics of the Greenland Ice Sheet*. CIRES Report, (Boulder).

Raynaud, D. & Whillans, I.M. (1982). Air content of the Byrd core and past changes of the West Antarctic ice sheet. *Ann. Glaciol.*, **3**, 269–73.

Reeh, N., Clausen, H.B., Gundestrup, N., Johnson, S.J. & Stauffer, B. (1978). Secular trends of accumulation rates at three Greenland stations. *J. Glaciol.*, **20**, 27–30.

Robin, G. de Q. (1986). Changing sea level. In *The Greenhouse Effect, Climatic Change and Ecosystems*, B. Bolin, B.R. Döös, J. Jäger & R.A. Warrick (eds.). John Wiley and Sons, New York, pp. 323–59.

Thomas, R.H. (1976). Thickening of the Ross ice shelf and equilibrium state of the West Antarctic ice sheet. *Nature*, **259**, 180–3.

Thomas, R.H. & Bentley, C.R. (1978). A model for Holocene retreat of the West Antarctic ice sheet. *Quat. Res.*, **10**, 150–70.

Thomas, R.H., Sanderson, T.J.D. & Rose, K.E. (1979). Effects of a climatic warming on the West Antarctic ice sheet. *Nature*, **227**, 355–8.

van der Veen, C.J. (1987). Longitudinal stresses and basal sliding: a comparative study. In *Dynamics of the West Antarctic Ice Sheet*, C.J. van der Veen & J. Oerlemans (eds.). Reidel, pp. 223–48.

van der Veen, C.J. (1986). *Ice Sheet, Atmospheric CO_2 and Sea Level*. Thesis, University of Utrecht (The Netherlands).

Weertman, J. (1961). Stability of ice-age ice sheets. *J. Geophys. Res.*, **66**, 3783–92.

Weidick, A. (1967). Observations on some Holocene glacier fluctuations in West Greenland. *Meddr Gronland*, **165**, 202 pp.

Whillans, I.M. (1981). Reaction of the accumulation zone portions of glaciers to climate change. *J. Geophys. Res.*, **86**, 4274–82.

Wilhelm, F. (1975). *Schnee- und Gletscherkunde*. Walter de Gruyter, Berlin, New York, 434 pp.

10

The Antarctic Peninsula contribution to future sea level rise

J.G. Paren, C.S.M. Doake and D.A. Peel

Abstract

It is not clear whether the Antarctic Peninsula, the warmest sector of Antarctica, is today contributing positively or negatively to sea level rise. In the last 40 years the Antarctic Peninsula has witnessed a 2 °C rise in temperature and precipitation (obtained from ice cores) has increased also. At sea level, where surface snow and ice melts each summer, there has been a dramatic disintegration of the Wordie Ice Shelf and an expansion of summer snow-free ground beside coastal Antarctic stations. At a few sites repeated surveys of the height of the ice sheet surface between benchmarks on nunataks show only minor changes of elevation, implying that there the ice sheet is stable. The absence of a strong link between Antarctic Peninsula temperature and mass balance makes uncertain a future prediction of the impact of the region on sea level. The best estimate is that over the next 100 years an additional contribution to sea level rise should not lie outside -3 cm to $+1.5$ cm.

Introduction

The total volume of ice in the Antarctic Peninsula has been estimated to be about 180×10^3 km³ (Doake, 1985). If all the ice melted, global sea level would rise on average by about 50 cm, roughly equivalent to the effect of melting all of the world's small glaciers. The Antarctic Peninsula region, therefore, has the potential to add significantly to global mean sea level rise. Whether this is an important factor, however, depends on how rapidly the net ice balance of the Peninsula might respond to global-scale warming and on the direction of the response. The purpose of this paper is to provide some quantitative estimates of the range of possible effects over the next century or so.

At present, it is not clear whether the net mass balance of the whole area is positive or negative. In some areas there appears to be an overall increase in ice volume, while in other areas there is evidence for mass loss. The rugged nature of the terrain makes it difficult to extrapolate sparse data with much degree of confidence.

The annual rate of snow accumulation for the Peninsula is more than three times the average for the whole of Antarctica. At 226 km³/yr of ice, it is equivalent to a change in sea level of 0.6 mm/yr. The rapid throughput of ice arises because the Antarctic Peninsula stretches further north than any other region of the continent. The climate is warmer, with greater potential for snow precipitation than areas further south. However, there is relatively little surface melting during the summer. Most of the ice loss is by calving of the ice sheet at sea level. There is a distinct difference between the climatic regime on the east and west coasts, the east coast being affected by the Weddell Sea and experiencing temperatures about 7 degrees lower than on the west coast at similar latitudes (Reynolds, 1981). On the western side, the interannual variability of temperature is higher than in any other sector of the Antarctic or in the Southern Ocean (Limbert, 1984).

If the climate warmed, one of the immediate effects could be an increase in snow accumulation. This would not be balanced by an increase in ice loss for a number of years. The time lag before the Antarctic Peninsula ice sheet reached a steady state with a new accumulation regime would be of the order of 1,000 years. However, it is unlikely that the accumulation rate is ever constant and therefore the ice sheet is always attempting to adjust its size to follow a climatically determined pattern.

To understand the effects of a changing climate, it is necessary to understand what factors determine the accumulation rate and how the loss and discharge of ice is controlled.

Accumulation

Past precipitation rates have been measured in ice cores dating back about 200 years from sites on James Ross Island on the northernmost tip of the Antarctic Peninsula, Dolleman Island, abutting the Larsen Ice Shelf and the Weddell Sea, Dyer Plateau in central Palmer Land, and Gomez Nunatak in Ellsworth Land/southern Palmer Land (see Fig. 10.1). These ice cores show considerable interannual to interdecadal variability in accumulation rate (see Fig. 10.2, from Peel, 1992). Since the 1950s, there has been an upward trend. At James Ross Island, values appear to have increased generally since about 1850 (Fig. 10.2). The generally rising temperature detected in the Antarctic Peninsula since about 1960 correlates with increasing rates of accumulation at these drill sites and with an increasing frequency of days of precipitation recorded at Faraday (Peel *et al.*, 1988).

Repeated measurement of elevation profiles across an ice surface give a direct measure of ice volume changes. Six level lines were established in the Antarctic Peninsula between 1972 and 1976 and were re-levelled in the 1985–6 summer (Paren and Richardson, 1988). Changes in four profiles on Alexander Island ranged from a thickening of 66 mm/yr to a thinning of 83 mm/yr: on average, the sites showed a thickening of 6 mm/yr. A thickening rate of 5 mm/yr was found for a cold site on the spine of Palmer Land at 1,600 m elevation. The average accumulation rate is around 500 mm/yr at all the sites. In a snow field in the Batterbee Mountains, thickening averaged 165 mm/yr. At all sites investigated, except in the Batterbee Mountains, glacier flow is in close balance (to 1%) with the present climate, despite the warming trends occurring in the region.

Ice loss

A noticeable feature of the west coast of the Antarctic Peninsula is the way in which land ice terminates at the shoreline. However, there are some areas of bare ground at

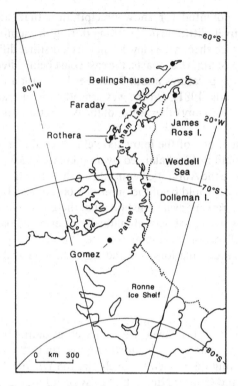

Fig. 10.1 The Antarctic Peninsula with the locations of chosen meteorological stations and ice core drilling sites.

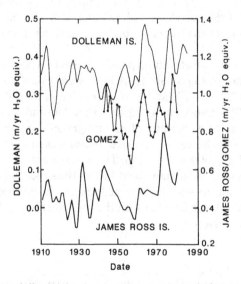

Fig. 10.2 Five-year binomially weighted average snow accumulation rate in m/yr water equivalent from Gomez, Dolleman Island and James Ross Island.

the coastline providing ideal sites for Antarctic stations. There is evidence in the vicinity of stations of a local retreat of the land margin over the last 20 years, which in one case returned it to a position occupied 50 years ago. Few ice shelves form on the west coast and this has been attributed to the existence of a climatic limit for ice shelf formation which is reached at about 70°S. The Wordie Ice Shelf, just to the north of this limit, has undergone considerable decay in recent years (Doake and Vaughan, 1991) and the George VI Ice Shelf to the south is also retreating at its northern ice front. A study of archival material could allow more quantitative and longer-term assessments to be made. Relatively rapid, short time-scale fluctuations of the ice sheet and the land ice are expected. Consequently, it is difficult to separate out the long-term climatic signal.

On the east coast, the Larsen Ice Shelf dominates the ice loss regime. Periodic and frequently spectacular changes in the ice front occur as icebergs calve, but these changes are not in themselves indicators of climatic change nor do they help in understanding how a change in climate would affect the mass balance of the area. A continuous retreat in the northernmost part of the ice shelf mirrors that of Wordie and George VI ice shelves.

Recent evidence for climatic change

Faraday, Rothera (Marguerite Bay) and Bellingshausen stations, with 40-year climate records, are climatically representative of the west coast of the Antarctic Peninsula. Since 1947, there has been a general warming trend. The Antarctic Peninsula average temperature has risen 2.2 °C in this interval (see Fig. 10.3 and Table 10.1). The three years 1988–90 were exceptionally warm, especially 1989 which was the warmest year at all three sites. The 40-year warming trend is mirrored in temperature profiles from drill holes; these data suggest that surface ice temperatures have warmed by 1 °C on the Palmer Land plateau (Nicholls and Paren, 1992) and between 2–3 °C on George VI Ice Shelf during the same period (Nicholls *et al.*, unpublished). The temperature trend for the Antarctic Peninsula is large compared with the Southern Hemisphere 40–60°S (0.53 °C, 1947–90) and Antarctica as a whole (0.72 °C, 1957–90). Because of the greater interannual variability over the Peninsula, however, the trends here are less significant than those over 40–60°S (Table 10.1).

A correlation between the warming trend and increased accumulation was noted above. Accumulation rates at the three ice-core sites are poorly correlated (see Table 10.2), but all three sites show a positive correlation with Peninsula mean temperature (Table 10.2, row E). The strength of the relationship is, however, weak ($r = 0.25$ between mean temperature and mean accumulation rate), insufficient to allow any quantitative estimate of future accumulation rate changes for realistic warming scenarios.

Estimates of contribution to future sea level rise

As we do not know whether or not the ice sheet over the Antarctic Peninsula is in mass balance equilibrium at present, and since the temperature/accumulation link is weak, one can only give a broad range of possible future sea level effects. The cases considered below give the additional changes to sea level expected for future potential changes to accumulation rates or ice loss.

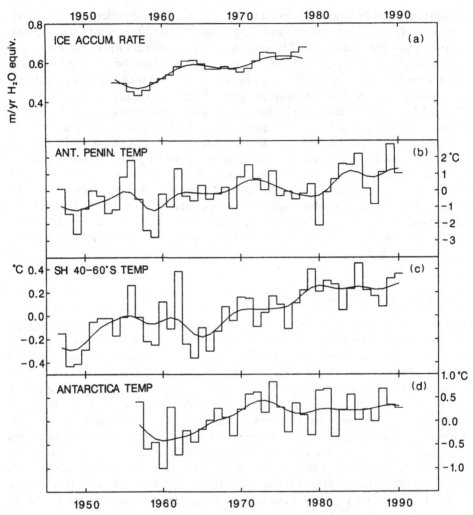

Fig. 10.3 Precipitation and temperature for the Antarctic Peninsula region from 1947 onwards compared to temperatures from 40–60°S and for Antarctica. The precipitation data (*a*) are the mean of the ice accumulation rates from Gomez, James Ross Island and Dolleman Island in m/yr water equivalent, from Peel (1991); panel (*b*) shows Antarctic Peninsula mean temperature changes (average of Rothera, Faraday and Bellingshausen); panel (*c*) shows temperature averaged over 40–60°S (mainly sea surface temperatures, data from Jones *et al.*, 1991); and panel (*d*) shows the Antarctic average (updated from Raper *et al.*, 1984).

Table 10.1. *Trends in annual mean temperature (total trend in °C). t-values (shown in brackets) indicate that these trends are highly statistically significant*

	1947–1990	1957–1990
Faraday	2.68 (3.55)	2.69 (3.24)
Rothera	2.91 (2.74)	3.25 (3.09)
Bellingshausen	1.24 (2.39)	1.40 (2.54)
Ant. Penin[1]	2.20 (3.41)	2,32 (3.26)
SH4060[2]	0.53 (6.01)	0.48 (4.92)
Antarctica[3]		0.72 (3.07)

Notes:
[1] Average of Faraday, Rothera and Bellingshausen.
[2] Average for the zone 40°–60°S (P.D. Jones, personal communication).
[3] Area-average for Antarctica (Raper *et al.*, 1984, updated)

Table 10.2. *Correlations between site and average accumulation rates (acc.) and various temperatures (T) (lengths of overlapping records shown in brackets)*

		A	B	C	D	E	F
A	Dolleman (acc.)						
B	James Ross (acc.)	− 0.06(25)					
C	Gomez (acc.)	− 0.03(34)	0.26(25)				
D	Average (acc.)	0.44(25)	0.74(25)	0.79(25)			
E	Ant. Pen. (*T*)	0.12(39)	0.35(25)	0.20(34)	0.25(25)		
F	SH4060 (*T*)	0.10(39)	0.27(25)	0.27(34)	0.14(25)	0.68(44)	
G	Antarctica	− 0.19(29)	0.56(22)	0.56(24)	0.32(22)	0.48(34)	0.23(34)

1. *Minimum case.* If the accumulation rate increased immediately by 50% and were sustained with no increase in ice losses, sea level would fall by 0.3 mm/yr, giving a total fall of 3 cm over the next 100 years.
2. *Intermediate case.* If the accumulation rate increases by 50% but is balanced by ice loss, there would be no change in sea level.
3. *Maximum case.* If the accumulation rate stays the same as present, but discharge increased linearly by 50% over a hundred years, sea level would rise on average by 0.15 mm/yr, a total of 1.5 cm after 100 years.

This range, − 3 cm to + 1.5 cm per century, is small compared with changes likely to occur from other sources (Wigley and Raper, this volume). On the century time-scale, therefore, although the Peninsula area represents an important 'reservoir' of land-based ice, climatic changes in this region are unlikely to significantly influence future sea level.

Acknowledgement

Recent temperature data and some of the analyses shown here in Tables 10.1 and 10.2 and Figure 10.3 were provided by Dr. P.D. Jones, Climatic Research Unit, University of East Anglia.

References

Doake, C.S.M. (1985). Antarctic mass balance: glaciological evidence from Antarctic Peninsula and Weddell Sea sector. In *Glaciers, Ice Sheets and Sea Level: Effect of a CO_2-induced Climatic Change*. Report of a workshop held in Seattle, Washington, September 13–15, 1984. United States Department of Energy, DOE/EV/60235-1, pp. 197–209.

Doake, C.S.M. & Vaughan, D.G. (1991). Rapid disintegration of the Wordie Ice Shelf in response to atmospheric warming. *Nature*, **350**, 328–9.

Jones, P.D. & Limbert, D.W.S. (1987). *A Data Bank of Antarctic Surface Temperature and Pressure Data*. US DOE TR038, Office of Energy Research, US Department of Energy, Washington, DC.

Jones, P.D., Wigley, T.M.L. & Farmer, G. (1991). Marine and land temperature data sets: a comparison and look at recent trends. In *Greenhouse-Gas-Induced Climatic Changes: A Critical Appraisal of Simulations and Observations*. M.E. Schlesinger (ed.). Elsevier Scientific Publishers B.V., Amsterdam, pp. 153–72.

Limbert, D.W.S. (1984). West Antarctic temperatures, regional differences and the nominal length of the summer and winter season. In *Environment of West Antarctica: Potential CO_2-induced Changes*. Report of a workshop held in Madison, Wisconsin, 5–7 July 1983, National Academy Press, pp. 116–39.

Nicholls, K.W. & Paren, J.G. (1992). Extending the Antarctic meteorological record using ice sheet temperature profiles. *J. Clim.* (in press).

Nicholls, K.W., Paren, J.G. & Cooper, S. (unpublished). Recent warming of the Antarctic Peninsula: analysis of ice temperature profiles and meteorological data.

Paren, J.G. & Richardson, N.A. (1988). Glacier fluctuations in the Antarctic Peninsula: the last decade. *Ann. Glaciol.*, **11**, 206–7.

Peel, D.A. (1992). Ice-core evidence from the Antarctic Peninsula region. In *Climate since AD1500*, R.S. Bradley & P.D. Jones (eds.). Routledge, London, pp. 549–571.

Peel, D.A., Mulvaney, R. & Davison, B.M. (1988). Stable-isotope/air-temperature relationships in ice cores from Dolleman Island and the Palmer Land plateau, Antarctic Peninsula. *Ann. Glaciol.*, **10**, 130–6.

Raper, S.C.B., Wigley, T.M.L., Mayes, P.R., Jones, P.D. & Salinger, M.J. (1984). Variations in surface air temperatures. Part 3: The Antarctic, 1957–1982. *Monthly Weather Review*, **112**(7), 1341–53.

Reynolds, J.M. (1981). The distribution of mean annual temperatures in the Antarctic Peninsula. *British Antarctic Survey Bulletin*, **54**, 123–33.

11

Model projections of CO_2-induced equilibrium climate change†

M.E. Schlesinger

Abstract

The changes in the equilibrium climate induced by a doubling of the CO_2 concentration have been simulated by energy-balance, radiative–convective and general circulation climate models. These models have given a surface air temperature warming of 0.2–10 °C, 0.5–4.2 °C and 2.8–5.2 °C, respectively. The warming simulated by the general circulation models (GCMs) increases from the tropics toward the winter pole, but the regional changes show significant differences among the models. The changes in soil water simulated by the GCMs show a moistening of the soil over much of Eurasia and North America in winter and decreased soil water in summer. However, the geographical extent of the summertime desiccation varies among the models. The CO_2-induced changes in the annual snow-accumulation rate simulated by one GCM show increases above and decreases below 1,500 m and 2,000 m over Antarctica and Greenland, respectively. The $2 \times CO_2 - 1 \times CO_2$ change in the annual, area-integrated snow-accumulation rates would raise sea level by 8.6 cm per century. The net change in sea level between now and 2030, due to this effect alone, would be small, less than 1 cm. Other models, however, suggest that total precipitation over both Greenland and Antarctica would increase, leading to a negative contribution to sea level rise. A combination of a GCM-simulated warming with a theoretical tropical cyclone model indicates that the intensity of tropical cyclones would increase. Testing of this hypothesis, however, must await simulations of CO_2-induced climate change with GCMs having a horizontal resolution about twice that of the current models.

Introduction

If the Earth's atmosphere were composed of only its two major constituents, nitrogen (N_2, 78% by volume) and oxygen (O_2, 21%), the Earth's surface temperature would be close to the -18 °C radiative-equilibrium value necessary to balance the approximately 240 W/m² of solar radiation absorbed by the Earth–atmosphere system. The fact that the Earth's global mean surface temperature is a life-supporting 15 °C is a consequence of the greenhouse effect of the atmosphere's minor constituents, mainly water vapour (H_2O, 0.2%) and carbon dioxide (CO_2, 0.03%). Measurements taken at Mauna Loa, Hawaii, show that the CO_2 concentration has increased from about 315ppmv in 1958 to 353ppmv in 1990 (Watson *et al.*, 1990), a 12% increase in 32 years. Measurements of the air entrapped within the ice sheet of Antarctica indicate that the pre-industrial CO_2

† This chapter was written in 1989 and therefore represents a review of the science as of that date.

concentration increased from about 280ppmv in 1750 to 290ppmv in 1880 (Siegen-thaler and Oeschger, 1987). Rotty and Masters (1985) report that, with the exception of the periods of the Depression and World Wars I and II, CO_2 emissions increased from 1860 to 1949 due to a 4.2%/yr growth in the consumption of fossil fuels (gas, oil, coal). Subsequently, the growth rate of CO_2 emissions was a steady 4.4%/yr from 1950 to 1973, and then decreased to 1.5%/yr from 1973 to 1982 at least partly as a result of the rise in the price of oil. A probabilistic scenario analysis of the future usage of fossil fuels predicts about an 80% chance that the CO_2 concentration will reach twice the pre-industrial value by 2100 (Nordhaus and Yohe, 1983).

Computer simulations of the equilibrium climatic change induced by a doubling of the CO_2 concentration have been made with a hierarchy of mathematical climate models, one of the most recent of which has given a warming as large as 5.2 °C in the global mean surface air temperature (Wilson and Mitchell, 1987). Since such a global warming is comparable to that which is estimated to have occurred during the transition from the last ice age to the present interglacial (Gates, 1976a, b; Imbrie and Imbrie, 1979), a transition which took thousands of years as a result of non-anthropo-genic factors, there is considerable interest in the detection of a CO_2-induced climatic change and in the potential impacts of such a change on the spectrum of human endeavours.

The majority of the simulations of CO_2-induced climatic change have been per-formed to determine the change in the equilibrium climate of the Earth resulting from an abrupt increase in CO_2 such as a doubling from 300 to 600ppmv. The goal of these equilibrium simulations has been to estimate the magnitude of the eventual climatic change which may occur as a consequence of the doubled CO_2 concentration projected to occur in the next century or the even earlier equivalent doubling likely to result when one accounts for other anthropogenic greenhouse gases.

In the following section the types of climate model that have been used to simulate CO_2-induced equilibrium climate change are described. The results from the physically most comprehensive models are then reviewed in terms of changes in surface air temperature, soil water and snow accumulation. Although the horizontal resolutions of contemporary climate models are not sufficient to simulate tropical cyclones well, a theoretical model suggests a CO_2-induced intensification of tropical cyclones; this is discussed later in the text.

Types of climate model

Three types of climate model have been used to simulate CO_2-induced equilibrium climate change: energy-balance models (EBMs), radiative–convective models (RCMs) and general circulation models (GCMs).

Energy-balance models

Energy-balance models predict the CO_2-induced change in temperature at the Earth's surface by imposing the condition that the change of energy at either the Earth's surface or the top of the atmosphere is zero. For the former case, the models may be called surface energy-balance models (SEBMs). In the latter case, the models may be referred to as planetary energy-balance models (PEBMs). SEBMs and PEBMs have given a warming induced by a CO_2 doubling of 0.2–10 °C and 0.6–3.3 °C, respectively. These wide ranges are due in part to the requirement of specifying the behaviour of the atmosphere entirely in terms of the surface temperature in EBMs, and on the inherent difficulty in doing so in a physically based manner (see Schlesinger, 1985, 1988a).

Radiative–convective models

Radiative–convective models determine the equilibrium vertical temperature distribution for an atmospheric column and its underlying surface for given solar insolation and prescribed atmospheric composition and surface albedo. An RCM includes submodels for the transfer of solar and terrestrial (long-wave) radiation, the turbulent heat transfer between the Earth's surface and atmosphere, the vertical redistribution of heat within the atmosphere by dry or moist convection, and the atmospheric water vapour content and clouds. For RCMs, estimates of the surface temperature change (ΔT_s) for a CO_2 doubling range from 0.5–4.2 °C. This response can be characterized by $\Delta T_s = G_f \Delta R_T$, where ΔR_T is the radiative forcing at the tropopause due to the CO_2 doubling (~ 4 W/m²), $G_f = G_o/(1-f)$ is the gain of the climate system with feedback f, and G_o is the gain of the climate system without feedback [~ 0.3 °C/(W/m²)] (see Schlesinger, 1985, 1988a).

Several feedback mechanisms have been included in RCMs; in particular, in order of increasing ΔT_s: (1) the increase in the amount of water vapour in the atmosphere as a consequence of the near-constancy of the relative humidity; (2) the change in the temperature lapse rate; (3) the increase in the cloud altitude as the clouds maintain their temperature; (4) the change in cloud amount; (5) the change in the cloud optical depth; and (6) the decrease in surface albedo. An analysis of these feedback mechanisms (Schlesinger, 1985, 1988a) shows that: (1) the water vapour, cloud altitude and surface albedo feedbacks are positive (i.e., they enhance the warming), with values that decrease in that order; (2) the cloud optical depth feedback is negative for low- and middle-level clouds, but is positive for high-level (cirrus) clouds; (3) the temperature lapse rate feedback is either positive or negative, depending on whether the lapse rate is controlled by baroclinic or convective processes; and (4) the cloud cover feedback is unknown. These feedbacks can be predicted credibly only by physically based models that include the essential dynamics and thermodynamics of the feedback processes. Such physically-based models are the general circulation models.

General circulation models

While EBMs and RCMs calculate only the surface temperature and the vertical temperature profile, respectively, GCMs calculate the geographical distributions of an ensemble of climatic quantities which includes the vertical profiles of atmospheric temperature, water vapour and velocity, and the surface pressure, precipitation, soil water and snow. In addition, GCMs used for climate change simulations must calculate the geographical distributions of the sea surface temperature (SST) and sea ice extent. (See Schlesinger, 1984 and 1988b for further information on GCMs.)

To study CO_2-induced equilibrium climatic changes with a GCM requires two simulations, a $1 \times CO_2$ simulation with a CO_2 concentration generally taken to be 300–330 ppmv, and an $N \times CO_2$ simulation with N generally taken to be 2 or 4. Each of the simulations is begun from some prescribed initial conditions and is run until these conditions are forgotten and the simulated climate reaches its statistically stationary equilibrium state. The differences between the $N \times CO_2$ and $1 \times CO_2$ equilibrium climates are then taken to be the CO_2-induced equilibrium climatic changes.

The earliest GCM simulations of CO_2-induced climatic change were performed with so-called 'swamp models' of the ocean. In such a swamp ocean model, the ocean has zero heat capacity and no horizontal or vertical heat transports. Therefore, the swamp ocean is like perpetually wet land and is always in thermodynamic equilibrium. Consequently, the time for the Earth–atmosphere system to reach its equilibrium climate with an atmospheric GCM/swamp ocean model depends only on the atmosphere and land surface, and generally requires about 300 days. While such models are economical of computer time, in order to prevent the freezing of the ocean in the latitudes of the polar night, they must be run without the seasonal insolation cycle. Consequently, these models simulate the CO_2-induced change only for a surrogate of the annual mean climate, namely, that which is obtained for the annual mean solar insolation.

More recently, atmospheric GCMs have been coupled to prescribed-depth mixed-layer ocean models which have finite heat capacity and sometimes a prescribed additional oceanic heating which includes the effects of oceanic heat transport. These atmospheric GCM/mixed-layer ocean models are run with the seasonal insolation cycle and generally require about 20 simulated years to reach their equilibrium climate. An additional 10 years is generally simulated to obtain estimates of the means and other statistics of the equilibrium climate.

The annual cycles of the equilibrium climatic changes induced by a doubling of the CO_2 concentration have been simulated by the atmospheric GCM/mixed-layer ocean (AGC/MLO) models of: (1) the Geophysical Fluid Dynamics Laboratory (GFDL; Wetherald and Manabe, 1986), (2) the Goddard Institute for Space Studies (GISS; Hansen *et al.*, 1984), (3) the National Center for Atmospheric Research (NCAR; Washington and Meehl, 1984); (4) Oregon State University (OSU; Schlesinger and Zhao, 1989), and (5) the United Kingdom Meteorological Office (UKMO; Wilson and Mitchell, 1987). (For a more complete list see Cubasch and Cess, 1990). In the following section the CO_2-induced changes in the surface air temperature, soil water and snow amount simulated by these models are presented.

Results from general circulation models

In this section we present results for the equilibrium climate changes simulated for a CO_2 doubling by the five AGC/MLO models above, in particular for the surface air temperature, soil water and snow amount.

Surface air temperature

The changes in the global mean equilibrium surface air temperature, ΔT_s, are presented in Table 11.1 together with the global mean precipitation changes, ΔP. This table shows that the simulated CO_2-induced changes in the annual mean, global mean surface air temperature range from 2.8–5.2 °C, and the corresponding changes in precipitation from 7.1-15% of their $1 \times CO_2$ values. Furthermore, this table shows that the size of ΔP

Table 11.1. *Changes in the global mean surface air temperature (ΔT_s) and precipitation rate (ΔP) simulated by atmospheric GCM/mixed-layer ocean models for a CO_2 doubling*

Model/Study	ΔT_s (°C)	ΔP (% of $1 \times CO_2$ value)
GFDL/Wetherald & Manabe (1986)	4.0	8.7
GISS/Hansen et al., (1984)	4.2	11.0
NCAR/Washington & Meehl (1984)	3.5	7.1
OSU/Schlesinger & Zhao (1989)	2.8	7.8
UKMO/Wilson & Mitchell (1987)	5.2	15.0

is positively correlated with the size of ΔT_s; this occurs as a result of the Clausius-Clapeyron relation between the saturation vapour pressure of water vapour and temperature.

A partial explanation for the range of values of ΔT_s and thus of ΔP is provided by Fig. 11.1 which displays the CO_2-induced warming plotted versus the annual mean, global mean surface air temperature of the simulated $1 \times CO_2$ climate. This figure shows that the warmer the $1 \times CO_2$ control climate, the smaller the CO_2-induced warming. This is due in part to the fact that there is a decrease in the positive ice-albedo feedback with increasing $1 \times CO_2$ temperature, as a result of there being less sea ice and snow. From this it appears that the CO_2-induced changes in temperature and precipitation simulated by these models would be in better agreement if their $1 \times CO_2$ global mean surface air temperatures were in better agreement. Furthermore, it is tempting to conclude that the resulting common model temperature and precipitation sensitivities would be correct if the common $1 \times CO_2$ surface air temperature were in agreement with the observed temperature. However, while this is a necessary condition for the common sensitivities of the models to be in agreement with the sensitivities of nature, it is not a sufficient condition. That there are considerable residual uncertainties in the climate sensitivity is indicated by the vertical arrows labelled sensitivity uncertainty in Fig. 11.1. In fact, the potential for such a disparity is evidenced by the fact that although the GISS and GFDL models simulate similar values of ΔT_s (Fig. 11.1), they do so with feedbacks that differ in both magnitude and sign despite the approximate agreement of their simulated values of T_s for the $1 \times CO_2$ climate (Schlesinger, 1989).

To establish the correctness of the models' temperature and precipitation sensitivities, even if they simulated the present climate correctly, is not a simple task[†]. Validation of climate sensitivity requires the simulation of at least one climate different from that of the present, for example, that of the last glacial maximum 18,000 years before the present, and the comparison of such a simulated paleoclimate with observations. This requirement for model validation is a fundamental and inherently difficult problem in climate modelling and simulation (Schlesinger and Mitchell, 1985, 1987). This notwithstanding, it is of interest here to present and compare further results of the CO_2-induced climatic changes simulated by these models.

[†] Editors' note: the correct simulation of the present climate requires far more than just simulating the correct global mean temperature!

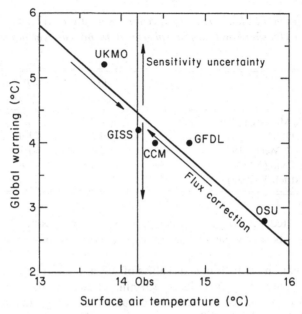

Fig. 11.1 The global mean surface air temperature warming (ΔT_s) simulated for a CO_2 doubling by five GCMs versus their simulated $1 \times CO_2$ global mean surface air temperature (T_s). 'Obs' indicates the observed global mean surface air temperature based on the data of Jenne (1975). (Adapted from Cess and Potter, 1988. Copyright American Geophysical Union).

The time–latitude distributions of the zonal mean surface air temperature changes simulated by the five models for doubled CO_2 are presented in Fig. 11.2. This figure shows that the CO_2-induced temperature changes increase from the tropics, where the values range from about 2 °C for the NCAR and OSU models to about 4 °C for the GISS and UKMO models, toward the poles, and that the seasonal variations of the CO_2-induced surface temperature changes are small between 50°S and 30°N and are large in the regions poleward of 50° latitude in both hemispheres. In the Northern Hemisphere a warming minimum of about 2 °C is simulated near the pole by all five models in summer. The models also simulate a warming maximum in the autumn with values that range from 8 °C for the NCAR model to 16 °C for the GFDL and UKMO models. This maximum extends into winter in all the simulations except that of NCAR, which instead exhibits a warming minimum near the pole. The NCAR model also simulates another polar warming minimum in spring that is not found in the other simulations. In the Southern Hemisphere all five models simulate a maximum warming in winter and a minimum warming in summer. The summer warming maximum occurs near the Antarctic coast in all five simulations and ranges from 8 °C in the GFDL and OSU simulations to 14 °C in the NCAR and UKMO simulations.

The geographical distributions of the $2 \times CO_2 - 1 \times CO_2$ surface air temperature changes simulated for December–January–February (DJF) and June–July–August (JJA) by the five models are presented in Figs. 11.3 and 11.4, respectively. These figures show that all five models simulate a CO_2-induced surface air temperature warming virtually everywhere. In general, the warming is a minimum in the tropics during both seasons, at least over the ocean, and increases toward the winter pole. The tropical

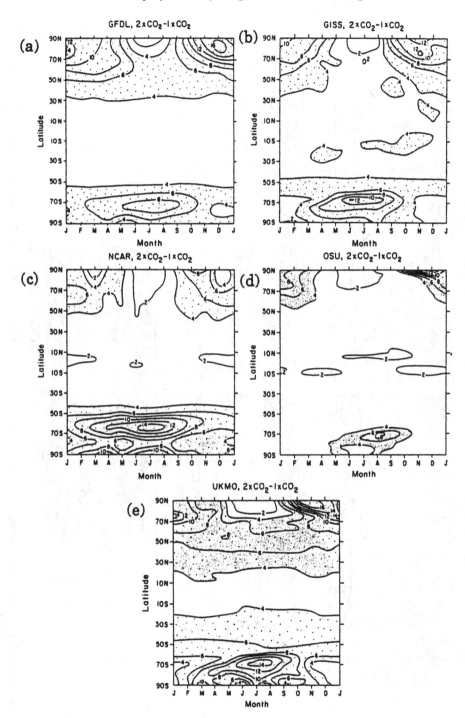

Fig. 11.2 Time–latitude distribution of the zonal mean surface air temperature change (°C), $2 \times CO_2 - 1 \times CO_2$, simulated by: (*a*) GFDL model; (*b*) the GISS model; (*c*) the NCAR model; (*d*) the OSU model, and (*e*) the UKMO model. Stipple indicates temperature increases larger than 4 °C.

TEMPERATURE DIFFERENCES FOR DJF

Fig. 11.3 Geographical distribution of the surface air temperature change (°C), $2 \times CO_2 - 1 \times CO_2$, for DJF simulated by (a) GFDL model; (b) the GISS model; (c) the NCAR model; (d) the OSU model, and (e) the UKMO model. Stipple indicates temperature increases larger than 4 °C.

TEMPERATURE DIFFERENCES FOR JJA

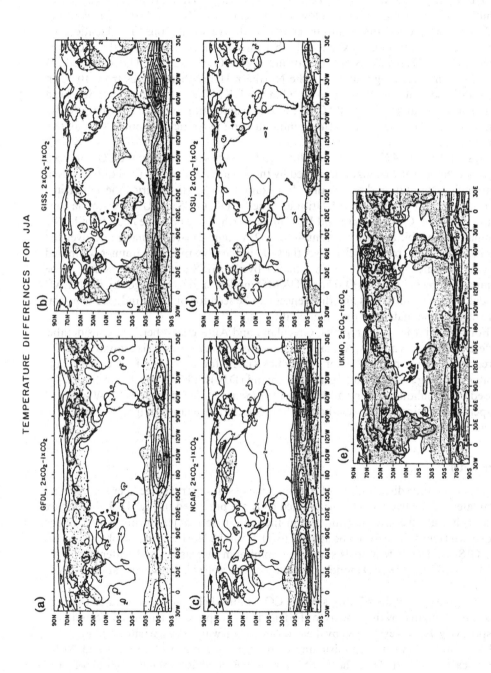

Fig. 11.4 Geographical distribution of the surface air temperature change (°C), $2 \times CO_2 - 1 \times CO_2$, for JJA simulated by (a) GFDL model; (b) the GISS model; (c) the NCAR model; (d) the OSU model, and (e) the UKMO model. Stipple indicates temperature increases larger than 4 °C.

maritime warming minimum ranges from about 2 °C in the NCAR and OSU simulations to about 4 °C in the GFDL, GISS and UKMO simulations. Maximum warming in DJF occurs in the Arctic in all the simulations except that of NCAR, which instead exhibits a warming maximum at a lower latitude, near 65°N. The maximum warming in JJA occurs around the Antarctic coasts in all five simulations. The locations of the wintertime warming maxima in both hemispheres coincide with the locations where the $1 \times CO_2$ sea ice extent retreats in the $2 \times CO_2$ simulation. The magnitude of the wintertime warming maxima in the Northern Hemisphere ranges from 10 °C in the GISS and OSU simulations to 20 °C in the UKMO simulation. In the Southern Hemisphere it ranges from 10 °C in the OSU simulation to 20 °C in the UKMO simulation. In JJA there is a warming minimum in the Arctic of about 2 °C in all five simulations.

Figs. 11.3 and 11.4 also show that although there are similarities in the CO_2-induced regional temperature changes simulated by the models there are marked differences in both their magnitude and seasonality. For example, in North America the wintertime warming generally increases with latitude in the GFDL, GISS and UKMO simulations with values of 4 °C in the south and 10 °C in the north, while both the NCAR and OSU simulations exhibit a warming minimum of 2 °C centred over Canada and over the Pacific Northwest, respectively. Also, the CO_2-induced warming in summer compared to that in winter is simulated to be smaller by the GISS and NCAR models, to be comparable by the GFDL, OSU and UKMO models, and to be larger by the NCAR model. Further similarities and differences can be seen for the CO_2-induced temperature changes simulated for the other continents.

For the major ice sheet regions, Greenland and Antarctica, there are considerable differences between models. Average equilibrium summer warming is around 2.5 °C for Greenland (JJA) and 3.5 °C for Antarctica (DJF) for $2 \times CO_2 - 1 \times CO_2$. Average winter warming is around 6.5 °C for Greenland (DJF) and 7 °C for Antarctica (JJA). Based on greenhouse forcing and transient response projections given by Wigley and Raper (this volume), about one-half of these equilibrium changes might be expected to occur between now and 2030.

Soil water

The time–latitude distributions of the zonal mean soil water change over ice-free land simulated by the five models for doubled CO_2 are presented in Fig. 11.5. This figure shows that all five models simulate an increase in soil water during winter in the Northern Hemisphere from about 45°N to 70°N. In the Northern Hemisphere summer the GISS and NCAR models simulate a minimum increase in soil water, while the GFDL, OSU and UKMO models simulate a decrease in soil water from about 30°N to 70°N.

The geographical distributions of the CO_2-induced changes in soil water over ice-free land simulated by the models for DJF and JJA are presented in Figs. 11.6 and 11.7, respectively. As was evidenced by the zonal mean soil water changes shown in Fig. 11.5, all five models simulate a moistening of the soil over much of Eurasia and North America in DJF. In JJA, the GISS and NCAR models simulate regions of both increased and decreased soil water over Eurasia and North America, while the GFDL, OSU and UKMO models simulate a desiccation virtually everywhere in the Northern

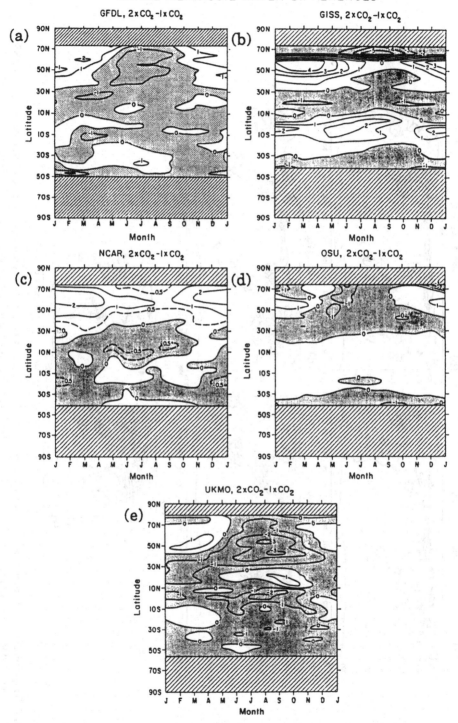

Fig. 11.5 Time–latitude distribution of the zonal mean soil water change (cm), $2 \times CO_2 - 1 \times CO_2$, over ice-free land only, simulated by: (a) GFDL model; (b) the GISS model; (c) the NCAR model; (d) the OSU model, and (e) the UKMO model. Stipple indicates a decrease in soil water, hatching indicates the latitudes where there is no ice free land.

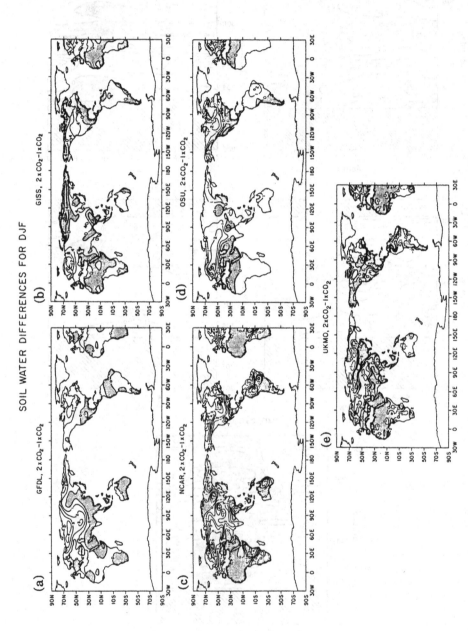

SOIL WATER DIFFERENCES FOR DJF

(a) GFDL, 2 × CO₂ − 1 × CO₂

(b) GISS, 2 × CO₂ − 1 × CO₂

(c) NCAR, 2 × CO₂ − 1 × CO₂

(d) OSU, 2 × CO₂ − 1 × CO₂

(e) UKMO, 2 × CO₂ − 1 × CO₂

Fig. 11.6 Geographical distribution of the soil water change (cm), $2 \times CO_2 - 1 \times CO_2$, for DJF simulated by (*a*) GFDL model; (*b*) the GISS model; (*c*) the NCAR model; (*d*) the OSU model, and (*e*) the UKMO model. Stipple indicates a decrease in soil water.

SOIL WATER DIFFERENCES FOR JJA

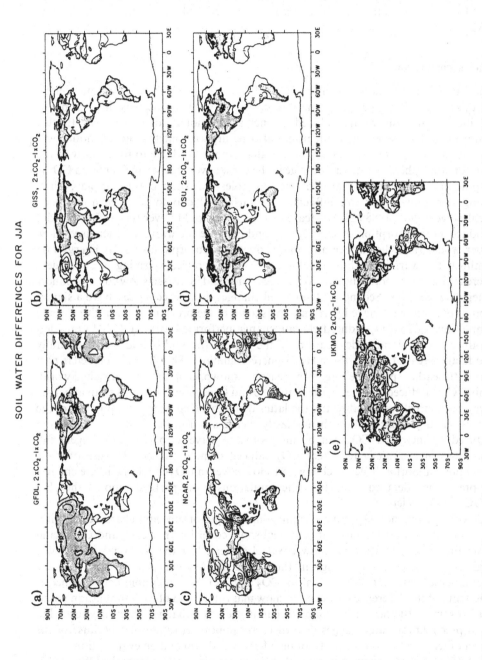

Fig. 11.7 Geographical distribution of the soil water change (cm), $2 \times CO_2 - 1 \times CO_2$, for JJA simulated by (a) GFDL model; (b) the GISS model; (c) the NCAR model; (d) the OSU model, and (e) the UKMO model. Stipple indicates a decrease in soil water.

Hemisphere during summer. A similar desiccation of the Northern Hemisphere soil in summer was first obtained in the 'annual mean' simulations with atmospheric GCM/ swamp ocean models and with the first annual-cycle simulation with an atmospheric GCM/prescribed-depth mixed-layer ocean model (see Schlesinger and Mitchell, 1985, 1987).

Snow accumulation

One aspect of a CO_2-induced change in climate is the change in the snow amount on the Earth's surface. It can be argued that the amount of snow will decrease in response to the increase in temperature induced by increased amounts of CO_2 in the Earth's atmosphere. On the other hand, it can also be argued that the snow amount will increase as a result of there being more water vapour available to form snow in the warmer atmosphere. The actual change in the snow cover induced by increased CO_2 is of potential importance because of the four roles that snow cover can play. First, changes in snow cover can influence the surface albedo to produce a feedback (Schlesinger, 1985, 1988a,b). Decreased snow cover would lower the surface albedo, produce more absorption of solar energy and enhance the warming. Increased snow cover would raise the surface albedo, produce less absorption of solar energy and diminish the warming. Thus, insofar as the CO_2-induced warming is concerned, decreased snow cover produces a positive feedback and increased snow cover produces a negative feedback. Second, CO_2-induced changes in snow cover can have a significant impact on the surface hydrology. As shown in the previous section, three of the simulations of $2 \times CO_2$-induced climate change performed with AGC/MLO models produce a considerable desiccation of the soil in the mid-latitude, agriculturally productive areas of the Northern Hemisphere in summer. This drying occurred in part due to the earlier spring melting of the seasonal snowpack in the CO_2-enriched world. Third, CO_2-induced changes in the Antarctic and Greenland snowpacks can affect sea level, both directly and indirectly, the latter through a change in the equilibria of the corresponding ice sheets. Fourth, and lastly, changes in the rate of snow accumulation induced by increased CO_2 can be monitored to serve as one of the 'fingerprint' quantities in a strategy to detect CO_2-induced climate change. Because of these potentially important roles of changes in snow cover in CO_2-induced climate change, we present here detailed results from the simulations by one particular model, the OSU AGC/MLO model.

   In considering the CO_2-induced changes in the equilibrium snow characteristics it is useful to separate the regions of perennial snow cover in Antarctica and Greenland from the regions of seasonal snow cover elsewhere. For the perennial snow cover regions we consider the change in the annual snow-accumulation rate, with the accumulation rate for the individual $1 \times CO_2$ and $2 \times CO_2$ simulations being defined as one-tenth of the difference between the snow mass at the end of each simulation minus the snow mass 10 years earlier. Snow-accumulation rate changes are clearly related to total precipitation rate changes, but there are important differences because of the effect of temperature on the partitioning of snow and rain and on evaporation.

   The CO_2-induced changes in annual snow-accumulation rate from the OSU AGC/ MLO model are presented in Fig. 11.8. This figure shows that over both Antarctica and Greenland there are areas where the annual snow-accumulation rate both increases

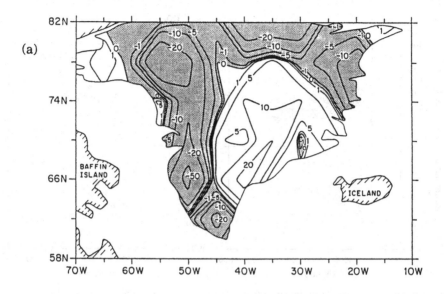

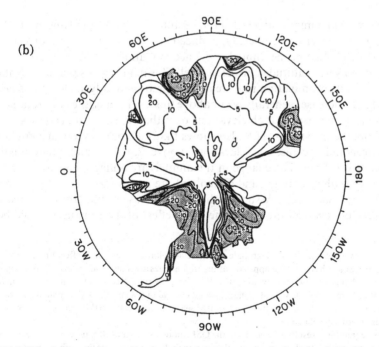

Fig. 11.8 CO$_2$-induced changes in the annual snow-accumulation rate (g/cm^2/yr) for (*a*) Greenland, and (*b*) Antarctica. Stipple indicates a decrease in snow accumulation.

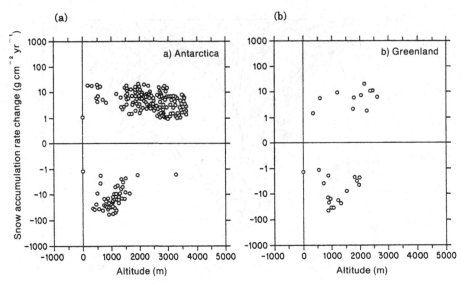

Fig. 11.9 Annual snow-accumulation rate change versus elevation above sea level for
(*a*) Antarctica, and (*b*) Greenland.

and decreases, with maximum values of about 20 and -50 g/cm²/yr. Thus, both of the
speculations given earlier about the CO_2-induced changes on snow are correct. Fig.
11.9 presents the change in the annual snow-accumulation rate for Antarctica and
Greenland plotted versus altitude above sea level. This figure shows that over Antarctica the decreases in accumulation rate are located predominantly at elevations below
1,500 m, while the increases generally occur above this level. Thus, the Antarctic snow-
accumulation rate increases at high elevations over the Plateau and decreases at low
elevations along the coast. Fig. 11.9 also shows that the annual snow-accumulation
rate over Greenland increases at elevations above 2,000 m and predominantly
decreases below this level. The change in the annual, area-integrated snow-accumulation rates over both Antarctica and Greenland are negative, -1.7×10^{17} g/yr and
-1.4×10^{17} g/yr, respectively. Assuming that this decrease results from a decrease in
evaporation exclusively over the ocean, the direct effect of these changes would be to

Editors' note: to give some idea of the magnitude of this accumulation-rate-change effect on sea level over
future decades, we assume that, as a first approximation, the changes are linear and scale according to global
mean temperature changes. Between now and 2030, therefore, the rate of sea level change would increase
from zero to about 4 cm per century, giving an integrated change of 0.8 cm. If $2 \times CO_2$ climate conditions
relative to today were reached by, say, 2070, the integrated sea level change due to this process alone would be
3.4 cm. These are relatively small effects.
  One should note that these results are specific to the OSU model. Insofar as they point to a net *decrease* in
snow-accumulation rate over the large ice sheets, they are probably unrepresentative. When results from all
of the five models listed in Table 11.1 are combined, the 'consensus' result points towards a net *increase* in
total precipitation over both Greenland and Antarctica. The analysis by Wigley *et al.* (1990) suggests that in
both summer and winter and for both regions, the probability of a precipitation increase exceeds 0.75. This
would lead to a decrease in sea level.

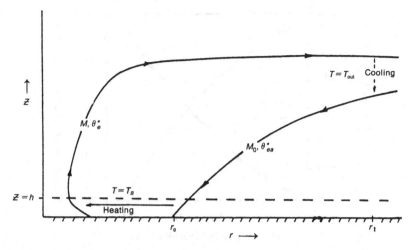

Fig. 11.10 The Emanuel (1986, 1987) Carnot-cycle model of a tropical cyclone. Reprinted by permission of the American Meteorological Society.

raise sea level by 8.6 cm per century. The indirect effects (in particular, those due to changes in ablation and ice discharge) can be estimated only by models of the Antarctic and Greenland ice sheets.

Potential changes in tropical cyclone intensities

The published results for the CO_2-induced climate changes simulated by the five AGC/ MLO models described above have not included changes in the characteristics of tropical cyclones, that is, disturbances in the tropics with wind speeds of at least 34 m/s. Nevertheless, Emanuel (1987) combined the CO_2-induced sea surface temperature changes simulated by the GISS model (Hansen *et al.*, 1984) with a theoretical tropical cyclone model to estimate the potential change in the intensity of tropical cyclones.

The theoretical model was developed by Emanuel (1986) and treats the tropical cyclone as a Carnot engine in which the hot reservoir is the sea surface and the cold reservoir is the lower stratosphere or upper troposphere. As shown schematically in Fig. 11.10, air in the boundary layer is heated isothermally by its receipt of latent heat as it moves radially toward the cyclone centre, during which it loses angular momentum due to friction. At the eyewall of the cyclone the boundary layer air is caused to ascend in an adiabatic process in which the moist entropy θ_e^* and angular momentum M are conserved. The ascent ceases at the level of neutral buoyancy in the lower stratosphere or upper troposphere where the air is cooled isothermally as it moves radially outward from the cyclone centre, during which it acquires the angular momentum of the environment, M_a. The Carnot cycle is completed by the adiabatic descent of the air, as required by mass conservation, during which the moist entropy of the environment θ_{ea}^* and angular momentum M_a are conserved. In the steady-state the work available from the Carnot cycle equals the frictional dissipation. From this balance and the thermo-dynamic efficiency of the Carnot cycle ϵ, Emanuel (1986, 1987) obtained an expression

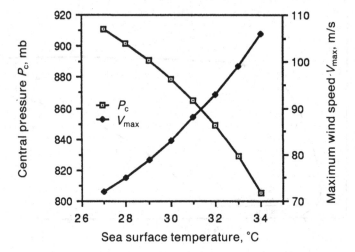

Fig. 11.11 Minimum sustainable central pressure P_c and maximum wind speed V_{max} as a function of sea surface temperature for RH = 0.78 and ϵ = 0.33. (Based on Emanuel, 1987. Reprinted by permission from *Nature*, **326**, copyright © 1987 Macmillan Magazines Ltd).

for the minimum sustainable central pressure for a tropical cyclone, P_c, as a function of the sea surface temperature (SST), ambient relative humidity (RH) and ϵ. Fig. 11.11 presents the result for RH = 0.78 and ϵ = 0.33, together with the corresponding SST-dependence of the maximum wind speed V_{max} based on empirical relations between wind and pressure. This figure shows that as the SST increases from 27 °C to 34 °C, the central pressure of the tropical cyclone decreases non-linearly from 911 mb to 805 mb and the maximum wind speed increases nonlinearly from 72 to 106 m/s. These non-linear relationships are the result of the non-linear dependence of the saturation vapour pressure on temperature as expressed by the Clausius-Clapeyron relation.

Using this model, Emanuel (1987) estimated CO_2-induced changes in the intensity of tropical cyclones based on the AGC/MLO results from the GISS model. The CO_2-induced SST changes simulated by the AGC/MLO model, which range between 2.3 °C and 4.8 °C in the tropics, were added to the observed August SSTs to obtain the estimates of $1 \times CO_2$ P_c (Fig. 11.12(*a*)) and $2 \times CO_2$ P_c (Fig. 11.12(*b*)). CO_2-induced decreases in P_c are as large as 50 mb. According to Emanuel (1987): 'Were these sea surface temperature increases realized, the maximum destructive potential of tropical cyclones would be substantially increased, in some places by as much as 60%.'

Considering Emanuel's hypothesis that the intensity of tropical cyclones will be increased as a result of greenhouse gas induced warming of the sea surface temperature, and the potential deleterious impact this would have on coastal regions, particularly in conjunction with a greenhouse gas induced rise in sea level, why have the five AGC/MLO models not reported results for changes in tropical cyclone intensity? The answer is that these models cannot simulate tropical cyclones well, if at all, because their horizontal resolution is too coarse. In particular, the horizontal scale of a tropical cyclone is less than 3,000 km, while the horizontal resolution of the AGC/MLO models ranges from 556 km at the equator for the OSU model (Schlesinger and Zhao, 1989) to 1,112 km for the GISS model (Hansen *et al.*, 1984). Thus, the number of grid boxes

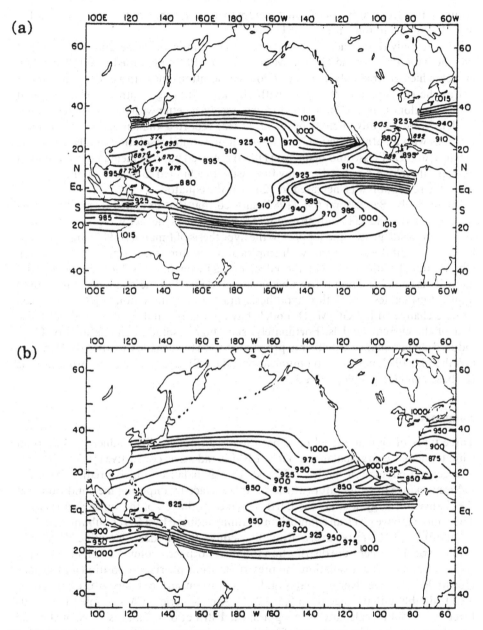

Fig. 11.12 Minimum sustainable central pressure (mb) for (*a*) present September mean climatological conditions, and (*b*) 2 × CO$_2$ August conditions simulated by the GISS AGC/MLO model. (Emanuel, 1987. Reprinted by permission from *Nature*, **326**, copyright © 1987 Macmillan Magazines Ltd).

occupied by a tropical cyclone from its centre to its periphery would range from one for the GISS model to three for the OSU model.

To date, only three studies have reported the spontaneous generation of a tropical cyclone with atmospheric GCMs. Manabe *et al.* (1970) presented results from two models which differed only in their horizontal resolution, one with a grid spacing of 417 km at the equator, and the other with 209 km. Although both models simulated tropical cyclones, the cyclones' structure was resolved better with the higher-resolution model. More recently, Bengtsson *et al.* (1982) reported the simulation of tropical cyclones with an atmospheric GCM having a horizontal resolution of 209 km at the equator, and Dell'Osso and Bengtsson (1985) reported improvement in the simulation of tropical cyclone Tip in a three-day forecast with a limited-area version of the model when the resolution was increased to 52 km at the equator.

Interestingly, Bengtsson *et al.* (1982) showed that the simulated tropical cyclones were sensitive to the SST, with no cyclones forming unless the SST was greater than 28–29 °C. This result gives some support to the hypothesis of Emanuel (1987), but it should be noted that it does not agree with empirical observations (see Raper, this volume). Observational data show that the effect of SSTs on both cyclone frequency and intensity is complex, and far from being as well-defined as these model results would imply. Further testing of this hypothesis must await simulations of CO_2-induced climate change with AGC/MLO models having a horizontal resolution about twice that of the current models. Fortunately, such high-resolution simulations of CO_2-induced climate change are either in progress or soon to be made with the GFDL model (Wetherald, 1987; personal communication), the OSU model and a few other models (Cubasch and Cess, 1990, Table 3.2a).

Summary

Three types of climate model have been used to simulate CO_2-induced equilibrium climate change: energy-balance models (EBMs), radiative–convective models (RCMs) and general circulation models (GCMs). Energy-balance models predict the CO_2-induced change in temperature at the Earth's surface by imposing the condition that the change of energy at either the Earth's surface or the top of the atmosphere is zero. These models have given an equilibrium warming induced by a CO_2 doubling of 0.2–10 °C and 0.6–3.3 °C, respectively. Radiative–convective models determine the equilibrium vertical temperature distribution for an atmospheric column and its underlying surface for given solar insolation and prescribed atmospheric composition and surface albedo. RCMs have given estimates of the equilibrium surface temperature change for a CO_2 doubling that range from 0.5–4.2 °C. Atmospheric GCMs coupled to mixed-layer ocean models calculate the geographical distributions of an ensemble of climatic quantities including the vertical profiles of atmospheric temperature, water vapour and velocity, and the surface pressure, precipitation, soil water, snow, sea surface temperature and sea ice extent. These AGC/MLO models have simulated CO_2-induced changes in the annual mean, global mean surface air temperature which range from 2.8–5.2 °C, with the size of the warming being negatively correlated with the simulated $1 \times CO_2$ temperature, this in part as a result the ice-albedo feedback mechanism.

The geographical distributions of the $2 \times CO_2 - 1 \times CO_2$ surface air temperature changes simulated for December–January–February (DJF) and June–July–August

(JJA) show a warming virtually everywhere. In general, the warming is a minimum in the tropics during both seasons and increases toward the winter pole. The locations of the wintertime warming maxima in both hemispheres coincide with the locations where the $1 \times CO_2$ sea ice extent retreats in the $2 \times CO_2$ simulation. Although there are similarities in the CO_2-induced regional temperature changes simulated by the models, there are significant differences in both their magnitude and seasonality. Because of these differences, and because these models generally perform poorly in their simulations of the regional details of today's climate, projections of future climate at the regional scale must be treated with caution.

The geographical distributions of the simulated CO_2-changes in soil water show a moistening of the soil over much of Eurasia and North America in DJF and decreased soil water over Eurasia and North America in JJA. The geographical extent of the summertime desiccation of the Northern Hemisphere continents varies among the models.

The CO_2-induced changes in the annual snow-accumulation rate simulated by the OSU model show both increases and decreases over both Antarctica and Greenland. Over Antarctica the snow accumulation rate increases at elevations above 1,500 m and decreases below this level, while over Greenland this threshold elevation, while less well defined, is about 2,000 m. The change in the annual, area-integrated snow-accumulation rates over Antarctica and Greenland would raise sea level by 8.6 cm per century.

Emanuel (1987) combined the CO_2-induced sea surface temperature changes simulated by the GISS model (Hansen *et al.*, 1984) with a theoretical tropical cyclone model and estimated that the intensity of tropical cyclones would increase as a result of the warmer tropical SSTs. However, empirical evidence confirming the realism of Emanuel's model in this regard is lacking. Current AGC/MLO models cannot simulate tropical cyclones well, if at all, because their horizontal resolution is too coarse. Consequently, proper testing of Emanuel's hypothesis must await simulations of CO_2-induced climate change with AGC/MLO models having a horizontal resolution about twice that of the current models.

Acknowledgement

This study was supported by the National Science Foundation and the Carbon Dioxide Research Division, Office of Basic Energy Sciences of the Department of Energy under grant ATM-8712033.

References

Bengtsson, L., Böttger, H. & Kanamitsu, M. (1982). Simulation of hurricane-type vortices in a general circulation model. *Tellus*, **34**, 440–57.

Cess, R.D. & Potter, G.L. (1988). A methodology for understanding and intercomparing atmospheric climate feedback processes within general circulation models. *J. Geophys. Res.*, **93**, 8305–14.

Cubasch, U. & Cess, R.D. (1990). Processes and modelling. In *Climate Change: The IPCC Scientific Assessment*, J.T. Houghton, G.J. Jenkins and J.J. Ephraums (eds.). Cambridge University Press, pp. 69–91.

Dell'Osso, L. & Bengtsson, L. (1985). Prediction of a typhoon using a fine-mesh NWP model. *Tellus*, **37A**, 97–105.

Emanuel, K.A. (1986). An air-sea interaction theory for tropical cyclones. Part I: Steady-state maintenance. *J. Atmos. Sci.*, **43**, 585–604.

Emanuel, K.A. (1987). The dependence of hurricane intensity on climate. *Nature*, **326**, 483–5.

Gates, W.L. (1976a). Modeling the ice-age climate. *Science*, **191**, 1138–44.

Gates, W.L. (1976b). The numerical simulation of ice-age climate with a global general circulation model. *J. Atmos. Sci.*, **33**, 1844–73.

Hansen, J., Lacis, A., Rind, D., Russell, G., Stone, P., Fung, I., Ruedy, R. & Lerner, J. (1984). Climate sensitivity: Analysis of feedback mechanisms. In *Climate Processes and Climate Sensitivity, Maurice Ewing Series, 5*, J.E. Hansen & T. Takahashi (eds.). American Geophysical Union, Washington, DC, pp. 130–163.

Imbrie, J. & Imbrie, K.P. (1979). *Ice Ages, Solving the Mystery.* Enslow Publishers, Short Hill, NJ, 224 pp.

Jenne, R. (1975). *Data Sets for Meteorological Research.* NCAR Tech. Note TN/1A-III, National Center for Atmospheric Research, Boulder, CO, 194 pp.

Manabe, S., Holloway, J.L. Jr. & Stone, H.M. (1970). Tropical circulation in a time-integration of a global model of the atmosphere. *J. Atmos. Sci.*, **27**, 580–613.

Nordhaus, W.D. & Yohe, G.W. (1983). Future paths of energy and carbon dioxide emissions. In *Changing Climate.* National Academy of Sciences, Washington, DC, pp. 87–153.

Rotty, R.M. & Masters, C.D. (1985). Carbon dioxide from fossil fuel combustion: trends, resources, and technological implications. In *Atmospheric Carbon Dioxide and the Global Carbon Cycle*, J.R. Trabalka (ed.). US Department of Energy, DOE/ER-0239, pp. 63–80.

Schlesinger, M.E. (1984). Climate model simulations of CO_2-induced climatic change. In *Advances in Geophysics*, **26**, B. Saltzman (ed.). Academic Press, New York, pp. 141–235.

Schlesinger, M.E. (1985). Feedback analysis of results from energy-balance and radiative–convective models. In *The Potential Climatic Effects of Increasing Carbon Dioxide*, M.C. MacCracken & F.M. Luther (eds.). US Department of Energy, DOE/ER-0237, pp. 280–319. (Available from NTIS, Springfield, Virginia.)

Schlesinger, M.E. (1988a). Quantitative analysis of feedbacks in climate model simulations of CO_2-induced warming. In *Physically-Based Modelling and Simulation of Climate and Climatic Change*, M.E. Schlesinger (ed.). NATO Advanced Study Institute Series, Reidel, Dordrecht, pp. 653–736.

Schlesinger, M.E. (1988b). *Physically-Based Modelling and Simulation of Climate and Climatic Change*, M.E. Schlesinger (ed.). NATO Advanced Study Institute Series, Reidel, Dordrecht, 1084 pp.

Schlesinger, M.E. (1989). Model projections of the climatic changes induced by increased atmospheric CO_2. In *Climate and the Geo-Sciences*, A. Berger, S. Schneider & J. Cl. Duplessy (eds.). Kluwer Academic Press, Dordrecht, pp. 374–415.

Schlesinger, M.E. & Mitchell, J.F.B. (1985). Model projections of the equilibrium climatic response to increased CO_2. In *The Potential Climatic Effects of Increasing Carbon Dioxide*, M.C. MacCracken & F.M. Luther (eds.). US Department of Energy, Washington, DC, DOE/ER-0237, pp. 81–147. (Available from NTIS, Springfield, Virginia.)

Schlesinger, M.E. & Mitchell, J.F.B. (1987). Climate model simulations of the equilibrium climatic response to increased carbon dioxide. *Rev. of Geophys.*, **25**, 760–98.

Schlesinger, M.E. & Zhao, Z.-C. (1989). Seasonal climate changes induced by doubled CO_2 as simulated by the OSU atmospheric GCM/mixed-layer ocean model. *J. Climate*, **2**, 459–95.

Schlesinger, M.E., Gates, W.L. & Han, Y.-J. (1985). The role of the ocean in CO_2-induced climate warming: Preliminary results from the OSU coupled atmosphere–ocean GCM. In *Coupled Ocean–Atmosphere Models*, J.C.J. Nihoul (ed.). Elsevier, Amsterdam, pp. 447–78.

Siegenthaler, U. & Oeschger, H. (1987). Biospheric CO_2 emissions during the past 200 years reconstructed by deconvolution of ice core data. *Tellus*, **39B**, 140–54.

Washington, W.M. & Meehl, G.A. (1984). Seasonal cycle experiment on the climate sensitivity due to a doubling of CO_2 with an atmospheric general circulation model coupled to a simple mixed-layer ocean model. *J. Geophys. Res.*, **89**, 9475–9503.

Watson, R.T., Rodhe, H., Oeschger, H. & Siegenthaler, U. (1990). Greenhouse gases and aerosols. In *Climate Change: The IPCC Scientific Assessment*, J.T. Houghton, G.J. Jenkins, J.J. Ephraums (eds.). Cambridge University Press, Cambridge, pp. 1–40.

Wetherald, R.T. & Manabe, S. (1986). An investigation of cloud cover change in response to thermal forcing. *Climatic Change*, **8**, 5–23.

Wigley, T.M.L., Santer, B.D., Schlesinger, M.E. & Mitchell, J.F.B. (1990). Developing climate scenarios from equilibrium GCM results. *J. Geophys. Res.* (submitted).

Wilson, C.A. & Mitchell, J.F.B. (1987). A doubled CO_2 climate sensitivity experiment with a global climate model including a simple ocean. *J. Geophys. Res.*, **92**, 13315–43.

12

Observational data on the relationships between climatic change and the frequency and magnitude of severe tropical storms

S.C.B. Raper

Introduction

Tropical cyclones are perhaps the most important of the natural hazards that coastal regions in the tropics and subtropics have to contend with. Danger to life and damage to property is caused by both the hurricane force winds and attendant storm surges. There is concern that an increase in tropical cyclone activity might occur in the future, associated with greenhouse gas induced global warming. The concern arises primarily from the fact that tropical cyclones derive their energy from the warm tropical oceans. An increase in sea surface temperatures could lead to increases in the sizes of the regions in which tropical cyclones form, as well as increases in frequency and intensity. Apart from the possible change in tropical cyclone activity, any sea level rise accompanying a global warming will clearly exacerbate the problems of coastal inundation by storm surges.

In this paper, the US National Oceanic and Atmospheric Administration tropical cyclone data set (NOAA, 1988), which spans the six tropical cyclogenesis regions, is examined and a review is made of the factors affecting tropical cyclone formation. Relationships between tropical cyclone frequency and parameters of the general circulation are then discussed. These relationships are necessarily determined largely by variations on the interannual time-scale because we do not know whether the longer-term trends in the data are real. Some new analyses of sea surface temperature (SST) data for all six regions are made with a view to speculating on the effect of a 1 °C increase in local sea surface temperature.

Tropical cyclone data

A *tropical cyclone* is defined here as a storm with maximum wind speeds > 17.5 m/s (34 knots). *Hurricanes* or *typhoons* have maximum wind speeds > 33 m/s (64 knots). Fig. 12.1 (from Gray, 1975) shows the regions in which storms are formed, defined by the first reported position of tropical cyclones for the 20–year period 1952–71. Superimposed on Fig. 12.1 are the regions which contain the tropical cyclone track data of NOAA (1988). Seasonal cycles of tropical cyclone frequencies are shown in Fig. 12.2.

The Indian Ocean (Bay of Bengal and Arabian Sea) is exceptional in that it shows a bimodal pattern of the annual cycle of tropical cyclones. The mid-season minimum is a consequence of the geography of the region. Between July and September, the monsoon trough is in its northernmost position and there is generally insufficient track

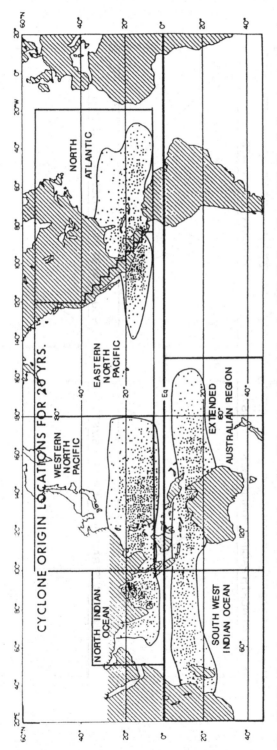

Fig. 12.1 The dots show tropical cyclone (wind speeds > 17.5 m/s) origin locations for 1952–71 (from Gray, 1975, Fig. 6). Superimposed are the NOAA (1988) tropical cyclone regions.

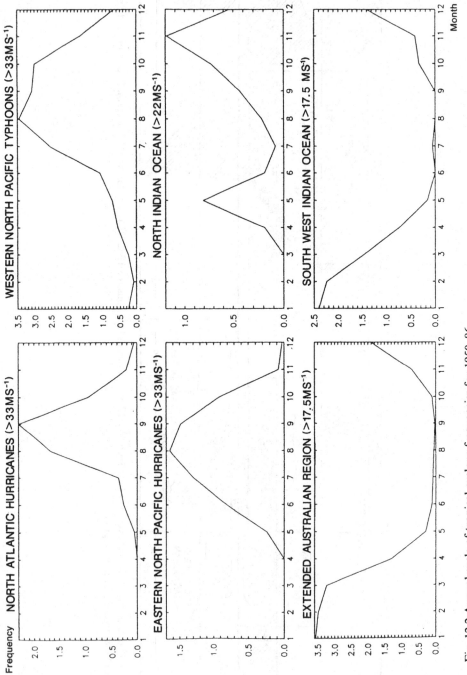

Fig. 12.2 Annual cycle of tropical cyclone frequencies for 1950–86.

over water for storms to intensify before landfall. In fact, compared with the North Atlantic and North Pacific, storms in the North Indian Ocean are generally less intense. For this reason, North Indian Ocean storms with maximum winds > 22 m/s are used for this study rather than hurricanes. In the Southern Hemisphere, tropical cyclones instead of hurricanes are shown purely because of data availability. The results shown in Fig. 12.2 are almost certainly representative of more severe storms because, in all regions except the North Indian Ocean, the seasonal mean tropical cyclone and hurricane frequencies are highly correlated in recent years.

Time series of tropical cyclone frequencies are shown in Fig. 12.3 (data from NOAA, 1988). The region denoted 'extended Australian region' is referred to as 'Southwest Pacific/Australian region' by NOAA. It is more extensive than the region which Nicholls (1985) refers to as the Australian region (viz. 105–165°E only). Fig. 12.3 also shows results for this more restricted region. Several of the series show an increase in the observed frequency over the period of record. These trends are most likely due to incomplete data in the early years. Changes in data coverage are a serious problem in all regions except (probably) the North Atlantic region where the overall trend in frequencies is not large. In the western North Pacific, coverage was restricted to west of 150°E until the late 1930s; frequencies before then are probably underestimated (Raper, 1978). The large jump in the number of recorded storms in the Eastern North Pacific during the 1960s (Fig. 12.3(*c*)) has been attributed to the recent use of satellite surveillance.

The Southern Hemisphere series (Fig. 12.3(*e*)-(*g*)) show marked increases during the twentieth century. In the extended Australian region, many of the recorded storms in the early years are classified as being of 'unknown intensity'. These have not been included in Fig. 12.3(*e*), so the time series is clearly non-homogeneous. Nicholls (1985) has compiled a more complete series for the Australian region (105–165°E; Fig. 12.3(*g*)). This also shows a trend, which Nicholls (1984, 1985) attributed to improving observational systems. He also suggests that the large number of recorded cyclones in the 1962–3 season may be due to some storms being counted more than once. To support this contention, he notes that, in this season, Lourenz (1981) recorded three storms located within 150 miles of each other at the same time, an unlikely circumstance.

Because of these data problems, it is not known whether any long-term (decadal) trends exist in the real frequencies (except perhaps for the North Atlantic). Thus, little confidence can be placed in derived linkages with long-term trends in other climate parameters. Any such associations, therefore, must rely largely on analyses of interannual variability.

Factors affecting tropical cyclone formation

Since Gray's (1968) seminal paper on the subject, there has been a great deal written on the factors affecting tropical cyclone formation. Tropical cyclones always develop from identifiable but far less severe tropical weather systems. Such systems are numerous, but relatively few intensify into fully developed cyclones. Development into a tropical cyclone is a result of large-scale influences rather than the characteristic of the individual weather systems (McBride and Zehr, 1981). The main factors appear to be

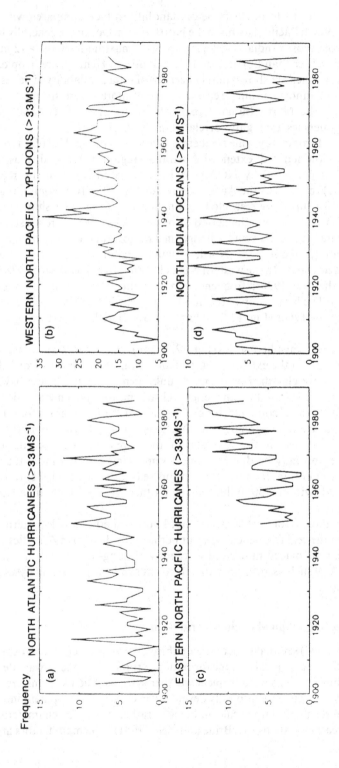

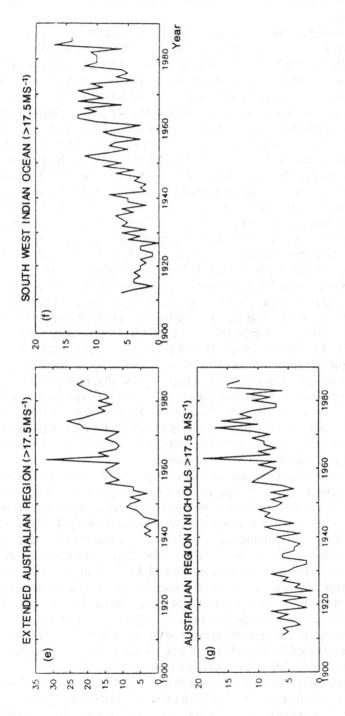

Fig. 12.3 Tropical cyclone frequencies for the six cyclogenetic regions extracted from the NOAA (1988) except for the Australian region which is from Nicholls (1985).

mixed-layer ocean temperature and the horizontal and vertical wind structure in the vicinity of the storm.

As a general rule, for tropical cyclone development and maintenance, an ocean temperature of at least 26 °C is needed down to a depth of about 60 m. Tropical cyclones derive their energy from this heat source. The relationship is demonstrated by the coincidence of the annual cycles of tropical cyclone frequency and SST. In the North Atlantic, winter ocean temperatures are too cold for storms to develop. In the Western North Pacific, where average temperatures are 1–2° C higher, tropical cyclones do occasionally develop and intensify in the winter months (see Fig. 12.2).

For the North Atlantic, Wendland (1977) correlated monthly tropical cyclone and hurricane frequencies for 1962–71 with monthly SST data. For his SST parameter, he used the area enclosed by the 28.6 °C isotherm between 7.5°N and 30°N. The relatively high positive correlations obtained using monthly data (r = 0.59 for tropical storms and r = 0.55 for hurricanes) are due to the strong annual cycles in both data sets. The relationship, however, appears to be non-linear. Wendland proposed that there is an upper limit to the tropical cyclone frequency associated with a given area of SST > 28.6 °C, with the actual number of storms that form in a month lying anywhere between zero and that limit. Hastenrath and Wendland (1979) obtained similar results for the North Atlantic using monthly data for 1911–72 and area-average SST from 10–30°N, west of 30°W. They also found a (weaker) SST-frequency correlation in the Eastern North Pacific using data for 1947–72 and area-average SST east of 180°W and north of 10°N. These results are probably dominated by the effects of intra-annual rather than interannual variability. More recently, Merrill (1987, 1988) has carried out a spatial analysis comparing tropical cyclone intensity and climatological monthly mean sea surface temperatures. He finds a threshold relationship, similar to Wendland, between the greatest storm intensities and SST. While there is some evidence of increasing maximum intensity with increasing SST, the use of climatological mean SSTs in this analysis and the strong annual cycle in both data sets makes the interpretation of these results obscure. Overall, these analyses support the idea that warm water is necessary, though not sufficient, for tropical cyclone development.

Despite the coincidence of the annual cycles in tropical cyclone frequency and SSTs, McBride (1981) reports that sea temperature did not emerge as a sensitive parameter when conditions associated with developing and non-developing disturbances were compared. This work is based on data published by Gray et al. (1982), which are separated into the North Atlantic and the Western North Pacific and consist of averaged daily data for all developing and non-developing disturbances. Average patterns for both disturbance types were formed by centring the data over the disturbance centres. McBride's (1981) result is consistent with some recent findings of Graham and Barnett (1987). They report that SSTs in excess of 27.5 °C are needed for deep tropical convection to occur but that, above this temperature, there is no relationship between the SSTs and the intensity of the convection. This threshold result is similar to that claimed by Merrill (1987); but Merrill (1988) interprets his results differently. Further SST analyses are discussed below.

The second controlling factor for cyclone development is the wind configuration. The conditions that are favourable for storm development on the day-to-day time-scale involve large-scale low-level cyclonic inflow overlain by upper-level anticyclonic outflow. In these conditions, there are two adjoining regions of strong 900–200 mb vertical wind shear on either side of the developing storm, with a change in sign and hence very small vertical wind shear near the centre of the storm. The wind configuration can be expressed as the superposition of low-level positive relative vorticity and upper-level negative relative vorticity. McBride and Zehr (1981) have demonstrated the importance for intensification of the wind configuration around the disturbance centre, using the data set compiled by Gray *et al.* (1982).

When daily weather conditions are averaged over months and seasons, larger-scale features of the prevailing winds reveal themselves. The conditions in which tropical cyclones occur are perturbations from the mean features. The nature of the link between tropical cyclone genesis and atmospheric conditions on this time-scale is therefore probabilistic rather than deterministic (Gray, 1975).

A number of authors have investigated the influences of various climatic parameters by averaging and comparing years with many and few tropical cyclones (Ballenzweig, 1957; Raper, 1978; Ding and Reiter, 1983; Aoki and Yoshino, 1984). Although these composite analyses are instructive, they do not tell us whether the anomalies are independent perturbations of the mean background circulation or whether they reflect the actual presence of the storms themselves. This is a point which needs to be borne in mind in all studies which attempt to relate changes in tropical cyclone activity to the general circulation.

Of the composite analyses cited above, only Ding and Reiter studied more than one region (viz. the North Atlantic and the Western North Pacific). Their results illustrate some important differences between the two regions.

For the North Atlantic, a northward shift in the position of the subtropical anticyclone is noted during years with many hurricanes and a broader than usual zone of surface easterlies is found to the south of the anticyclone. Gray (1968) had earlier noted that tropical cyclones in the North Atlantic form from disturbances in this broad zone of easterlies. Unlike systems in the other regions, they do not form in association with the intertropical convergence zone (ITCZ). In this genesis region, the long-term mean summertime relative vorticity is *negative* at 850 mb (-1.5×10^6/s) (McBride and Zehr, 1981).

In the Western North Pacific, during seasons with many typhoons, stronger Westerlies are found equatorward of 20°N with stronger Easterlies to the north. The associated seasonal mean position of the ITCZ is near 20°N and extends eastward to 160°E. In the years with few typhoons, the ITCZ is near 10°N and extends only to about 130°E. The reason for this is because typhoons form in association with, and just poleward of, the monsoon equatorial trough in a region of strong mean background *positive* relative vorticity (3.0×10^{-6}/s) at 850 mb (McBride, 1981).

In both the North Atlantic and Western North Pacific, the 200 mb flow analyses for years with many intense tropical cyclones show anomalous anticyclonic circulations over the genesis regions. In the few-cyclone cases, anomalous upper-level cyclonic circulations are found.

Interannual changes in tropical cyclone frequency related to the general circulation

The two general circulation parameters which have been studied most are sea level pressure (SLP) and sea surface temperature (SST). Studies using 500 mb and 200 mb data have been limited because of the relatively short records available, though some interesting results consistent with the concept of the steering of tropical cyclone tracks by the mid-tropospheric flow have been obtained (Shapiro, 1982). It should be noted that SLP and SST data are not independent. On a seasonal time-scale, changes in SSTs are related to the surface winds, which are in turn influenced by the SSTs. Surface pressure gradients are therefore related to SSTs (Davis, 1976; Shapiro, 1982).

Sea level pressure

Most studies of SLP relationships have been for the North Atlantic region. For this region, Raper (1978) related indices of the position and intensity of the subtropical anticyclone derived from SLP data to tropical cyclone frequency for June–November ($r = 0.44$ and -0.37 respectively for 1900–69). The indicated northward shift and weaker intensity of the subtropical anticyclone during years with high tropical cyclone frequency is consistent with the composite analyses described in the previous section. Raper also analysed daily SLP data to show that these relationships are probably *not* due to the presence or absence of the storms themselves. The storms appear to have 'slipped through' the observing network on most occasions and so not had much influence on the mean SLP values.

These relationships between SLP and North Atlantic tropical cyclone activity have been confirmed by Shapiro (1982). He has mapped correlation coefficients across the North Atlantic from 20–50°N between his measure of August–October hurricane activity and average SLP for the same months using data for 1899–1978. To answer the question of causality, Shapiro also computed the correlations between August–October hurricane activity and May-July SLP. (According to Shapiro, the correlations between May–July and August–October hurricane activity are negligible.) Both maps show relatively large negative correlations, the largest of which are in the trade wind region between 20 and 30°N (see Fig. 12.4(*a*),(*b*)). There is a shift in the region of maximum correlation from the west of the Atlantic for August–October ($r > -0.6$) to the mid-Atlantic for May–July ($r = -0.5$).

Shapiro has used the first empirical orthogonal function (EOF) of the May–July SLP as a predictor for August–October North Atlantic hurricane activity. Both series are first detrended. The correlation coefficient between the first EOF of SLP and hurricane activity is $r = 0.44$. Following Davis (1976, 1977) he estimates that the amount of variance that can be accounted for in a predictive sense is 17%. The inclusion of further EOFs does not add to the predictive skill. Gray (1984b) has noted that some reduction in predictive skill is likely to occur when a forecast is made using independent data. On the other hand, Shapiro's estimate may be conservative because he detrended the data; there is evidence that there is a real increasing trend in hurricane activity associated with decreases in SLP.

Sea surface temperature

Several studies have reported positive correlations between North Atlantic tropical cyclone frequencies and SSTs across the North Atlantic between about 10°N and 30°N (Riehl, 1956; Raper, 1978; Shapiro, 1982). However, the SST data used in these studies were not homogeneous (Jones *et al.*, 1986); the pre-1940s SSTs are too low because of

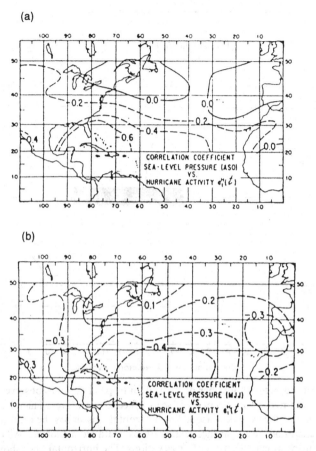

Fig. 12.4 Correlations between (*a*) August–October SLP and (*b*) May–July SLP and August-October hurricane activity 1899–1978. (Shapiro, 1982. Reprinted by permission of the American Meteorological Society).

the transition from bucket measurements to engine intake measurements. This inhomogeneity is present in the SST data world-wide, the necessary corrections depending on the region and time of year.

Studies in the Western North Pacific have used post-1940s data and are therefore not affected by these data inhomogeneities. In contrast to the North Atlantic, the largest correlations between Western North Pacific tropical cyclones and SSTs west of 150°E are negative (Raper, 1978; Aoki and Yoshino, 1984). The reason for this negative relationship is uncertain, and the relationship itself appears to breakdown for earlier periods (see below).

The only other tropical cyclone region which has been extensively examined with regard to SSTs is the Australian region. Nicholls (1984) investigated the relationship between tropical cyclone generation and SST in a 'North Australian SST area' (5–15°S, 120–160°E) for 1964–82. He correlated cyclone-season frequencies with overlapping three-month means of area-average SST, starting with the period January–March preceding the cyclone season and finishing with the period October–December follow-

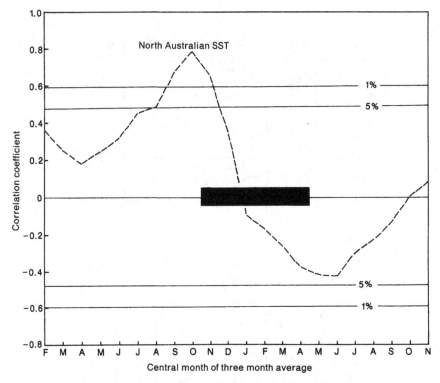

Fig. 12.5 Correlations between number of Australian tropical cyclones observed in total cyclone season and three-month averages of sea surface temperatures. Thick horizontal bar indicates the extent of the cyclone season. The horizontal axis shows the central month of three-month averages of the sea surface temperatures. Data from 1964–82. All correlations were calculated using at least 18 pairs of observations. The horizontal lines show the 1% and 5% significance levels for the correlation coefficients (Nicholls, 1984. Reprinted by permission of the Royal Meteorological Society).

ing the cyclone season (see Fig. 12.5). The correlation with SST in September–November, near the start of the cyclone season, is 0.78. This high positive correlation drops off rapidly using SSTs after the start of the season and the relationship becomes negative in the second half of the season. Nicholls confirmed these results using cyclone and SST data for 1913–31 (SST data were for 7°S, 118°E, just west of the 'North Australian SST area'). As an explanation, Nicholls (1981, 1984) proposed that anomalously warm SSTs are associated with the early commencement of a stronger than normal northwest monsoon, allowing earlier than usual tropical cyclone formation. He attributes the rapid change in sign of the SST correlation just after the start of the tropical cyclone season to increased ocean mixing caused by the stronger monsoon and possibly also by the tropical cyclones themselves.

The relationship between pre-season SSTs and tropical cyclone frequencies using both the 1964–82 and the 1913–31 data is shown in Fig. 12.6(*a*) (from Nicholls, 1984). The relationship looks impressive, but there are problems with both data sets. The bucket model results of Farmer *et al.* (1989) suggests that, for September–November, the pre-1940s SSTs for this region should be increased by 0.32 °C. When the data are

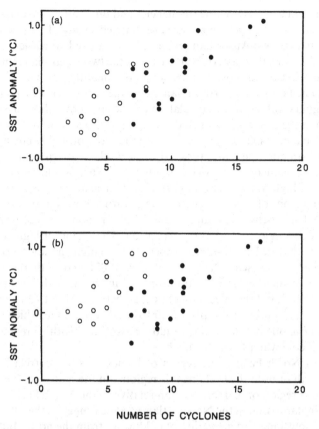

Fig. 12.6(*a*) Scatter diagram of cyclone numbers versus north Australian SST anomaly averaged over September–November just prior to the start of the cyclone season, data for 1964–82 (full circles) and 1913–31 (open circles). (Nicholls, 1984. Reprinted by permission of the Royal Meteorological Society). (*b*) Same as (*a*) but with a correction of 0.32 °C added to the data for 1913–31.

replotted with this correction, the overall linear relationship largely disappears, but the two different time periods individually still show positive relationships (Fig. 12.6(*b*)). This may indicate that the earlier cyclone frequency data are in error (as suggested by Nicholls to explain the apparent increasing trend in Fig. 12.3(*g*)). If tropical cyclone frequencies were consistently underestimated in the period 1913–31, an estimate of the average number of storms each year which went undetected during 1913–1931 can be made simply by comparing the positions of the regression lines through the two data sets. This gives a value of 4.5 undetected storms, compared with the 'observed' average increase of 5.7 storms between the two periods 1913–31 and 1964–82.

Re-analysis of SST data

In this section, the relationship between seasonal mean tropical cyclone frequency and SST is examined region by region. The SST data used are those described by Farmer *et al.* (1989) and Jones *et al.* (1991), based on COADS. Using data for 1950–86, the correlations between tropical cyclone frequencies and three-month-average grid-point SSTs near the start of the cyclone season are mapped in Fig. 12.7 (see caption for details). (Note that the two Australian regions are considered, as earlier.)

In the North Atlantic, the Eastern North Pacific, the two Australian regions and the Southwest Indian Ocean, the correlations are predominantly positive, whereas in the Western North Pacific and the North Indian Ocean the correlations are predominantly negative. As a guide, values of the correlation coefficient above about 0.3 are statistically significant at the 5% level, but this does not take into account the multiplicity of correlations calculated. Isolated large values are therefore probably not significant.

In the North Atlantic, the correlations show a similar pattern to those shown by Shapiro (1982), using uncorrected SST data for 1899–1967, but the values are slightly lower. The area of highest correlation, off the West African coast, is to the north and east of the main regions of cyclone genesis. Some storms, however, do originate in the eastern North Atlantic between about 5°N and 15°N, a region which coincides with the southern half of the region of positive correlation. The region of high correlation is consistent both with the SLP anomalies associated with high and low tropical cyclone frequencies already discussed (Ding and Reiter, 1983) and with the correlations between cyclone frequencies and the position and intensity of the subtropical anti-cyclone (Raper, 1978). Off the West African coast, between about 10°N and 25°N, an anomalously intense, southward displaced subtropical anticyclone, which is unfavourable for tropical cyclone development, would be associated with increased upwelling and advection of cold water from the north.

In the Eastern North Pacific, the region of highest positive correlation off Baja, California, is north of both the cyclogenesis region and the main cyclone tracks. The positioning of this region of high correlation relative to the North Pacific subtropical anticyclone is similar to that in the North Atlantic, which suggests that it might also be associated with upwelling and advection of cold water from the north, but composite analyses for low and high cyclone frequencies are not available to confirm this. The relatively high positive correlations found in the Australian cyclogenesis region reproduces the results of Nicholls (1984) (see Fig. 12.5). When cyclone data for the extended Australian region are used, the correlations are slightly lower, implying a smaller SST influence for storms in the eastern part of the extended region. In the Southwest Indian Ocean positive correlations are found, coincident with the region of storm formation and intensification.

Mainly negative correlations are found in the other two regions. It is possible that the relatively high negative correlations found in the South China Sea are, in part, due to the passage of the typhoons themselves causing near-surface ocean mixing in this land-restricted sea. However, early season SSTs were chosen to reduce this effect, and Aoki and Yoshino (1984) found that in the Western North Pacific, to the east of the Philippines, lag correlations with SSTs up to two years preceding the typhoon season are also negative. The North Indian Ocean negative correlations occur in the cyclone track region. The three months May–July, used for the SSTs, coincide with the second half of the first seasonal peak in North Indian Ocean cyclone frequency so, again, the negative correlation may be related to the passage of the storms.

Some of the correlations noted above are strong enough to be used, albeit tentatively, to estimate possible changes in storm frequency in a warmer world. But first we need to test the stability of the relationships. The ephemeral nature of many climatic correla-

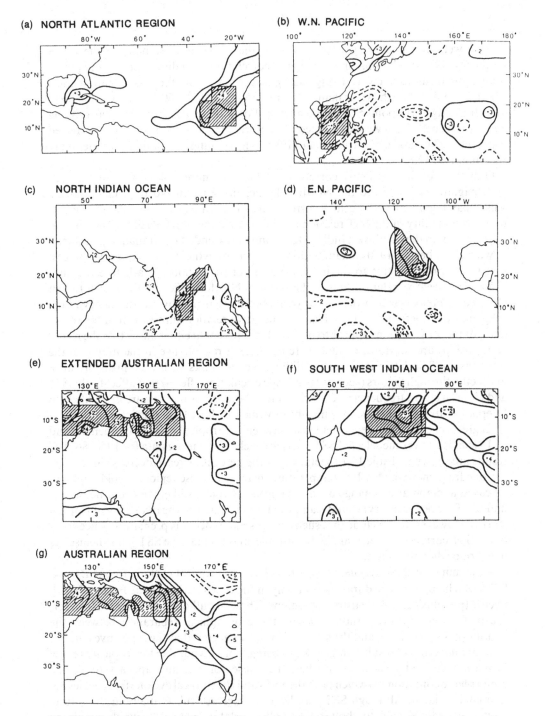

Fig. 12.7 Correlation maps of regional tropical cyclone frequency versus three-month-average SST. The SST months used were May–July for the Northern Hemisphere regions, September–November for the Australian and extended Australian regions and October–December for the South West Indian Ocean. Shaded areas show selected regions of highest correlation (see text).

tions is well-known (Ramage, 1983). To test the longer-term stability of the storm frequency/SST correlations, areas of highest correlation extending over 5–8 grid points were chosen subjectively (see Fig. 12.7), except in both the Australian regions where Nicholls' 'North Australian SST area' (5–15°S, 120–160°E) was used. Three-month-average grid-point SSTs were averaged over these areas and 15–year running correlation coefficients were calculated. As before, pre-1942 SSTs were first corrected using the bucket model of Farmer *et al.* (1989). The running correlation coefficients are shown in Fig. 12.8.

Only three of the areas have correlations which are more-or-less of consistent sign and magnitude. These are the North Atlantic, the Eastern North Pacific and the Australian region, all with positive correlations. (There are insufficient data to be able to test the stability of the SST relationship in the Eastern North Pacific.) The positive relationships in the Southwest Indian Ocean and the extended Australian regions break down during the mid-1950s. Examination of the trends in the storm frequencies suggest that this is probably due to the incompleteness of the cyclone data before the mid-1950s. The negative relationships in the Western North Pacific and the North Indian Ocean also break down in the earlier years, but there is no obvious deficiency in the frequency data before this time. Thus, the negative relationships shown in the recent data for these two regions appear to be unstable. This could either imply that the recent relationships are spurious, or that there has been a real change in the nature of the relationship. The latter is a real possibility because links with SST probably arise indirectly, through factors (such as large-scale circulation effects) that affect both SSTs and tropical storm frequencies. Complex interrelationships like this are more likely to be ephemeral. Nevertheless, they may be of value for short-term predictions.

Regression equations were calculated between the tropical cyclone frequencies and SSTs averaged over the selected SST areas for each region using data for 1950–86. The results are shown in Table 12.1 together with the percent increase/decrease in tropical cyclone frequency which a 1 °C temperature increase in these regions would imply. In all cases, a temperature change of this magnitude (which is likely to occur before the middle of the next century) would lead to a marked change in the frequency of tropical storms. However, to conclude that such changes might occur is premature, since some of the SST correlations are manifestly unstable, and because the SST links themselves may not be directly causal.

To summarize these results, correlations between tropical cyclone frequency and SSTs which appear to be durable occur only in the North Atlantic, the Eastern North Pacific (possibly) and the Australian regions. These are all of positive sign. However, except for the Southwest Indian Ocean, the geographical separation between the regions of SST influence and the storms themselves argues that these positive correlations are not a direct result of higher SSTs being more favourable for tropical cyclone formation and intensification. Rather, the cause for the link appears to be the atmospheric conditions to which both the SSTs and the tropical cyclone frequencies are intimately related. Although SSTs are likely to increase substantially in the future, model predictions tend to show only small circulation changes, usually within or similar to past natural variability (Santer *et al.*, 1991). If circulation changes are the primary reason for the observed SST relationships, then future cyclone frequency changes may well be within their past natural variability.

The results presented have only used SSTs averaged over three months near the start

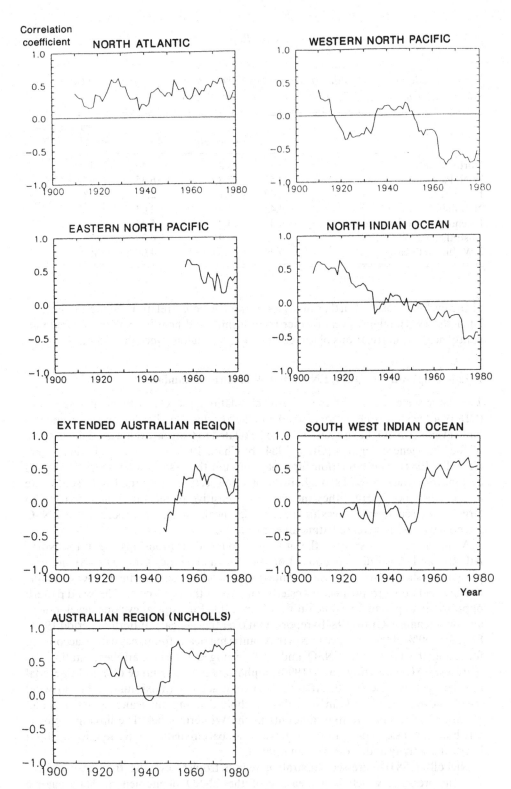

Fig. 12.8 15-year running correlation coefficients between tropical cyclone frequencies and three-month-average SSTs for the shaded areas indicated in Fig. 12.7.

Table 12.1. *Regression results for tropical cyclone frequencies in each region and selected SST areas (see Fig. 12.7), using data for 1950–86*

	r	Constant	Slope	Increase/decrease for $\Delta T = +1°C$
North Atlantic	0.48	5.82	2.40	41%
W.N. Pacific	−0.53	17.4	−6.45	−37%
E.N. Pacific	0.49	6.18	2.35	38%
N. Indian Ocean	−0.42	4.48	−2.03	−45%
Extended Australian region	0.37	14.9	5.48	37%
Australian region (Nicholls)	0.73	9.91	6.20	63%
S.W. Indian Ocean	0.49	9.05	4.08	45%

of the cyclone season. Further analyses, using lagged correlations along the lines of Nicholls (1984), might reveal further relationships and help to explain the predominantly negative correlations observed in two of the regions since the 1950s.

Relationships with the QBO, ENSO and West African Rainfall

Two large-scale factors of the general circulation, the Quasi-Biennial Oscillation (QBO) of the stratospheric zonal wind and the Southern Oscillation and associated El Niño events (ENSO) have been shown to be related to tropical cyclone activity in some of the cyclogenesis regions (Gray, 1984a,b; Chan, 1985; Revell and Goulter, 1986; Nicholls, 1984). These relationships may obscure the associations between tropical cyclone frequency and SSTs sought in the previous section. QBO and ENSO studies are also of relevance because they indicate how sensitive tropical cyclone activity is to large-scale circulation changes and because significant changes may occur in the ENSO phenomenon with increased atmospheric CO_2.

As in the previous section, the most well-founded relationships are in the North Atlantic and Australian regions. For the North Atlantic, Gray (1984a) presents evidence to show that suppression of hurricane activity occurs during ENSO years due to enhanced upper tropospheric westerly winds over the Caribbean. This wind pattern opposes the upper anticyclonic outflow favourable for tropical cyclone development and maintenance. Gray (1984b) reports that over the 33-year calibration period 1950–82, about 60% of the variance in North Atlantic hurricane frequency can be accounted for using the QBO, the ENSO and SLP averaged for six Caribbean stations as regressors. More recently Gray (1990) emphasizes the importance of a multi-decadal link between intense North Atlantic hurricane activity (maximum sustained wind speed above 50 m/s, cf. 33 m/s for the threshold defining hurricanes in general) and variations of summer rainfall amounts in the Western Sahel. The linking factor is attributed to basic changes in the larger-scale global circulation patterns affecting both hurricane activity and West African rainfall.

Nicholls (1984) regressed Australian region tropical cyclone frequency against Darwin pressure, which is a measure of the ENSO phenomenon. The negative correlation was strongest between July–September average Darwin pressure and the number of cyclone days; 46% of the variance was explained over the period 1958/9 to 1981/2.

Conclusions

Increasing atmospheric greenhouse gas concentrations may affect tropical cyclones in three ways. Firstly, increasing SSTs may cause tropical storms to be, on average, more intense (Emanuel, 1987; see also, Schlesinger, this volume). Secondly, greenhouse gas induced changes in the general circulation and surface ocean temperatures may change the frequency, tracks and regional occurrence of the storms. Thirdly, sea level rise will add to the already present problems of coastal inundation. Only the second of these has been considered in depth here.

With reference to the first point, it should be noted that there is no convincing empirical evidence to support a relationship between SSTs and cyclone intensities. Nevertheless, there is some evidence of a threshold effect that could lead to increases in the intensity of the most severe storms with an increase in SSTs (Merrill, 1987, 1988). The issue is still open, however. Graham and Barnett (1987) have shown that, while SSTs in excess of 27.5 °C are required for large-scale deep tropical convection, further increases in SSTs have little effect on the intensity of convection. Thus, an increase in SSTs would not necessarily lead to a general increase in storm intensity. (For further discussion, see Schlesinger, this volume).

Relationships have been sought between tropical cyclone activity and SSTs with a view to speculating on the effect of a local warming of 1 °C on cyclone frequency. The statistical analysis presented here used area-mean SSTs over prescribed areas. Nevertheless, the results may well reflect the effects of changes in the areas of cyclogenesis and/ or areas with warm SSTs, although these aspects have not been singled out specifically. For the former, while the observed changes in tropical cyclone frequencies could be due to changes in the number of storms developing in a fixed cyclogenetic area, they may also be due to changes in the area of the cyclogenesis regions (e.g., the maximum area at height of season, seasonal duration, etc. For the latter, because of the spatial and temporal (month-to-month) autocorrelation in the SST data, the preseasonal SST data used in this analysis will also reflect changes in the area and duration of SSTs > 27.5 °C. No simple relationships have emerged from the analyses. The fact that a warm ocean surface is needed for tropical cyclone development is well-known, but based on empirical evidence SST does not seem to be the primary variable in determining whether incipient storms develop or not (McBride, 1981). Empirical analyses presented in this paper found that, on the interannual time-scale, four of the six cyclone regions show predominantly positive correlations between cyclone frequency and early cyclone season SSTs. However, in the North Atlantic and Eastern North Pacific, the areas of highest correlation are not coincident with the regions of cyclone origins and tracks. In three of the four cases there is evidence that the link between tropical cyclone frequency and SSTs is indirect, both being associated with changes in the atmospheric circulation. The reason for the predominantly negative correlations found in the Western North Pacific and North Indian Ocean from 1950–86 is uncertain. These negative relationships are not evident in earlier years.

In the regions of highest positive correlation (shaded regions in Fig. 12.7), simple regression of frequency on SST indicated that a 1° C increase in SST would be associated with an increase in average tropical cyclone activity ranging up to 63% (see Table 12.1). However, it would be unwise to extrapolate these results to the case of future greenhouse gas induced global warming until the causes of the relationships are more firmly established.

A possible effect of SSTs may arise through changes in atmospheric stability. If near-surface warming were greater than that in the upper troposphere, enhanced instability may lead to conditions that were generally more favourable for cyclogenesis and development. However, equilibrium GCM studies give no evidence of any increases in instability in the 30°S to 30°N region (see Schlesinger and Mitchell, 1987); indeed, some models show more stable conditions in a $2 \times CO_2$ world, which, as noted by Holland *et al.* (1988) could lead to fewer tropical storms. As with the direct SST results, however, it would be unwise to place too much weight on this conclusion. Model-derived stability changes depend critically on tropical cloud parameterizations and on model fidelity in general, and considerable uncertainties still exist with regard to these items. Further-more, the role of large-scale stability in tropical cyclogenesis is uncertain; McBride and Zehr's (1981) empirical analysis found that stability was not a major factor in determining cyclogenesis.

Besides SSTs, two other factors important for tropical cyclone activity are the Coriolis force and the large-scale three-dimensional wind configuration around the storms. Little can be said about the effect of possible wind changes, because current models are unable to simulate tropical climates well enough to trust their projections of wind field changes. The Coriolis effect, will, of course, not change, but it involves an interaction with SSTs that could be important. As cyclones move poleward, increasing f (Coriolis parameter) necessitates either a drop in the central pressure or a decrease in wind speed (Ivanov and Khain, 1983). The intensity of poleward tracking cyclones may therefore be limited by energy supply effects, which in turn may be limited by subtropical SSTs. With a warming world, increasing SSTs might therefore allow poleward tracking cyclones to be more intense, a suggestion made by Pittock (this volume). However, poleward motion is also dependent on the prevailing large-scale atmospheric circulation, and the effects of circulation changes on the cyclone tracks may be more important than SST effects.

The main point emerging from this analysis is that changes in tropical cyclone activity in a warmer world will depend crucially on the changes that occur in the regional-scale atmospheric circulation. Until models allow these changes to be pre-dicted, while there is some evidence for an increase in the frequency and geographical extent of storms in some regions, any projections must be considered speculative.

References

Angell, J., Korshover, J. & Cotton, G. (1969). Quasi-biennial variations in the 'centres of action'. *Monthly Weather Review*, **97**, 867–72.

Aoki, T. & Yoshino, M.M. (1984). Relation between the frequency of typhoon formations and sea surface temperature. *Journal of the Meteorological Society of Japan*, **62**, 172–5.

Ballenzweig, E.M. (1957). *Seasonal Variations in the Frequency of North Atlantic Cyclones Related to the General Circulation*. National Research Project, Report No. 9, Washington, DC.

Chan, J.C.L. (1985). Tropical cyclone activity in the Northwest Pacific in relation to the El Niño/

Southern Oscillation phenomenon. *Monthly Weather Review,* **113**(4), 599–606.

Davis, R. (1977). Techniques for statistical analysis and prediction of geophysical fluid systems. *Geophysics, Astrophysics and Fluid Dynamics,* **8**, 245–77.

Davis, R. (1976). Predictability of sea-surface temperature and sea-level pressure anomalies over the North Pacific Ocean. *Journal of Physical Oceanography,* **6**, 249–66.

Dickson, R.R. (1975). A preliminary analysis of factors affecting the frequency of August southwestern North Pacific tropical storms and hurricanes since the advent of Satellite observation. *Monthly Weather Review,* **103**, 926–8.

Ding, Yi-Hui. & Reiter, E.R. (1983). Large-scale hemispheric teleconnections with the frequency of tropical cyclone formation over the Northwest Pacific and North Atlantic Oceans. *Arch. Met. Geoph. Biocl.,* **A 32**, 311–37.

Emanuel, K.A. (1987). The dependence of hurricane intensity on climate. *Nature,* **326**, 483–5.

Farmer, G., Wigley, T.M.L., Jones, P.D. & Salmon, M. (1989). *Documenting and Explaining Recent Global-Mean Temperature Changes.* Report to the U.K. Natural Environment Research Council, 141 pp.

Graham, N.E. & Barnett, T.P. (1987). Sea surface temperature, surface wind divergence, and convection over tropical oceans. *Science,* **238**, 657–9.

Gray, W.M. (1990). Strong association between West African rainfall and US landfall of intense hurricanes. *Science,* **249**, 1251–6.

Gray, W.M. (1984a). Atlantic seasonal hurricane frequency Part I: El Niño and the 30 mb Quasi-Biennial Oscillation influences. *Monthly Weather Review,* **112**, 1649–68.

Gray, W.M. (1984b). Atlantic seasonal hurricane frequency Part II: forecasting its variability. *Monthly Weather Review,* **112**, 1669–83.

Gray, W.M. (1975). *Tropical Cyclone Genesis.* Atmospheric Science Paper No. 24, Department of Atmospheric Science, Colorado State University, Fort Collins, Colorado.

Gray, W.M. (1968). Global view of the origin of tropical disturbances and storms. *Monthly Weather Review,* **96**, 669–700.

Gray, W.M., Buzzell, E. & Burton, G. (1982). *Tropical Cyclone and Related Meteorological Data Sets Available at CSU and their Utilization.* Department of Atmospheric Science, Colorado State University, Fort Collins, Colorado.

Hastenrath, S. & Wendland, W.M. (1979). On the secular variation of storms in the tropical North Atlantic and Eastern Pacific. *Tellus,* **31**, 28–38.

Holland, G.J., McBride, J.L. & Nicholls, N. (1988). Australian region tropical cyclones and the greenhouse effect. In *Greenhouse Planning for Climate Change,* G.I. Pearman (ed.). CSIRO, Australia, pp. 438–55.

Ivanov, V.N. & Khain, A.P. (1983). Parameters determining the frequency of tropical cyclone genesis. *Atmospheric and Oceanic Physics,* **19**(8), 593–9.

Jones, P. D., Wigley, T.M.L. & Farmer, G. (1991). Marine and land temperature data sets: a comparison and a look at recent trends. In *Greenhouse-Gas-Induced Climatic Change: A Critical Appraisal of Simulations and Observations,* M.E. Schlesinger (ed.). Elsevier Science Publishers B.V., Amsterdam, pp. 153–72.

Jones, P.D., Wigley, T.M.L. & Wright, P.B. (1986). Global temperature variations between 1861 and 1984. *Nature,* **322**, 430–4.

Lamb, H.H. (1972). *Climate: Present, Past and Future: Vol. 1., Fundamentals and Climate Now.* Methuen, London.

Lourenz, R.S. (1981). *Tropical Cyclones in the Australian Region, July 1909 to June 1980.* Bureau of Meteorology, Melbourne, Australia, 94 pp.

McBride, J. (1981). Observational analysis of tropical cyclone formation. Part 1: Basic description of data sets. *J. Atmos. Sci.,* **38**, 1117–31.

McBride, J.L. & Zehr, R. (1981). Observational analyses of tropical cyclone formation. Part II:

Comparison of non-developing versus developing systems. *J. Atmos. Sci.*, **38**, 1132–51.

Merrill, R.T. (1988). Environmental influences on hurricane intensification. *J. Atmos. Sci.*, **45**, 1678–87.

Merrill, R.T. (1987). An experiment in the statistical prediction of tropical cyclone intensity change. *Volume of Extended Abstracts, 17th Conference on Hurricanes and Tropical Meteorology*. AMS, Boston, pp. 302–4.

Nicholls, N. (1985). Predictability of interannual variations of Australian seasonal tropical cyclone activity. *Monthly Weather Review*, **113**, 1144–9.

Nicholls, N. (1984). The Southern Oscillation, sea-surface-temperature, and interannual fluctuations in Australian tropical cyclone activity. *J. Clim.*, **4**, 661–70.

Nicholls, N. (1981). Air-sea interaction and the possibility of long-range weather prediction in the Indonesian Archipelago. *Monthly Weather Review*, **109**, 2435–43.

NOAA (1988). *Constructed World-wide Tropical Cyclones 1871–1988*. (TD-9636).

Ramage, C.S. (1983). The teleconnections rush – bonanza or fool's gold? *Proceedings of the Seventh Annual Climate Diagnostics Workshop*, NCAR, Boulder, Colorado. US Department of Commerce, NOAA, pp. 207–10.

Ramage, C.S. (1974). Monsoonal influences on the annual variation of tropical cyclone frequency development over the Indian and Pacific Oceans. *Monthly Weather Review*, **102**, 745–53.

Raper, S.C.B. (1978). *Variations in the Incidence of Tropical Cyclones and Relationships with the Global Circulation*. Ph.D. Thesis, University of East Anglia, Norwich. UK.

Rasmusson, E.M. & Carpenter, T.H. (1982). Variations in tropical sea surface temperature and surface wind fields associated with the Southern Oscillation/El Niño. *Monthly Weather Review*, **110**, 354–84.

Revell, C.G. & Goulter, S.W. (1986). South Pacific tropical cyclones and the Southern Oscillation. *Monthly Weather Review*, **114**, 1138–45.

Riehl, H. (1956). *Sea Surface Temperatures of the North Atlantic, 1887–1936*. Project NR 082–120, Office of Naval Research, US.

Santer, B.D., Wigley, T.M.L., Jones, P.D. & Schlesinger, M.E. (1991). Multivariate methods for the detection of greenhouse-gas-induced climate change. In *Greenhouse-Gas-Induced Climatic Change: A Critical Appraisal of Simulations and Observations*, M.E. Schlesinger (ed.). Elsevier Science Publishers B.V., Amsterdam, pp. 511–36.

Schlesinger, M.E. & Mitchell, J.F.B. (1987). Climate model simulations of the equilibrium climatic response to increased CO_2. *Rev. Geophys.*, **25**, 760–98.

Schregardus, W.M.F. (1938). *Sea Surface Temperature on some Steamer Routes in Netherlands India*. Verhadelingen No. 28, Koninklijk Magnetisch en Meteorologisch Observatorium te Batavia, Batavia, Indonesia.

Shapiro, L.J. (1982). Hurricane climatic fluctuations. Part II: Relation to large-scale circulation. *Monthly Weather Review*, **110**, 1014–23.

Weare, B.C., Navato, A.R. & Newell, R.E. (1976). Empirical orthogonal analysis of Pacific sea surface temperatures. *J. Phys. Oceanogr.*, **6**, 671–8.

Wendland, W.M. (1977). Tropical storm frequencies related to sea surface temperatures. *J. Applied Met.*, **16**, 477–81.

IV

IMPACTS AND CASE STUDIES

13

Geographic information systems and future sea level rise

I. Shennan

Abstract

Geographic information systems (GIS) are potentially valuable in assessing the impact of future sea level rise on coastal lowlands. A GIS is a computer-based system, no longer dependent on large mainframe computing facilities, for integrating the collection, processing, storage, retrieval, analysis and display of geographic data. The development of such a system is not an easy task as data are rarely in a usable, i.e., digital, form. In order to illustrate how a simple GIS can be built up from a low level of complexity into a form usable by decision makers two contrasting coastal areas of the UK are discussed: the Tees and the Fenlands. These two cases also illustrate the application of the GIS approach at different scales. Input includes census data, land-use data, topographic map data, sea level scenarios, current flood defence data and tide data. From this input a GIS can produce an impact map for specified sea level rise scenarios.

Introduction

For assessing the impacts of future sea level rise, many different types of data over a wide range of spatial and temporal scales are potentially useful. For example, case studies of the impacts of sea level rise on the coastal urban areas of Charleston, South Carolina, and Galveston Bay, Texas, carried out with support from the US Environmental Protection Agency (EPA), illustrate the data requirements for detailed, local impact assessment. At the other extreme, Henderson-Sellers and McGuffie (1986) have calculated the global impact of a 10 m rise in sea level. Using a data archive with a resolution of 50 km they showed that more than 10×10^6 km^2 of land are at risk. Ideally, it would be desirable to adopt a single approach that embraces the global perspective while retaining the richness of local detail, but clearly this is impracticable. Financial costs have to be considered, along with data quality, quantity and availability. Thus, a major problem in assessing the local impacts of global sea level rise is how to select data from a wide range of scales in order to achieve a manageable balance.

Another problem common to such impact assessments pertains to the collection, integration, analysis and display of many different types of data. These data are spatially, or geographically, referenced and include natural resources, descriptions of infrastructure, patterns of land use, health, wealth, housing, employment and other socio-economic data, as well as the range of topographic data usually presented on maps. The following statement summarizes well the requirements for manipulating such geographic data:

'Large sums of money are spent by Government, commerce and industry, the utilities, the armed forces, and others in collecting and using it [geographic information]. Much human activity depends upon the effective handling of such information – its collection, processing, storage and retrieval – but it is the ability of computer systems (Geographic Information Systems or GIS) to integrate these functions and to deal with the locational character of geographic information that makes them such potentially powerful tools.'

(Her Majesty's Stationery Office,(HMSO), 1987, p. 2).

This brief review emphasizes the practical considerations in the use of GIS in assessing the impacts of future sea level rise.

Geographic Information Systems (GIS)

In its most basic form a GIS is a computer-based system for integrating the collection, processing, storage, retrieval, analysis and display of geographic data. The specific configuration of a GIS may involve any combination, and not necessarily all, of these functions. Fig. 13.1 shows in a simplistic fashion (but adequate for illustrative purposes) a GIS for integrating a range of geographic data. The different data are overlaid, based on their geographic attributes, according to the individual require-ments of the user. The organization of a GIS in the specific context of sea level rise impact assessment is illustrated in Fig. 13.2. Whilst these schemata may suggest a simple concept, in practice the development of an operational GIS is not a trivial task. (For further discussion of the general problems, of international as well as national relevance, see HMSO, 1987).

Availability of data

Much relevant data already exist, but rarely are the data available in the correct, i.e., digital, form. The type of raw data required and their presentation in specialist maps varies from country to country. National policies towards unrestricted use of basic topographic data, land-use data and other relevant GIS/impact assessment data vary tremendously.

These points are highlighted in the results of an international survey carried out under the auspices of IGCP/UNESCO Project 200: *Late Quaternary Sea level Changes; Measurement, Correlation and Future Applications.* Each national Project leader was sent a questionnaire requesting information concerning the range of relevant data available for their coastal lowlands in traditional paper map form and in digital form. Thirty-five national leaders were contacted, of whom one-third responded. The response was not sufficient to make a firm statement regarding the proportion of the world's coastal lowlands for which specific data are available, but some general points are worth noting. Two major obstacles to the immediate development of GISs on a national scale are the unavailability of digital data, and the restrictions on access to conventional map data sources. Currently only one of the responding countries has the relevant basic topographic data for all their coastal lowlands available in digital form; a number of other countries are still developing procedures and standards for converting

(a)

(b)

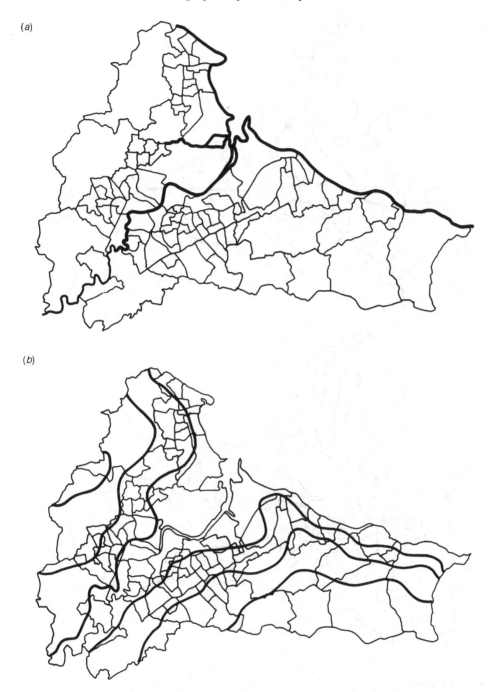

Fig. 13.1 Components of a basic GIS.
(a) Definition of spatial areas, in this example administrative 'ward' areas, and edge conditions where appropriate, i.e., coastline + tidal rivers.
(b) Topographic data in vector form, i.e., contours.

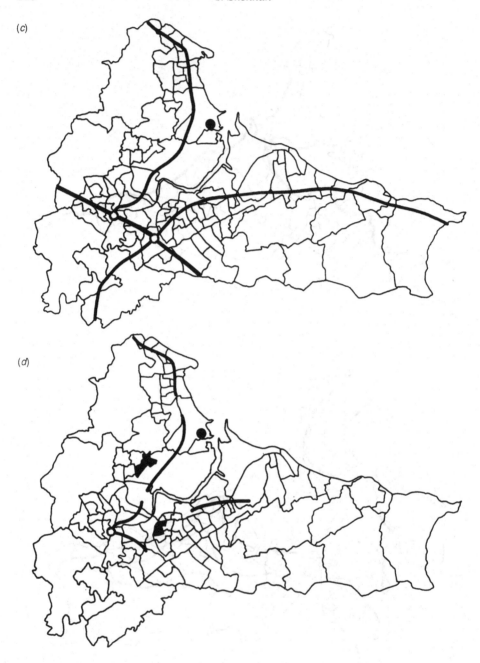

Fig. 13.1 (*cont.*)
(*c*) Vector data: major roads and point data: location of nuclear power station (shaded circle).
(*d*) Definition of the flooding zone hazards based on a defined water level, and the information relevant to, for example, the evacuation of large numbers of old people (shaded areas, data obtained by census data references for each ward), and access along flood-prone main roads (thick lines) to the nuclear power station (shaded circle).

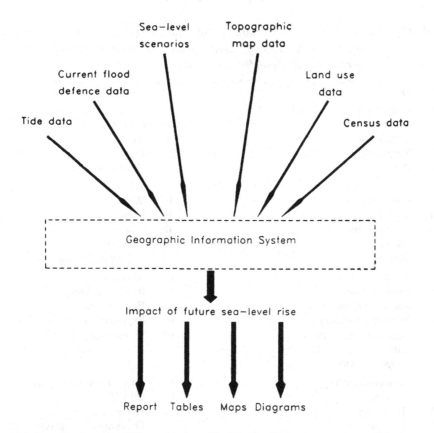

Fig. 13.2 Organization of a GIS used to provide data for future management of the impact of sea level rise (Shennan and Tooley, 1987. Reprinted by permission of Basil Blackwell).

the data. Nearly one-third of the respondents indicated that there were government or military-imposed restrictions on the availability of large-scale maps of coastal low-lands. Regarding topographic information for maps at a scale of 1:25,000 or larger a contour interval of 2.5 or 5 m for coastal lowlands was most commonly reported, with a range of 0.5 m (supplemented by spot height data) to 10 m.

The range of thematic map data variously available is shown in Table 13.1. This range is based on a review by Perrotte (1986) of specialist maps and represents some of the output required for impact assessment. These data can be produced by a GIS. The USA National Ocean Service 1:62500 Storm Evacuation Map (in Perrotte, 1986) is a good example of the type of map data required for impact assessment. A global coverage of available maps is contained in Parry and Perkins (1987). The automated production of land-use maps for flood prone areas using a commercially available GIS has been demonstrated by Dangermond (1984) (for a non-coastal site but this is unimportant in the present context). One obvious source of additional digital data for GIS/impact assessment of sea level rise is multi-spectral remote sensing (discussed later in the text).

Table 13.1. *Range of thematic map data available for coastal zones. This list is based on Perrotte (1986) and is the range of thematic map data known to National Leaders of the IGCP Project 200 who responded to a questionnaire survey*

1	*Habitat*	2	*Oceanography*
1a	Land, marine, inter-tidal habitats	2a	Water composition
		2b	Currents
1b	Vegetation	2c	Ice and icebergs
1c	Sanctuaries	2d	Tides
1d	Refuges	2e	Hydrodynamics
1e	Soils		
3	*Climatology*	4	*Biology*
3a	Temperature	4a	Land fauna
3b	Precipitation	4b	Coastal birds
3c	Pressure	4c	Fisheries
3d	Winds	4d	Marine micro-organisms
5	*Socio-economic*	6	*Resources*
5a	Demography	6a	Inventory: living and non-living
5b	Social characteristics	6b	Exploration
5c	Economy, industry	6c	Exploitation
5d	Recreation & tourist sites	6d	Transportation
5e	Land use		
5f	Support services		
7	*Coastal hazards*	8	*Environmental problems*
7a	Flooding	8a	Oil spills
7b	Hurricanes	8b	Sewage outfalls
7c	Damages	8c	Dumping of dangerous material
7d	Lost value	8d	Air pollution
7e	Evacuation	8e	Land fills
7f	Protection strategies		
9	*Coastal zone management plans*	10	*Geophysical and geological*
9a	Existing government programs	10a	Solid geology
9b	Recommendations	10b	Surficial geology
9c	Strategies	10c	Geomorphology
9d	Results	10d	Gravity
		10e	Magnetic field
		10f	Seismic
		10g	Neotectonic
11	*Administrative*		
11a	National/local administrative zones		
11b	Maritime zone boundaries		
11c	Coastal zone planning boundaries		
11d	Other . . .		

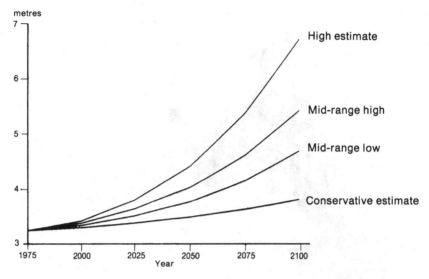

Fig. 13.3 An example of sea level rise scenarios (from Barth and Titus, 1984), adapted for local conditions in the Tees Estuary. All levels are related to Ordnance Datum (Shennan and Tooley, 1987. Reprinted by permission of Basil Blackwell).

Examples employing a GIS

Two coastal lowlands of the United Kingdom, the Tees and the Fenlands, are discussed briefly to illustrate how a simple GIS can be built up from a low level of complexity into a form usable by decision makers. The input data for the two sites is given in Fig. 13.2, except that no formal flood defence standard is stipulated for the Tees.

The Tees

The Tees estuary is heavily populated and industrialized, with a major concentration of oil refineries, chemical works and a nuclear power station. In order to identify the areas potentially at risk, scenarios of future sea level rise were required. In this case, the estimates of Hoffman (1984), modified for current tidal conditions and crustal movements (Shennan 1987, Shennan and Tooley 1987), were chosen (Fig. 13.3). Of course, other scenarios could have been used. These scenarios provided the basis for defining flood prone areas using altitudes measured to the national datum (Ordnance Datum, OD).

The spatial unit used in the analysis was the 'ward'. Wards were a basic unit of the 1981 census for the United Kingdom and the ward boundaries were available as digital files (as were all the census variables for each ward). A major drawback to the use of the ward is that the area within a ward will not necessarily be affected equally by a rise of sea level because of the variations in elevation found within the ward boundaries which are not distinguishable from the topographic data base. Ideally, digital contour and point data, and digitized boundaries for smaller spatial units (e.g., enumeration districts) would have been preferable to ward overlays, but this would have increased significantly the costs of data collection and verification. Nevertheless, in the case of the Tees

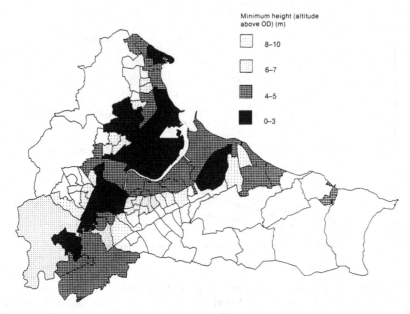

Fig. 13.4 Minimum height recorded for each ward within the County of Cleveland. The minimum height shown will either be the lowest contour or the lowest spot height within the ward, shown on the latest 1:25,000 Ordnance Survey maps. A range of altitudinal data was collected, e.g., all spot heights, the three lowest and the maximum contours, so that various measures of the proportion of a ward below a particular altitude could be used. This map illustrates those wards which will be dependent on protection from flooding along the River Tees for any particular sea level rise scenario. The use of minimum height indicates that the ward will be partly and not wholly affected. Unshaded areas are wards for which no detailed topographic data were collected since no part of the ward is below 15 m OD (Shennan and Tooley, 1987. Reprinted by permission of Basil Blackwell).

the presentation of data on a ward basis rather than another spatial unit (e.g., grid squares) has merit, in so far as the ward is a frequently used administrative unit upon which the implementation of many planning decisions are made. In general, if the output of a GIS is to be used by decision making authorities the user must be familiar with the spatial units used.

For assessing the impacts on the Tees, environmental and socio-economic data within each ward were required. However, most of the data sets were not readily available in digital form and manual acquisition was necessary (which significantly increased costs and time requirements). Topographic data were obtained manually from Ordnance Survey 1:25,000 scale maps. Outlines of the wards were overlain on the topographic maps, and contour and spot height altitude data were noted for each ward. The topographic data were used to indicate those wards which may require protection for any specified scenario (Fig. 13.4). An example (not a firm prediction) of the sort of impact map that can be produced is shown in Fig. 13.5. This shows on a ward-by-ward basis the population over retirement age (based on a recent census) potentially at risk from future flooding from sea level rise. The raw data required to create some of the inputs to the GIS are shown in Fig. 13.6. More complex maps using any of the census-

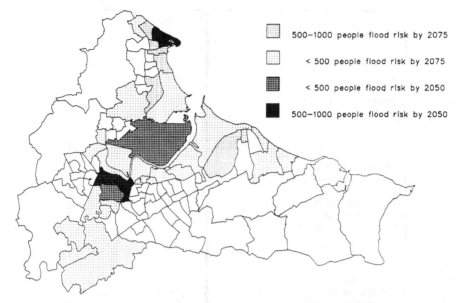

500–1000 people flood risk by 2075

< 500 people flood risk by 2075

< 500 people flood risk by 2050

500–1000 people flood risk by 2050

Fig. 13.5 An example of the type of map that could be produced if probability estimates were made for sea level rise scenarios which are then integrated with socio-economic variables. In this example the concentrations of people over retirement age which may require special flood relief or evacuation planning are illustrated. Note the constraints of using census wards as the spatial unit, since the different parts of the ward will be affected to different degrees yet the precise location of the population cannot be indicated.

derived data, locational data of flood prone utilities and transport networks could equally be produced. Such improvements are being made as more digital data become available.

The Fenland

The Fenland area of Eastern England differs in many ways from the Tees area. It is predominantly an agricultural area of low population density with a small number of towns and numerous villages and a history of organized flood protection and land reclamation from the sea (e.g., Godwin, 1978). Much of the land is already below contemporary high tide levels, and most areas already require major sea and river defences for protection. The specification of standards for raising sea defences is a statutory requirement of the regional water authority, Anglian Water.

Despite the differences between the Tees and the Fenland, the same GIS approach can be used. The purpose of the Fenland example (Figs. 13.7 and 13.8) is to illustrate the application of the approach at different scales. For the Tees case study a single administrative unit, one county consisting of 98 wards, was used; for the Fenland case study, four counties totalling 755 wards were integrated into one geomorphic unit, a coastal lowland zone. The latter scale of analysis would, for example, be appropriate for the Thames lowlands and all other coastal lowlands of the UK. This simple approach can be recommended to other countries with extensive coastal lowlands

Ward =	File number
MHW line = Contour 1 = Contour 2 = Contour 3 = Maximum contour =	
Minimum spot height = Maximum spot height = Mean spot height = Median spot height = Modal spot height = Number of spot heights =	Raw spot height data
Oil refinery = Power station = Chemical works = Other industry / factory = Dumps etc =	Tidal data calculations
MHWS = MTL =	
Area of ward	
Map sheet number Copyright date Imperial / 'metric' / true metric Data collected by date Computer storage by date Verified by date	Comments

Fig. 13.6 The data sheet used for the manual abstraction of map data used in the examples described in the text.

where sea level is likely to disrupt agriculture, drainage, settlement, industrial and service activities.

The immediate effects of sea level rise on the Fenland estuary will concern sea defences, land drainage and subsequent environmental changes. The sea defences front onto salt-marshes, and changes in these habitats require careful monitoring. Land drainage leads to changes in soil properties. Monitoring of such changes and the use of GIS are discussed in the next section.

New data requirements

Availability of data in digital form is a prerequisite for using computerized data handling techniques. For the UK, the Committee of Enquiry into the Handling of

Plate I A classified multi-spectral scanner (MSS) image (acquired during the NERC 1986 Airborne MSS campaign) of an area (~2 x 2 km) of Leverton Marsh, Eastern England. The reclaimed agricultural land, with rectangular fields, is seen on the west side of the image, with the seabank as the purple line. The intricate nature of the inter-tidal salt-marsh communities can be seen in the middle of the image, with high marsh communities as mauve and cyan, middle and low marsh communities yellow, blue and red, unvegetated mud and sand flats are green. Note how the upper marsh communities are absent from the south part of the image due to the recent enclosure of land for agriculture. A rise in sea level will cause a reduction in the area of valuable upper marsh ecosystems. Imagery can be used for detailed monitoring of the response of inter-tidal ecosystems to both management and sea level rise and can be easily integrated into a GIS. This image was produced by Danny Donoghue.

A colour version of this image is available for download from www.cambridge.org/9780521115032

Plate II A false-colour composite image (acquired during the NERC 1986 Airborne MSS campaign) of a part (~5 x 5 km) of Morton Fen, Eastern England. The soil marks and crop marks reveal ancient tidal channels, archaeological features and the variation in soil parent material. Due to changes in inland drainage requirements, influenced significantly by any future sea level rise, further soil changes, resulting from oxidation, wind erosion, compaction or waterlogging, can be monitored using these MSS data and used in impact assessment, allowing consideration of factors such as landscape and archaeological value as well as agricultural benefits. This image was produced by Danny Donoghue.

A colour version of this image is available for download from www.cambridge.org/9780521115032

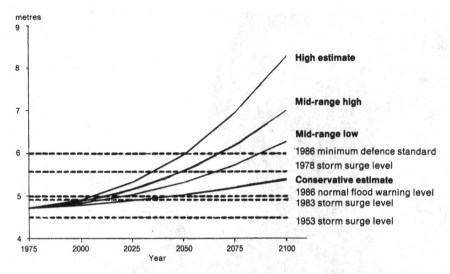

Fig. 13.7 Sea level rise scenarios (from Barth and Titus, 1984) related to local tidal conditions, recent storm surges and 1986 flood defence data for the Fenland. All levels are related to Ordnance Datum.

Geographic Information identified as one of four immediate tasks the rapid conversion of the basic scales map series to digital form. However, the Committee cautioned that the conversion of existing paper records into digital form is not easily automated and is costly (HMSO, 1987; further details are found in that report, with particular detail on spatial units and locational referencing given in an appendix).

The potential advantages of data being available in digital form can be highlighted by the example of remote sensing in the context of the Fenland case study. Two impacts of sea level rise in the Fenland will be, firstly, on the inter-tidal zone and salt-marshes which are found immediately in front of the sea banks, and, secondly, on soil properties in the low-lying areas where drainage is required to evacuate freshwater to the sea and where the removal of soilwater can lead to oxidation, wind erosion and sediment compaction. Plates I and II illustrate how multi-spectra, remotely sensed imagery can be used to evaluate such changes if multi-temporal data are collected and integrated to quantify rates of changes. Such data are very suitable for use within a GIS since they are in digital form and the different resolutions of the various satellite and airborne sensors can be integrated to monitor change. The two plates are for nominal ground resolutions of around 4 m and 10 m using an airborne scanner. Widely available satellite data of 30 m nominal ground resolution (LANDSAT Thematic Mapper (TM) imagery) can also be used (e.g., Donoghue & Shennan, 1987) while the 10 m panchromatic and 20 m multi-spectral data from the SPOT satellite have yet to be analysed for this application.

Installation

The computing requirements for GIS vary widely. No recommendation of a particular configuration is made here since it would soon become outdated with rapid developments in computing technology and GIS. Examples given in this paper reveal the range of possibilities: the maps were produced on a mainframe computer system using the

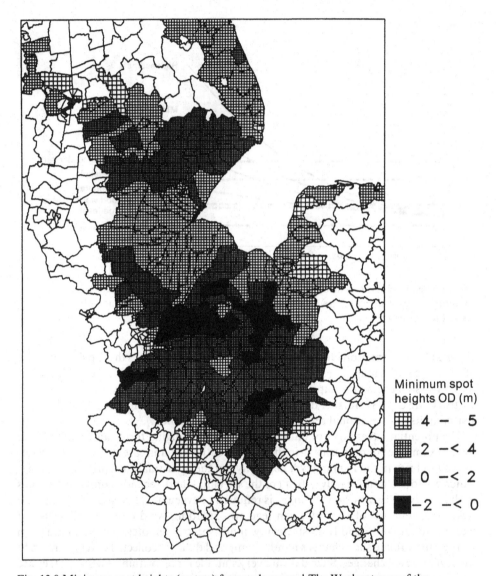

Fig. 13.8 Minimum spot heights (metres) for wards around The Wash estuary of the Fenland. This map was produced by merging the ward boundary files and topographic data files for four separate counties and selecting a portion of the total area to map and illustrate how the GIS can be used to integrate data collected and stored in one format, i.e., by administrative county units, to produce information for a different spatial grouping i.e., areas around an estuary.

GIS package GIMMS, the remotely sensed images were produced on a minicomputer system using an I²S system, and the GIS approach is being further developed on IBM personal computers using PC-ARC/INFO. Other alternatives are available. Geographic Information Systems are no longer dependent on large mainframe computing facilities and smaller systems are available to suit the particular requirements of users.

Impediments to the use of GIS in assessing the impacts of sea level rise include lack of awareness, and human and organizational problems:

'Many users and potential users are not sufficiently aware of the benefits of Geographic Information Systems. A major promotional exercise is required to maximize returns on the existing national investment in geographic information. There is also a need to increase substantially, and at all levels, the provision of trained personnel. The human and organizational problems of achieving change are also an important barrier to development. Information needs to be seen as a corporate resource and be more widely shared between departments and organisations' (HMSO, 1987, p. 3).

Conclusions

Geographic Information Systems offer large potential benefits in assessing the impacts of future sea level rise. The flexibility they offer in terms of data collection, storage, verification, analysis and display can be shown with relatively simple examples, and more complex uses can be easily envisaged. In the near future, the major constraints to the full development of GIS for this purpose are the lack of digital data, a lack of qualified personnel, and insufficient awareness on the part of users and potential users of the benefits. However, developments of new technology will ease the collection of digital data. The increase in national and international networks for data transfer, as well as the continued reduction in the cost of computer processing power and digital storage media, means that GIS will become increasingly affordable over the coming decade. Importantly, the efficient use of this technology is dependent on the training of operators and the education of users.

Acknowledgements

The Natural Environment Research Council (NERC) provided the multi-spectral scanner data (Plates I and II) as part of their 1986 MSS airborne campaign. These data were analysed by Danny Donoghue.

Note: This paper was written in 1987. Substantial advances have taken place since then regarding data availability and methods of analysis. I Shennan, 28 July 1992.

References

Barth, M.C. & Titus, J.G. (1984). *Greenhouse Effect and Sea Level Rise: a Challenge for this Generation*. Van Nostrand Reinhold Co., 325pp.

Dangermond, J. (1984). *ARC/INFO Maps*. Redlands, California : ESRI.

Donoghue, D.N.M. & Shennan, I. (1987). A preliminary assessment of Landsat TM imagery for mapping vegetation and sediment distribution in the Wash estuary. *Int. J. Remote Sensing*, **8**, 1101–8.

Godwin, H. (1978). *Fenland: its Ancient Past and Uncertain Future*. Cambridge University Press, Cambridge.

Henderson-Sellers, A. & McGuffie, K. (1986). The threat from melting ice caps. *New Sci.*, **1512**, 24–5.

HMSO (1987). *Handling Geographic Information*. Report to the Secretary of State for the Environment of the Committee of Enquiry into the Handling of Geographic Information, HMSO, London.

Hoffman, J.S. (1984). Estimates of future sea level rise. In *Greenhouse Effect and Sea Level Rise:*

A Challenge for This Generation, M.C. Barth & J.G. Titus (eds.). Van Nostrand Reinhold, New York, pp. 79–103.

Parry, T.B. & Perkins, C.R. (1987). *World Mapping Today*. Butterworths, London.

Perrotte, R. (1986). A review of coastal zone mapping. *Cartographica*, **23**, 3–71 (+ 24 maps).

Shennan, I. (1987). Holocene sea level changes in the North Sea Region. In *Sea Level Changes*, M.J. Tooley and I. Shennan (eds.). Blackwells, Oxford, pp. 109–51.

Shennan, I. & Tooley, M.J. (1987). Conspectus of fundamental and strategic research on sea level changes. In *Sea Level Changes*, M.J. Tooley and I. Shennan (eds.). Blackwells, Oxford, pp 371–92.

14

The storm surge problem and possible effects of sea level changes on coastal flooding in the Bay of Bengal

R.A. Flather and H. Khandker

Abstract

The effects of a change in mean sea level (MSL) on tides, storm surges and their combination are examined. The relevant dynamics are first reviewed in order to understand the possible influence of an increase in MSL on tide and surge elevations.

Using techniques developed at the Proudman Oceanographic Laboratory, a numerical model was established for the computation of tides and storm surges on the northern shelf of the Bay of Bengal, the coastal regions of which are particularly susceptible to inundation. The model was verified by using it to compute the distribution of the main (M_2) constituent of the tide and comparing model results with data from tide gauges.

The model was then used to estimate the effect of a 2 m increase in MSL on the M_2 tide; on a storm surge generated by the cyclone of 24–25 May 1985; and on the total (tide + surge) motion during this cyclone event. The estimates were produced by taking differences between model solutions with existing MSL and equivalent solutions assuming MSL to be 2 m higher.

For the M_2 tide, *increased* amplitudes are predicted in the NE corner of the Bay with *decreased* amplitudes in the NW corner. The magnitude of the change in each case is roughly 10% of the tidal amplitude, typically 10 cm. The differences are consistent with the anticipated changes in the response of the shelf due to the increased MSL.

For the storm surge of 24–25 May 1985, which raised sea levels in the NE corner of the Bay by up to 2.75 m, a reduction in the maximum computed surge elevation of 20–30 cm was predicted in the areas most affected. Note, however, that total water levels ~ 1.8 m above those expected with the present MSL would result.

For tide and surge combined, the maximum elevation being about 4 m above MSL near Cox's Bazaar, a *reduction* in level was predicted in the NE and NW corners of the Bay, with an increase in level in an area between. This more complex behaviour was attributed to variations in the relative timing of tidal high water and peak surge. The effect of the 2 m increase in MSL would therefore be to raise the highest flood level by about 1.8 m.

Introduction

Storm surges are changes in water level generated by storms passing over the sea and are a major cause of coastal floods affecting many parts of the world, none more severely than the Bay of Bengal. Surges occur together with, and are superimposed on, the normal astronomical tides and variations in mean sea level (MSL). It is usually an adverse combination of these components that leads to high total water levels and inundation. The separation of tide, surge and MSL, however, is not straightforward and raises questions of definition and approach, which we now address.

Observational data are usually comprised of hourly water level values measured relative to a suitable datum by a tide gauge. Harmonic analysis of these data provides estimates of the amplitude and phase of each tidal constituent resolvable in the period of data considered and a zero frequency component which is defined as MSL (relative to the benchmark) over that period. The resulting constants may then be used to predict, at hourly intervals, the contribution of tide and MSL to the total water level. Subtracting these values from the corresponding observed water levels leaves a time series of hourly residuals, which are defined as the storm surge elevations. On this basis MSL contains contributions arising from the tides and generated by storms.

From the viewpoint of dynamics, on the other hand, it is possible to separate the different contributions to the total water level by their distinct generating mechanisms. Thus tides (astronomically generated), surges (storm generated) and variations in MSL itself (perhaps due to changes in river flow in an estuary or to climate-related processes) can be examined independently or in combination. Because of non-linear effects in shallow coastal seas each dynamical component of the total water level depends on the others, so that a linear combination of

tide + surge + MSL,

in which each component is estimated independently of the others, may be substantially different from the total water level produced by tide, surge and MSL occurring together and interacting dynamically. In the context of climate change and the rise in sea level that may result, it is of particular interest to examine how changes in MSL may affect the tides, storm surges and their combination. This is the purpose of the present paper.

Initially, a brief account of storm surge and tidal dynamics is given. This is intended to identify the important processes and to describe some mechanisms that may be significant. Next, the numerical models for the Northern Bay of Bengal are described. Some preliminary results obtained from the models indicate the probable changes in tide, storm surge and combined tide + surge levels produced by an imposed increase in MSL. Finally, the conclusions are presented.

Dynamical considerations

Storm surges are meteorologically generated motions which typically last for a period of between a few hours and several days. Like the tides, they belong to the class of motions known as long gravity waves and are described by essentially the same dynamical equations.

Basic equations

In geographical co-ordinates, these equations may be written in the form

$$\frac{\partial \zeta}{\partial t} + \frac{1}{R\cos\phi}\left\{\frac{\partial}{\partial \chi}(Du) + \frac{\partial}{\partial \phi}(Dv\cos\phi)\right\} = 0 \qquad (1)$$

$$\frac{\partial u}{\partial t}+\frac{u}{R\cos\phi}\frac{\partial u}{\partial \chi}+\frac{v}{R}\frac{\partial u}{\partial \phi}-\frac{uv\tan\phi}{R}-fv=\frac{-g}{R\cos\phi}\frac{\partial \zeta}{\partial \chi}-\frac{1}{\rho R\cos\phi}\frac{\partial p_a}{\partial \chi}+\frac{1}{\rho D}(F_s-F_b) \quad (2)$$

$$\frac{\partial v}{\partial t}+\frac{u}{R\cos\phi}\frac{\partial v}{\partial \chi}+\frac{v}{R}\frac{\partial v}{\partial \phi}+\frac{u^2\tan\phi}{R}+fu=\frac{-g}{R}\frac{\partial \zeta}{\partial \phi}-\frac{1}{\rho R}\frac{\partial p_a}{\partial \phi}+\frac{1}{\rho D}(G_s-G_b) \quad (3)$$

where: t denotes time,
χ,ϕ east-longitude and latitude respectively
ζ elevation of the sea surface
u,v components of the depth mean current, q
F_s, G_s components of τ_s, the wind stress on the sea surface
F_b, G_b components of τ_b, the bottom stress
p_a atmospheric pressure on the sea surface
D the total water depth ($=h+\zeta$, where h is the undisturbed depth)
ρ the density of sea water, assumed uniform
R the radius of the Earth
g the acceleration due to gravity
f the Coriolis parameter ($f=2\omega\sin\phi$, where ω is the angular speed of rotation of the Earth)

Equation (1) is the continuity equation, expressing conservation of volume. In simple terms, with respect to a vertical column in the sea, changes in surface elevation, ζ, associated with the tide or a storm surge, are related to net fluxes of water in or out of the column.

Equations (2) and (3) equate, for each co-ordinate direction, the acceleration of the water (on the left side of the equation) to the forces acting on it (on the right). The meteorological forces that generate storm surges are thus wind stress and horizontal gradients of surface atmospheric pressure.

Mathematically, the problem is closed by relating bottom stress, τ_b, to the current, q, using a quadratic law

$$\tau_b=\kappa\rho q|q| \quad (4)$$

where κ is a friction parameter, and relating the wind stress, τ_s, to the surface wind velocity, w, also using a quadratic law

$$\tau_s=c\rho_a w|w| \quad (5)$$

where ρ_a is the density of air and c a drag coefficient. The consensus now is that c depends weakly on wind speed. For example Wu (1982) gives

$$c\times 10^3=0.8+0.065w,$$

for any wind speed from breeze to hurricane (this formulation assumes w is in m/s). Initial and boundary conditions must also be specified. These typically take the form

$$\zeta=u=v=0 \text{ at time } t=0; \quad (6)$$

on coastal boundaries,

$$q_n=0 \quad (7)$$

where q_n is the component of current along the outward-directed normal to the boundary; and either

$$\zeta = \hat{\zeta} \tag{8}$$

or

$$q_n = \hat{q}_n + \left(\frac{g}{h}\right)^{1/2}(\zeta - \hat{\zeta}) \tag{9}$$

on an open-sea boundary, where $\hat{\zeta}$ and $\hat{q}_n$ are specified functions of position and time.

Equations (1)–(3) with (4) and (5) must then be solved within the required domain, taking appropriate conditions from (6)–(9), and with p_a, **w**, defined as functions of time and position within the domain, and $\hat{\zeta}$ (both $\hat{\zeta}$ and $\hat{q}_n$ if (9) is adopted) as a function of time and position along its open-sea boundaries.

Tide and surge dynamics

Tides are generated in the deep ocean by gravitational forces associated with the Moon and Sun, and are dissipated on shallow continental shelves by friction. The tides nearshore are thus determined by the response of the waters on the continental shelf to the oscillation imposed at the shelf edge. This response depends on the size, shape and bathymetry of the shelf sea. Large tides occur near resonance, i.e., when a natural mode of oscillation of a part of the shelf occurs at a frequency close to that of one constituent of the ocean tide. The simplest case of this is when the shelf width, L, say, is approximately one-quarter of the wavelength, λ, of the tide. If tides are close to resonance, then a change in MSL could produce a larger change in the tides than would otherwise be the case. If a rise in MSL increases water depth, the wavelength of the tide would increase. For a shelf with $L > \lambda/4$, this could bring the regime closer to resonance, increasing the tidal range for certain constituents. If $L < \lambda/4$, increased MSL would shift the regime away from resonance and reduce the tidal range.

Storm surges are generated by wind stress acting over the sea surface and by variations of surface atmospheric pressure. It is important to note that the relevant forces in equations (2) and (3) are τ_s/D and ∇p_a. The pressure forcing is thus *independent* of water depth, whereas the wind stress forcing is *not*. An important consequence is that wind stress becomes increasingly effective in shallower water. Also, in view of equation (5), the strongest winds are most important since $\tau_s \propto w^2$. A kind of resonance is also possible for meteorologically generated disturbances when the speed of motion of the atmospheric forcing, V_c, matches that of free progressive waves in the sea, $(gh)^{1/2}$ (Proudman, 1929). However, this is likely to be relevant only for the most rapidly moving storms in very shallow water. In such cases, a change in MSL might affect the surge response to a greater extent than might otherwise be expected.

From the above discussion, the effect of an increase in MSL on surge magnitudes will depend on the resulting changes in water depth in the area of surge generation. We can propose two contrasting scenarios as follows:

a. If the shoreline comprises a vertical wall, high enough not to be overtopped by the raised levels giving a bathymetric profile with a step, then water depths will increase and as a consequence storm surges should be reduced in magnitude.
b. If the bathymetric profile in the nearshore region has a linear slope, then the

shoreline will advance inland as MSL rises so that the area of sea corresponding to a given water depth will remain broadly unchanged. Correspondingly, surge magnitudes may be unaltered.

In general, some intermediate situation is likely to exist: for example, much of the coast of Bangladesh in the Meghna estuary is protected by earth banks, high enough to prevent flooding by normal tides, so that the profile of bathymetry does include a step, as in (a). On the other hand, these banks are *not* capable of preventing inundation when cyclones generate extreme water levels and, under such circumstances, the shoreline will move inland, as in (b). A further consequence may be that levels at the original shoreline location are *reduced* as compared with those to be expected in (a), since without inundation the water would pile up at the coast giving still higher levels there. Perhaps the only way of resolving these complex questions is by investigations using a technique that allows for possible changes in water depth and inundation (i.e., the deformation of the coastline). Such an approach has been adopted here.

Numerical models

A complete analysis requires a means of solving the basic equations with real bathymetry and coastal configurations and of including the non-linear terms (i.e., those containing products or implied products of ζ, u and v). It is these non-linear terms, in particular the quadratic bottom stress, equation (4), and those containing the time-dependent total depth, D, which give rise to the dynamical interactions between different components of the motion mentioned in the introduction. Numerical methods have been used increasingly in recent years, providing a powerful approach for the simulation and prediction of tides and storm surges.

In the most widely used form, all derivatives in equations (1)–(3) are replaced by finite-differences between values defined at a regular array of points (the 'grid') covering the solution area, and at discrete but regular intervals of time. The equations are then rearranged such that values of ζ, u, v at each point at a given time can be computed directly from known values one timestep earlier. Repeated solution of these finite-difference equations for every grid point advances the solution timestep-by-timestep over the period of interest. The computer program to solve the equations, together with input data defining bathymetry boundary locations etc. for a specific region, constitutes a 'numerical model'. Many models can be applied to different regions simply by changing the relevant input data sets. One of the best-known and widely used surge prediction models is SPLASH, developed in the United States by C.P. Jelesnianski for hurricane generated surges on the open coasts bordering the Gulf of Mexico and the North Atlantic. In this region, tides are relatively unimportant and the bathymetry changes only slowly in the longshore direction. This means that the problem can be simplified and tides ignored in the surge calculation, leading to a very efficient modelling approach. Recognising the similarity between this region and parts of the Indian coast, Ghosh (1977) used SPLASH successfully for simulating cyclone generated surges on the open coasts bordering the west side of the Bay of Bengal. Indeed, SPLASH has been applied in many parts of the world.

A different kind of model has evolved for 'partially enclosed' sea areas, typified by complex coastlines with inlets, estuaries and islands, and correspondingly complex bathymetry. The greater shelf widths in these regions frequently produce large tides, so that dynamical interaction of tide and surge is important. The North Sea is an example of this type of sea area. The development of models for North Sea surges, pioneered amongst others by W. Hansen, H. Lauwerier and N.S. Heaps, is well reviewed in Heaps (1967, 1983).

The work described in the remainder of this paper employs numerical modelling techniques developed at the Proudman Oceanographic Laboratory (formerly the Institute of Oceanographic Sciences, Bidston) over many years. The techniques have been used successfully in a wide variety of applications, notably in operational prediction of storm surges in UK waters for flood warning, navigation and other purposes. The numerical scheme is essentially as described by Davies and Flather (1978), incorporating a simple moving boundary approximation to simulate coastal inundation (Flather and Heaps, 1975), combined with a method of approximating narrow one-dimensional channels (as in Flather, 1987).

Some preliminary results from a model of the Northern Bay of Bengal

The coastal zones of the Bay of Bengal, particularly the regions bordering the northern shelf, are especially susceptible to flooding from storm surges. A recent review (Murty, Flather and Henry, 1986) describes the problem and reviews the substantial literature on it.

The work described here was instigated following the cyclone on 24–25 May 1985, which caused some 11,000 deaths in the Meghna estuary of Bangladesh. A programme involving the development of numerical models and measurements of off-shore pressure was proposed. The scheme is illustrated in Fig. 14.1(*a*). Three models were planned: a 'large' model covering the region north of 17°N latitude, intended primarily for tidal computations; a 'shelf' model covering the region north of 20°N, intended for computation of tides and storm surges; and a 'small' model representing the NE corner of the Bay in greater detail. The grid size of the 'large' and 'shelf' models is 1/10° by 1/10° (equivalent to about 10 km) and that of the 'small' model is 1/30° by 1/30° (or about 3.3 km). All models include representations of the major channels of the Delta. The off-shore measurements along 17°N, together with data from coastal tide gauges were intended to define the tidal variations (and hence $\hat{\zeta}$) along the open-sea boundary of the 'large' model. Tidal simulations with the 'large' model, corrected as necessary by data from coastal and off-shore measurements further north, would subsequently provide input tides for the other models.

Only a limited amount of work has so far been carried out, aimed at establishing the 'large' and 'shelf' models. Early tests indicated that there was little to be gained from calculations using the 'large' model without the off-shore measurements because of uncertainties in the tidal variations on its open boundary, so most experiments have used the 'shelf' model alone, with open boundary tidal input derived by linear interpolation from the coastal gauges at Devi River Entrance (DR in Fig. 14.1(*a*)) and Akyab (AK). A preliminary set of experiments was carried out for the main (M_2) constituent of the tide to assess the sensitivity of the results to various aspects of the model design. These experiments indicated that the solutions were moderately sensitive to changes in bathymetry, as expected; that the presence of the channels within the Delta region tended to reduce tidal amplitudes on the open coast (providing a reservoir into which water that otherwise would augment levels on the coast could flow); and that

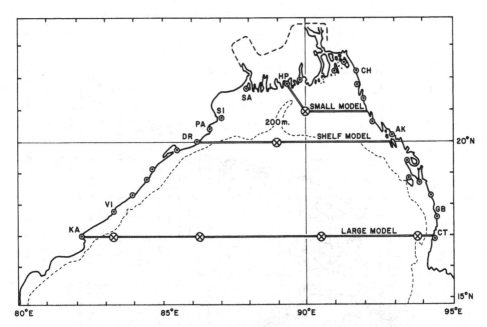

Fig. 14.1(*a*) Scheme of numerical models and off-shore pressure measurements for the northern Bay of Bengal. Also included are locations of some coastal tide gauges.

a smaller value of the bottom friction parameter ($\kappa = 0.0020$) than generally adopted for models of UK waters ($\kappa = 0.0025$) appeared appropriate. The experiments, carried out with elevations specified, equation (8), as the open boundary condition, also provided a distribution along the southern open boundary of $\hat{q}_n$ for the M_2 tide, thereby making possible the use of the (preferred) 'radiation' condition (equation (9)).

Fig. 14.1(*b*) shows the finite-difference grid for the 'shelf' model. The broken line indicates the limit of the computational area. Within this area, the position of the shoreline (approximated by the sides of grid boxes) is determined entirely by the bathymetry and sea level. Here, bathymetry includes both water depths and land elevations (negative depths relative to the model datum). In the absence of detailed information on land elevations, it was assumed that the land areas were *above* the high water level of normal tides, so that tidal amplitudes from observations and model experiments gave a lower bound for the height of the land above the model datum.

The 'shelf' model was used to estimate the influence of an increase in MSL (assumed to be 2 m) on the M_2 constituent of the tide, on a storm surge generated by the cyclone of 24–25 May 1985, and on total (tide + surge) motion during this cyclone event. The procedure adopted was to run the model for the existing regime, verifying as far as possible that the solution reproduced known conditions. A second solution was then computed, identical to the first in all respects other than the increased MSL. The difference between these solutions could then be attributed to the influence of the 2 m rise in MSL. The results of these experiments are described below. Note that the shoreline position *is* changed by the higher MSL. However, because the amplitude of normal tides exceeds the assumed 2 m MSL rise in much of the coastal region, these changes are not extensive. On the other hand, inundation with MSL at + 2 m obviously covers a larger area than with MSL at its present value.

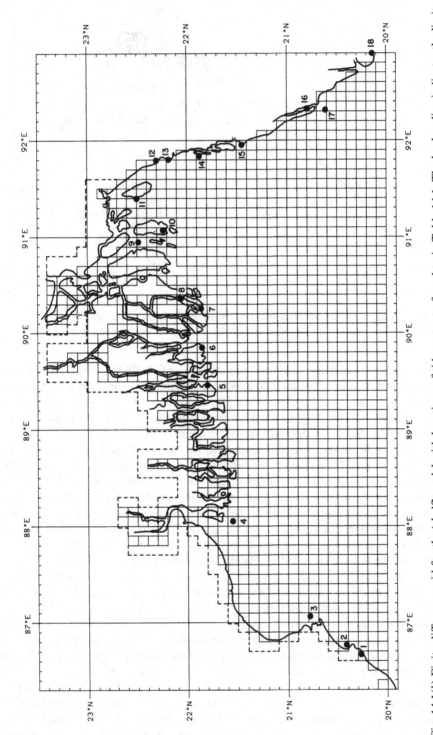

Fig. 14.1(*b*) Finite-difference grid for the 'shelf' model with locations of tide gauges referred to in Table 14.1. The broken line indicates the limit of the computational area, within which inundation can occur.

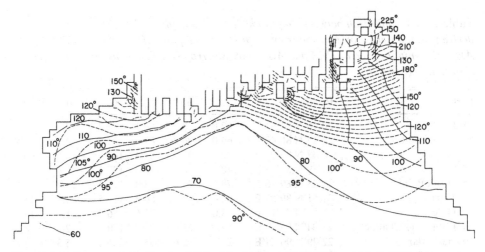

Fig. 14.2(*a*) Computed distribution of the M_2 tide from the 'shelf' model. Solid lines are contours of equal tidal amplitude; broken lines are contours of equal tidal phase.

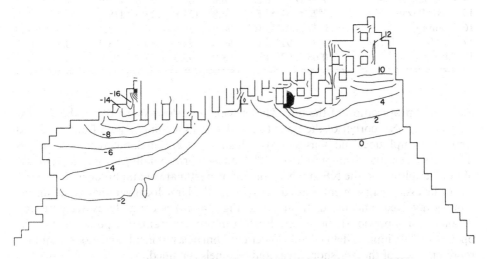

Fig. 14.2(*b*) Change (cm) in the amplitude of M_2 produced by a 2 m rise in MSL.

The M_2 constituent of the tide

Fig. 14.2(*a*) shows the computed chart for the M_2 tide, depicting contours of equal tidal amplitude by solid lines, and contours of equal tidal phase (along which tidal high water occurs simultaneously) by broken lines. Note that all times and tidal phases here refer to Greenwich Mean Time (GMT), not local time. The chart shows an amplification of the tide in the NE and NW corners of the Bay where the shelf is wide and a

Table 14.1. *Comparison between amplitude H (cm) and phase G (°) of the M₂ tide derived from observation (o) and from the nearest grid point of the 'shelf' model (c). ΔH is the amplitude error (%) and ΔG the phase error (°). Tide gauge locations are shown in Fig. 14.1(b)*

Port	Position	H_o	G_o	H_c	G_c	ΔH	ΔG
1 Pradip	20°16′N 86°41′E	62.0	8.2	60.1	87.7	−3.1	5.7
2 False Point	20°25′N 86°47′E	69.0	96.0	67.6	93.6	−2.1	−2.4
3 Short Island	20°47′N 87°04′E	92.0	103.0	82.4	102.5	−10.5	−0.5
4 Sager Roads	21°39′N 88°03′E	140.0	116.0	126.8	127.1	−9.4	11.1
5 Hiran Point	21°48′N 89°28′E	78.0	124.5	98.8	119.3	26.7	−5.2
6 Tiger Point	21°51′N 89°50′E	82.0	128.0	86.2	139.5	5.1	11.5
7 Duleswar	21°51′N 90°15′E	73.0	143.0	71.6	156.9	−1.9	13.9
8 Rabnabad Channel	22°04′N 90°22′E	72.0	162.0	73.8	170.6	2.5	8.6
9 Hatia Bar	22°29′N 90°57′E	129.0	255.0	135.8	251.5	5.3	−3.5
10 Char Changa	22°14′N 91°04′E	97.6	228.0	110.5	236.1	13.2	8.1
11 Sandwip Island	22°30′N 91°25′E	185.5	223.9	148.0	231.8	−20.2	7.9
12 Patenga-Chittagong	22°16′N 91°49′E	141.6	199.1	127.9	193.1	−9.7	−6.0
13 Norman Point	22°11′N 91°49′E	136.0	185.0	124.0	184.6	−8.8	−0.4
14 Kutubdia Island	21°52′N 91°50′E	126.0	164.0	123.8	152.9	−1.7	−11.1
15 Cox's Bazaar	21°26′N 91°59′E	96.8	131.8	113.9	119.0	17.7	−12.8
16 Shahpuri	20°47′N 92°20′E	96.9	91.0	94.7	95.3	−2.2	4.3
17 St. Martin's Island	20°37′N 92°19′E	90.0	82.0	90.4	93.2	0.5	11.2
18 Akyab	20°08′N 92°54′E	78.0	93.0	77.7	92.9	−0.4	−0.1

northward propagation of the tide. The propagation slows markedly in the shallow waters of the NE corner of the Bay. Table 14.1 gives a comparison of observed and computed amplitudes and phases of M_2. Mean errors are small but at some locations the errors reach 20% in amplitude and 10° in phase, corresponding to an underestimate of tidal amplitude in the NE and NW and an overestimate of amplitude in the centre. Nevertheless, the agreement is good considering the large local variability and uncertainties involved, and the distribution in Fig. 14.2(a) is certainly as accurate and detailed as any previously produced. Further improvement may be possible when the open boundary input is defined more precisely from measurements and models, and the representation of the nearshore areas and channels is refined.

Fig. 14.2(b) shows the change in amplitude of the M_2 tide resulting from a 2 m rise in MSL. Amplitudes increase by about 10 cm in the NE and decrease by a similar amount in the NW of the Bay. Given that the phase differences from the shelf edge to the coast at the head of the Bay are $\sim 120°$ in the NE corner, corresponding to $L \sim \lambda/3$, and $\sim 30°$ in the NW corner, corresponding to $L \sim \lambda/12$, these changes are consistent with those anticipated on the theoretical grounds described earlier.

The storm surge produced by the cyclone of 24–25 May 1985

The track and development of the cyclone, based on positions (indicated by dots) at various times in the days preceding landfall near Chittagong, are shown in Fig. 14.3.

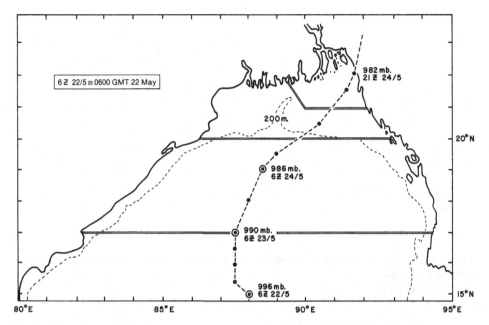

Fig. 14.3 Track and central pressure of the cyclone causing the surge in Bangladesh on 24–25 May 1985.

The limited available data were interpolated linearly in time, and the cyclone model of Holland (1980) was used to determine the associated distribution of surface atmospheric pressure. Surface winds were then computed using the gradient wind equations, the cyclone motion added, and the wind stress fields required for the surge calculations derived using the drag coefficient of Wu (1982), given earlier. This simple approach was satisfactory for the present purpose, in which *changes* in computed surges and levels are of interest, but could possibly be improved.

The first experiment was carried out by computing the surge alone (neglecting the tide) during the period 1200 GMT 22 May to 1200 GMT 25 May 1985. Fig. 14.4 shows the computed surge development. Elevations increased from small values at 1500 GMT 24 May and reached maximum values at about the time of landfall, 2100 GMT 24 May. Subsequently, elevations decreased rapidly, with negative elevations produced as winds veered to the NW. Fig. 14.5(*a*) shows the distribution of maximum computed elevations, indicating clearly that the surge was confined to the NE corner of the Bay. The computed peak surge, ~ 300 cm, occurs near Cox's Bazaar, with maximum surges in excess of 200 cm at Chittagong, Sandwip and Urichar, where most of the fatalities occurred. These results are in reasonable agreement with observed events (Ku, 1986), although the observational data required to make detailed comparisons were not available at the time of writing.

The above calculation was then repeated with MSL raised by 2 m. The resulting difference in maximum computed surge is plotted in Fig. 14.5(*b*). The maximum computed surges are reduced by 20–30 cm in the areas most affected. This is consistent with the anticipated effect of increased water depth in the region of surge generation, as discussed earlier, and suggests that increased depth is more important than shoreline

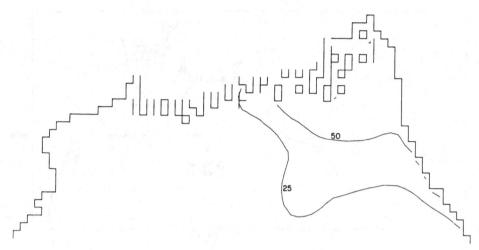

Fig. 14.4 Contours of computed surge elevation (cm) and current vectors (*d*) during the storm surge of 24–25 May 1985.
(*a*) 1500 GMT 24 May 1985

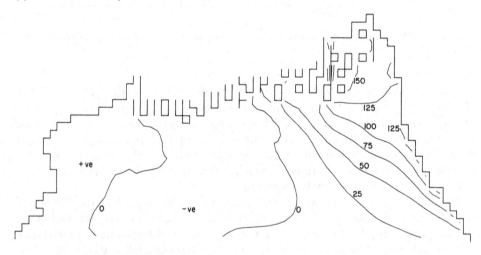

(*b*) 1800 GMT 24 May 1985

deformation in the present case. Note that the reduction in surge magnitude occurs together with the imposed 2 m rise in MSL, so that the highest levels reached are still 1.7–1.8 m *above* those resulting with the present MSL.

Tide and surge combined

In a final experiment, the calculations were repeated for tide and surge combined. The distribution of maximum computed water level is shown in Fig. 14.6(*a*). The highest level is again predicted to occur near Cox's Bazaar. Fig. 14.6(*b*) shows the change in the

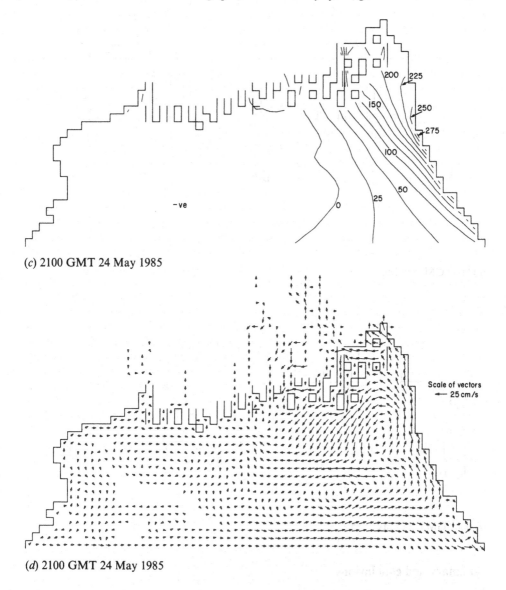

(c) 2100 GMT 24 May 1985

(d) 2100 GMT 24 May 1985

maximum computed tide + surge level resulting from an increase of 2 m in MSL. (Note again that these values exclude the 2 m rise in MSL itself, and this must be added to obtain the total change in maximum water level.) Levels are reduced in the NE and NW corners of the Bay, but increased in an area in between. These changes reflect not only the mechanisms affecting tides and surges independently, but also the possibility of a shift in the relative timing of tidal high water and the occurrence of peak surge. When tide and surge are both substantial (as in this case) their timing becomes critical in determining the maximum water level; the level may be large if tidal high water and peak surge occur at the same time, but much less if they occur at different times.

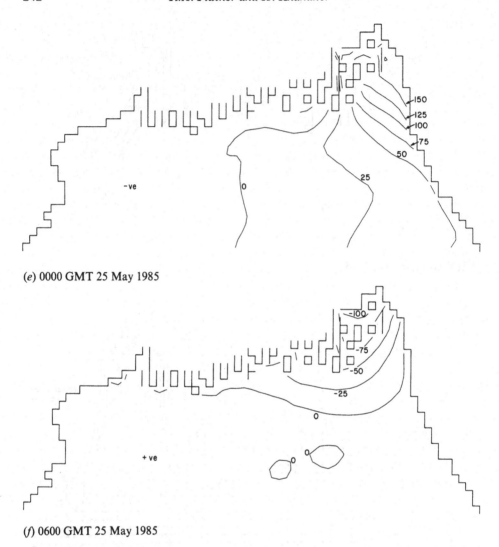

(*e*) 0000 GMT 25 May 1985

(*f*) 0600 GMT 25 May 1985

Summary and conclusions

Dynamical aspects of tides and storm surges have been reviewed, emphasising the processes and mechanisms that may help to understand the possible consequences of an increase in MSL on the tide and surge elevations. Experiments using a numerical model of the northern shelf of the Bay of Bengal were carried out to determine the probable influence of a 2 m rise in MSL on the tide, on the storm surge associated with the cyclone of 24–25 May 1985, and on the water levels due to tide and surge combined during this event.

The results of the model experiments were broadly in line with those anticipated on theoretical grounds. For tides, increased amplitudes were predicted in the NE corner of the Bay, with decreased amplitudes in the NW corner. These changes are consistent with a shift towards resonance in the NE and away from resonance in the NW of the

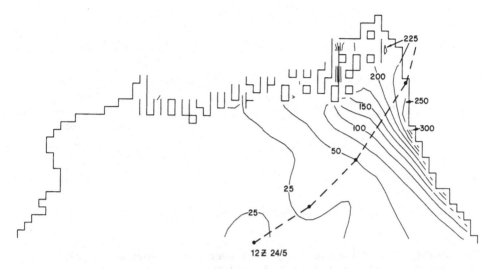

Fig. 14.5(*a*) Contours of maximum computed surge elevation (cm) in the period 1200 GMT 24 May to 1200 GMT 25 May 1985. The broken line indicates the track of the cyclone.

Fig. 14.5(*b*) Change (cm) in maximum computed surge elevation produced by a 2 m rise in MSL.

Bay produced by the increase in water depth. For surges, the dominant effect is a reduction in the main surge generating force, wind stress/water depth. The reduction is proportionally greatest in the shallowest water leading to a decrease in maximum computed surge in excess of 10%. The tide + surge response does not result solely from a combination of their individual responses, but is complicated by changes in the timing of surge and tide and by non-linear interactions between them. In the region affected by the highest water levels, a small reduction in elevation occurs when MSL is increased.

All of these changes are, of course, *small* compared with the assumed increase in

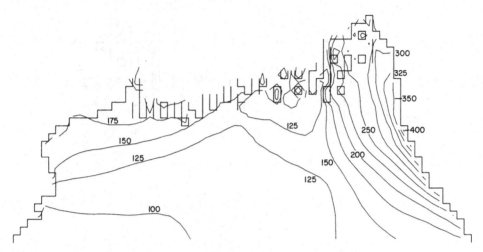

Fig. 14.6(*a*) Contours of maximum computed elevation (cm) due to tide and surge in the period 1200 GMT 24 May to 1200 GMT 25 May 1985.

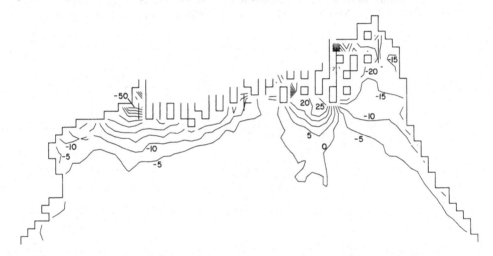

Fig. 14.6(*b*) Change (cm) in maximum computed tide + surge elevation produced by a 2 m rise in MSL.

MSL itself, which is the dominant effect and potentially the most serious. However, the predicted increase in tidal amplitude may be of some concern. Translated into the probable change in the extreme tidal range, an increase of up to 0.5 m could result.

Clearly, many questions have been ignored in the above analysis. An obvious omission is any consideration of river flow. If MSL rises, then presumably the surface slope and hence the pressure gradient driving the river flow will decrease, reducing the currents, increasing sedimentation rates, and so modifying the bathymetry. The extent of tidal penetration into the rivers and the estuarine circulation must also be affected. In an active region such as the Meghna estuary, these changes may be of major importance. Possible changes in cyclone tracks and intensity due to climate-related factors, which will modify surge elevations and hence maximum water levels, also need investigation. A better understanding of the effects of MSL rise on storm surges requires that these research issues be addressed in the future.

Acknowledgements

The authors are indebted to Jackie Huxley for typing the manuscript, to Robert Smith for work on the diagrams, and to Sian Broome for help in setting up the models.

References

Davies, A.M. & Flather, R.A. (1978). Application of numerical models of the Northwest European continental shelf and the North Sea to the storm surges of November to December 1973. *Dt. hydrogr. Z. Ergänzungsheft A*, No. 14, 72pp.

Flather, R.A. (1987). A tidal model of the northeast Pacific. *Atmosphere–Ocean*, **25**, 22–45.

Flather, R.A. & Heaps, N.S. (1975). Tidal computations for Morecambe Bay. *Geophys. J. Roy. Astron. Soc.*, **42**, 489–517.

Ghosh, S.K. (1977). Prediction of storm surges on the east coast of India. *Indian J. Met. Hydrol. Geophys.* **28**, 157–68.

Heaps, N.S. (1983). Storm surges, 1967–1982. *Geophys. J.R. Astron. Soc.*, **74**, 331–76.

Heaps, N.S. (1967). Storm surges. In *Oceanography and Marine Biology Annual Review*, **5**, H. Barnes (ed.). Allen and Unwin, London, pp. 11–47.

Holland, G.J. (1980). An analytical model of the wind and pressure profiles in hurricanes. *Monthly Weather Review*, **108**, 1212–18.

Ku, L.-F. (1986). *An Improvement to Tide Gauging for Cyclone Surge Studies in Bangladesh.* Paper presented at WMO/ESCAP Regional Cyclone Surge Workshop, Chittagong, Bangladesh, December 1985 (unpublished manuscript).

Murty, T.S., Flather, R.A. and Henry, R.F. (1986). The storm surge problem in the Bay of Bengal. *Prog. Oceanog.* **16**, 195–233.

Proudman, J. (1929). The effects on the sea of changes in atmospheric pressure. *Mon. Not. R. Astron. Soc. Geophys. Suppl.*, **2**, 197–209.

Wu, J. (1982). Wind stress over sea surface from breeze to hurricane. *J. Geophys. Res.*, **87**, C12, 9704–6.

15

Geographical complexities of detailed impact assessment for the Ganges–Brahmaputra–Meghna delta of Bangladesh

H. Brammer

Abstract

The Bangladesh delta is neither homogeneous nor static. Eight physiographic regions are recognized. Alluvial deposition and bank erosion cause significant annual changes along the lower Ganges and Meghna rivers, and major shifts in river courses have occurred in recent centuries. Land use is changing in response to increasing population pressure, spread of modern technology and environmental changes. Regional differences, both 'normal' and erratic geomorphic changes and unreliable official statistics create problems for sampling and for isolating the effects of a slow rise in sea level due to the 'greenhouse effect'.

A gradually rising sea level would be matched by a rise in levels of river levees and of tidal and estuarine floodplain land near the coast. Interior floodplain areas not subject to seasonal sedimentation would suffer increased seasonal flood levels, and increased areas in basins would stay wet in the dry season. The net effect on agricultural production would be negative. Protection works to counter increased flooding would be difficult and costly to provide, operate and maintain.

Introduction

Bangladesh provides a difficult subject for any kind of impact assessment, whether that be physical, social or economic. Both the physical and the socio-economic environments are complex; they are also dynamic. In addition, there are familiar Third World problems of difficult communications, weak institutions and dubious official statistics. Together, these factors make it extremely difficult to isolate and assess the specific consequences of any particular environmental trend or intervention.

That is especially so for relatively slow changes such as those predicted for sea level in response to the 'greenhouse effect'. Year-to-year variability and occasional catastrophic changes occurring in the Bangladesh delta are on a much greater scale than the gradual changes expected to result from a slowly rising sea level over the next half-century. It may therefore require considerable ingenuity to devise practical and reliable techniques for measuring and assessing the actual impact of global changes in sea level on Bangladesh's delta area.

This paper is divided into two main sections. The diversity and complexity of physical and socio-economic conditions in the delta are described first. This provides

246

the background for assessing possible impacts of the predicted rise in sea level, discussed in the second section. Conclusions and recommendations are summarized at the end. Fig. 15.1 shows the pre-1983 administrative districts of Bangladesh. Fig. 15.2 shows the major physiographic units referred to in the text.

Regional diversity

The first problem requiring recognition is that of locating 'representative' sites for monitoring. The Ganges-Brahmaputra-Meghna delta is far from homogeneous. It includes all or parts of eight major physiographic regions lying south of the Ganges-Jamuna confluence (excluding the eastern hills and piedmont fans; see Fig. 15.2). Subregions within some regions also need to be taken into consideration for detailed monitoring studies. Detailed descriptions of these regions and subregions are given in a recent Food and Agriculture Organization (FAO) report (FAO, 1988).

The two sides of the lower Meghna river and the Meghna estuary are quite different from each other. The eastern side has predominantly silty estuarine sediments on almost level land with few rivers and creeks. The western side has predominantly clay soils on ridge (levee) and basin landscapes of tidal and meander floodplains crossed by numerous rivers and creeks. These physical differences influence land use, population density, settlement patterns, modes of transport and accessibility. They also imply that a rising sea level might have different geomorphic and socio-economic consequences in different regions.

Instability

The Bangladesh delta is not static. Each year, large areas of land are formed by alluvial deposition and lost by bank erosion in the Meghna estuary (Fig. 15.2, units 12a,b) and along the active floodplains of the rivers Jamuna (unit 6a), Ganges (7a) and lower Meghna (southern part of 7a). Coleman (1969) describes lateral advances and recessions along sections of the Jamuna river banks of up to 850 m in a single year. On the other hand, the Meghna channel through the Middle Meghna Floodplain (11a) is virtually stable, for Meghna sediments are largely filtered out upstream in the Sylhet Basin (13b).

Historically, there have been several major shifts in the Brahmaputra, Ganges and other river courses through the delta (Fergusson, 1863; Ascoli, 1910). The most recent catastrophic changes occurred in and around the Meghna estuary following the Assam earthquake in August 1950. That earthquake released enormous quantities of sediments into the Brahmaputra river in Assam (Kingdon-Ward, 1955), raising river-bed levels by about 3 m (UN, 1957). This, in turn, raised flood levels on the adjoining land and in downstream areas in the subsequent 10–15 years, until the peak of the bed load had passed through. The sudden increase in sediment load created substantial areas of new land in the Meghna estuary; cross embankments built in 1957 and 1964 helped in its consolidation and encouraged further alluvial deposition to the southeast (Land Reclamation Project, 1987). Field evidence collected by the author during 1964–7 indicated that the formation of this new land had geomorphological repercussions elsewhere in the delta: namely, the erosion of the northern sides of the three major estuarine islands (Bhola, Hatiya, Sandwip), the impedance of drainage from the

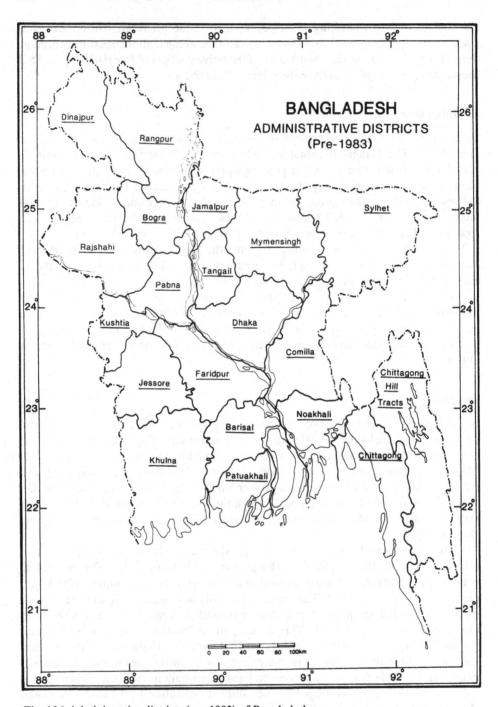

Fig. 15.1 Administrative districts (pre-1983) of Bangladesh.

southern part of the Old Meghna Estuarine Floodplain (12c) and the diversion of increased river flow into the Ganges Tidal Floodplain (8a,b). The latter caused erosion of much of the older river floodplain land north of Bhola Island, replacing it with new alluvial land as well as pushing back the saltwater limit.

Currently, the saline limit appears to be moving inland again. In the west of the Ganges Tidal Floodplain, this is a long-continued trend, associated with the natural deterioration of drainage through the so-called moribund Ganges delta (the High Ganges River Floodplain, unit 7b). This has been aggravated in recent decades by the damming of former Ganges distributaries upstream and the abstraction of water for dry-season irrigation. Further east, dry-season flow has been progressively reduced in recent years by the abstraction of water for irrigation in upstream parts of the Ganges-Brahmaputra-Meghna system and by the diversion of flow from the Ganges at the Farraka barrage to flush Calcutta port.

Flooding

Considerable differences in seasonal flooding characteristics exist within the delta. Flooding is mainly shallow (<90 cm) on the Young Meghna Estuarine Floodplain (12a,b), on the Ganges Tidal Floodplain (8a–c) and on much of the High Ganges River Floodplain (7b). Elsewhere, at its peak in July–September, flooding is mainly moderately deep (90–180 cm) or deep (>180 cm).

However, the floodplain land is not flat. Even within an individual village, differences in elevation occur between adjoining ridges and basins: up to 1 m on the Ganges Tidal and Young Meghna Estuarine Floodplains; up to 2 m on the Lower Meghna River Floodplain (11b) and the Old Meghna Estuarine Floodplain (12c); and up to 2–5 m on the Jamuna (6a,b), Ganges River (7a–c) and Middle Meghna (11a) Floodplains. Peak flood levels may vary by a metre or more and by a month or more between individual years, both of which differences may affect crop production. These year-to-year fluctuations, as well as the regional and local variations in flood levels, must obviously be taken into account in designing studies to assess the impact of a rising sea level.

Over much of the delta the flooding is by rainwater (or the raised groundwater-table) which is ponded on the land by high external river levels during the monsoon season. Flooding by silty water (with consequent alluvial deposition) occurs on the Active Jamuna and Ganges Floodplains (6a, 7a) and on parts of the Lower Meghna River Floodplain (11b), the Young Meghna Estuarine Floodplain (12b) and the Ganges Tidal Floodplain (8a–c). Inland parts of the young estuarine floodplain are mainly flooded by rainwater, as are empoldered areas on the tidal floodplain and within the Chandpur Irrigation Project area on the Lower Meghna River Floodplain. Alluvial sedimentation is nil or negligible on land flooded by rainwater or by the raised groundwater-table. The implications of these regional differences in sedimentation are discussed in a later section.

Salinity

Flood flow through the Meghna estuary is sufficient to carry freshwater to the coast during the rainy season, except in the western half of the Ganges Tidal Floodplain

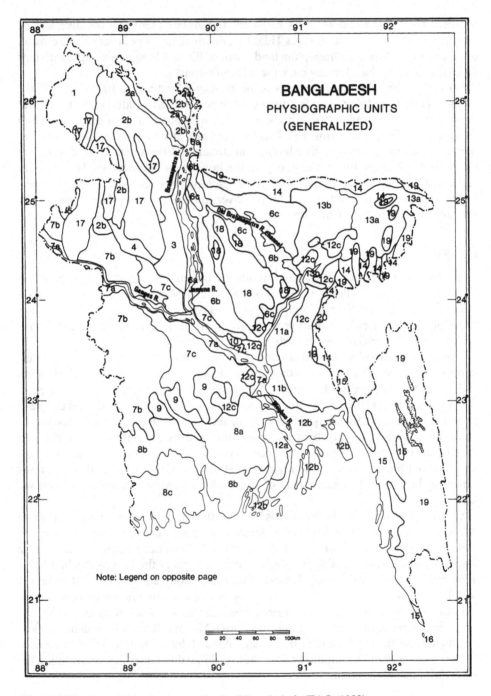

Fig. 15.2 Physiographic units (generalized) of Bangladesh (FAO, 1988).

Physiographic Units

1	Old Himalayan Piedmont Plain
2	Tista Floodplain
2a	Active Tista Floodplain
2b	Tista Meander Floodplain
3	Karatoya-Bangali Floodplain
4	Lower Atrai Basin
5	Lower Purnabhaba Floodplain
6	Brahmaputra Floodplain
6a	Active Floodplain
6b	Young Brahmaputra and Jamuna Floodplains
6c	Old Brahmaputra Floodplain
7	Ganges River Floodplain
7a	Active Floodplain
7b	High Ganges River Floodplain
7c	Low Ganges River Floodplain
8	Ganges Tidal Floodplain
8a	– non-saline
8b	– saline
8c	– Sunderbans mangrove forest
9	Gopalganj-Khulna Bils
10	Arial Bil
11	Meghna River Floodplain
11a	Middle Meghna River Floodplain
11b	Lower Meghna River Floodplain
12	Meghna Estuarine Floodplain
12a	Young Meghna Estuarine Floodplain (non-saline)
12b	Young Meghna Estuarine Floodplain (saline)
12c	Old Meghna Estuarine Floodplain
13	Surma-Kusiyara Floodplain
13a	Eastern Surma-Kusiyara Floodplain
13b	Sylhet Basin
14	Northern and Eastern Piedmont Plains
15	Chittagong Coastal Plain
16	St Martin's Coral Island
17	Barind Tract
18	Madhupur Tract
19	Northern and Eastern Hills
20	Akhaura Terrace

(8b,c). Even in the latter area, the heavy monsoon rainfall is sufficient to desalinize topsoils so that, as elsewhere in the delta, a rainy-season rice crop can be grown (except in the Sunderbans mangrove forest, unit 8c). (Mean annual rainfall increases from ~1500 mm in the west of the delta to >2500 mm in the south and east). In the dry season, the saltwater limit in the rivers gradually moves inland, and soils over substantial areas of the Young Meghna Estuarine Floodplain and the Ganges Tidal Floodplain (mainly 8b,c) become, to varying degrees, saline by the capillary movement of moisture to the surface from the saline groundwater-table.

The areal extent and degree of soil salinity vary from year to year, depending on the length of the dry season and whether or not the dry season is punctuated by rain showers. Storm surges flooding over coastal land when tropical cyclones occur may also increase soil and groundwater salinity for a subsequent period of weeks or years. The saltwater limit in rivers crossing the Ganges Tidal Floodplain also varies considerably from year to year due to variations in dry-season flow in the Ganges and Meghna rivers and in the date of onset of heavy pre-monsoon rainfall. This variability greatly complicates the choice of sites for impact studies and the interpretation of field data.

Settlement

The considerable regional differences in settlement patterns and in population density within the delta need to be taken into account in designing impact studies. Settlements generally are strongly nucleated, concentrated on floodplain ridges or high river banks, but they are dispersed on newly settled parts of the Young Meghna Estuarine Floodplain (12b).

The southern part of the Old Meghna River Floodplain (12c) and the neighbouring Lower Meghna River Floodplain (11b) have the highest population density in Bangladesh: 1,000/km^2 in 1981, with settlements occupying 20–40% of the land area. Population density is also relatively high over much of the High and Low Ganges River Floodplains (7b,c), the northern part of unit 8a on the Ganges Tidal Floodplain and older parts of the Young Meghna Estuarine Floodplain (12a,b), especially in areas where bank erosion has pushed people back. The lowest population densities occur on the youngest alluvial land in the Meghna estuary and on the Active Ganges Floodplain (7a), in saline southern and western parts of the Ganges Tidal Floodplain (8b,c), in the peat swamps of the Gopalganj-Khulna Bils (9), and in the low-lying Arial Bil (10) and the Middle Meghna Floodplain (11a). Nationally, population is presently increasing at around 2.5% annually, but there are appreciable regional differences and especially between rural and urban rates. The national rate of increase is projected to decline over the next 25 years to between an optimistic 1.24% and a pessimistic 2.1%.

Communications

Regional differences in geomorphology within the delta strongly influence modes of communication and ease of access. This must be considered in selecting sites for impact studies as well as in planning development measures to counter the effects of a rising sea level.

On both sides of the delta, the cost of building and maintaining roads and railways is high because of the high embankments needed to raise the beds above flood levels, the large numbers of bridges and culverts needed, the generally poor quality of the alluvial sediments as foundation materials (aggravated by the lack of rock for use in foundations), and the heavy monsoon rainfall. In most regions, communication is largely by boat in the rainy season and by foot or by bullock cart in the dry season. Communications become particularly difficult after severe cyclones, when roads may be made almost impassable because of muddy surfaces, breached embankments and broken bridges, and boats and ferries may be sunk.

Land use

Ultimately, the main object of studies to assess the implications of a rising sea level in the Bangladesh delta must be to measure and predict the effects of such a change on the population living in that area. Since that population is overwhelmingly rural and dependent on agriculture, such studies must aim to assess the implications for land use and, by extension, for the rural and national economies. That is a formidable task, given the intrinsic complexity of land use in Bangladesh and the dynamic changes taking place in geomorphology, population density and agricultural technology independent of the effects of a rising sea level.

Most of the delta is cultivated. Only the Sunderbans mangrove forest in the south–west (8c), residual areas of reed swamp within the Gopalganj-Khulna Bils (9), and some very young and exposed alluvial formations on the Active Ganges Floodplain (7a) and the Young Meghna Estuarine Floodplain (12b) remain under natural vegetation.

Rice is the principal crop grown. There are three seasonal rice groups; *aus*, pre-monsoon to mid-monsoon; *aman*, pre-monsoon (deep-water) or mid-monsoon (transplanted) to post-monsoon; *boro*, dry season. Cropping intensity and cropping patterns vary considerably between regions and within regions according to regional and local differences in the depth and duration of seasonal flooding, soil moisture properties, salinity and availability of irrigation. Single, double and triple cropping practices are used on different kinds of land: usually one or two rice crops, sometimes plus a dryland winter crop (especially pulses, oilseeds, wheat).

Cropping is most intensive on the densely populated Lower Meghna River Floodplain (11b) and the old Meghna Estuarine Floodplain (12c); also in parts of other regions where irrigation is available or where cash crops such as jute, sugar-cane, vegetables, potato or tobacco are grown. In such areas, modern crop varieties and fertilizers are widely used. Elsewhere, traditional crop varieties generally prevail, and there is little use of modern technology.

The division of rice into seasonal aus, aman and boro groups cloaks the existence of many thousands of traditional paddy cultivars, especially in the aman group. Also, the farmers' very fine adjustment of their crops and cultivars to local micro-environments means that there usually are significant differences in cropping patterns and practices within a single village and on the different kinds of land cultivated by an individual farmer; (an average-sized holding of 1 ha may comprise 10–20 scattered fields).

There are also year-to-year and longer-term changes in the cropping patterns and practices due to differences in the time of onset or withdrawal of the rains or seasonal floods, in seasonal flood levels and timings, in the extent of crop damage by floods, cyclones or unseasonal rains, as well as fluctuations in market prices for rice and cash crops. Longer-term changes may occur because of increasing population pressure on the land; the introduction of new crops or cultivars; improvement in land quality by the provision of flood protection and/or irrigation; or deterioration of land quality by impeded drainage, increased flood hazard, increased salinity, or the substitution of raw alluvium for older cultivated soils eroded by shifting river channels.

Even the areas of natural vegetation do not remain static. New alluvial land is usually brought under cultivation within a year or two, and reed-beds in the Gopalganj-Khulna Bils (9) are gradually being encroached upon to provide additional land for cultivation. In the Sunderbans (8c), the vigour and quality of sundri trees (*Heretiera minor*), the main mangrove species in the north of the forest, has deteriorated over the past 30 years, probably in response to increasing salinity as the flow of sweetwater through the delta has decreased.

Land tenure

No systematic regional studies of land tenure appear to have been made in Bangladesh. However, information gained informally by the author in the course of his work in the country indicates that significant differences occur between different parts of the delta and that the situation is far from static. Those monitoring land use as part of environmental impact studies need to be aware that such complexities exist.

Broadly, the Young Meghna Estuarine Floodplain (12a,b), the Active Ganges Floodplain (7a), parts of the High and Low Ganges River Floodplains (7b,c), southern and western parts of the Ganges Tidal Floodplain (8b) and the Gopalganj-Khulna Bils (9) are areas of relatively big land holdings (> 2 ha). Big landowners are often absentee, operating their land through short-term tenants or migrant labour. The common factor linking the areas where such practices occur seems to be the insecurity of crop production. Small farmers may be wiped out economically by recurring crop disasters. Only relatively bigger landowners have the resources to withstand such losses. They also have the resources – or the influence – to buy or grab the land of poor cultivators in times of distress.

In contrast with this pattern is the densely populated Old Meghna Estuarine Floodplain (12c) and the adjoining Lower Meghna River Floodplain (11b) where small owner–cultivators occupy most of the land. Areas of small owner–cultivators also occur in some densely settled parts in the north of the Young Meghna Estuarine Floodplain (mainly in 12a), the northeast of unit 8a on the Ganges Tidal Floodplain, and in other regions. In such areas, although rice remains the principal crop, cash crops often are grown as well to supplement farmers' incomes; rice may also be produced as a cash crop on irrigated land where modern high yielding varieties can be grown. Even in areas where small owner–cultivators predominate, land is often rented in or leased out. Land rental takes a variety of forms, with indications that leases for cash or a fixed quota of production are becoming increasingly important.

Implications and speculations

Detailed and complex though the above description of the diversity of conditions in the Bangladesh delta may appear, in reality it is very generalized and incomplete. Few detailed geomorphological studies have been made and the scattered social studies that have been carried out have generally not been related to the physical or regional environments. Much more detailed information on the physical, social and economic environments and their interrelationships needs to be obtained before reliable monitoring and assessment of change can be undertaken.

Statistics

Unfortunately, in this respect, official statistics – especially crop statistics – are often unreliable, though to what degree remains uncertain: (Pray, 1980; Montgomery, 1985; Boyce, 1987). Meteorological and hydrological data – obtained, in any case, from distantly separated locations – also appear to be dubious in some cases. The main causes of data unreliability probably are the general lack of supervision given to data collection – a consequence of poor communications and inadequate funds for travel and supervision – and uncritical acceptance (or, worse, manipulation) of field data by office-bound bureaucrats or researchers.

The data unreliability creates problems for determining historical trends (see especially Boyce, 1987). It also points to the sort of problems which may arise in collecting and using data specifically for impact studies relating to changes in sea level. Systematic statistical checks on the credibility of all data used in attempting to establish historical trends are a *sine qua non* for such studies. So are the provision of adequate funds and facilities for collecting new data and for supervising their collection and interpretation.

Benchmarks

In Bangladesh, one may even suspect the reliability of survey bench marks against which changes in sea level might be measured. These benchmarks were made many decades ago, and Bangladesh may not have stood still since they were established, due to its location within an active tectonic zone. The District Gazetteers have recorded a number of severe earthquakes in past centuries, with descriptions of land areas – including some coastal areas – being thrown up or foundering, and of river courses being changed.

Not recorded are the possibly slower changes in land level which may have occurred due to more gradual tectonic movements or to consolidation of the great thickness of alluvial sediments underlying the delta. Evidence of such changes exists in the buried organic layers revealed in excavations and boreholes at various distances below the present surface (and below present sea level) (Morgan and McIntire, 1959).

For future studies, the possibility also needs to be considered of local reductions in land level due to consolidation of alluvial sediments where water-tables are lowered by the increasing abstraction of water for irrigation or urban supplies. Misguided projects to drain the Gopalganj-Khulna Bils peat swamps (unit 9) could have a similar, and perhaps greater, effect on local land levels.

It is the responsibility of the Bangladesh Survey Department, the Bangladesh Water Development Board and naval, port and inland water transport authorities to operate the national network of survey benchmarks and tidal gauges, and to ensure that periodic adjustments are made in response to the millimetric annual rise in mean sea level which is expected to occur. Researchers and project planners studying the impact of sea level changes need to ensure that they obtain up-to-date information on such benchmarks and to adjust their own local benchmarks regularly.

Geomorphology and hydrology

Even without the anticipated change in sea level, one could expect considerable physical changes to occur in the delta over the next 50–100 years, as they have occurred in past centuries. The lower Meghna river, for instance, presumably will continue to eat away at its eastern bank; and this movement would be aggravated if the Ganges were to continue its historical trend by breaking through the low-lying Arial Bil (10) to join the Dhaleswari river south of Dhaka and thus join the Meghna 20–30 km upstream from the present junction.

Morphological change can be expected to continue in the Meghna estuary, too. The trend of tidal currents there is such as to force part of the river flow eastwards as it enters the Bay of Bengal. Thus there is a tendency for it to erode its northern bank on the Noakhali coast. Given time and relative stability, the Meghna could erode much or all of the new land it has deposited since 1950. A rising sea level would probably accelerate this process.

Opposing this trend, it may be expected that increasing quantities of sediments will be delivered to the estuary as degradation and erosion in the Himalayas and the Assam Hills continue to accelerate over the next several decades and as additional flood protection embankments along the major rivers channel increased flood flows through to the estuary. This is apart from any further catastrophic deliveries of sediments which might occur consequent upon any future earthquakes upstream on the scale of the 1950 Assam earthquake.

Given the magnitude of changes likely to continue (and perhaps increase) in the Meghna estuary, it might be impossible to isolate the specific geomorphic effects of a slow change in sea level in this environment. Similar difficulties seem likely to be experienced in other regions where dynamic changes in river courses and in the saltwater limit are taking place.

The author considers, in fact, that the greatest effects of the expected rise in sea level may not be seen near the coast but inland. On both the Young Meghna Estuarine Floodplain (12a,b) and the Ganges Tidal Floodplain (8a–c), land levels may be expected to increase concurrently with sea level as alluvium continues to be deposited on the land at high tide (except where this may be prevented by coastal embankments). Inland, however, a rise in sea level might have more drastic consequences. In the first place, the rise in sea level in the Meghna estuary would raise river levels upstream in the delta. In turn, the levels of the levees alongside active river channels would rise to match the increased height of seasonal river floods. Associated with this would be a corresponding increase in the depth of seasonal flooding over most of the Ganges River Floodplain (7b,c), the Gopalganj-Khulna Bils (9) and the Old Meghna Estuarine Floodplain (12c). These areas are flooded mainly by rainwater and so are not subject, like the river levees and coastal areas, to build-up by regular alluvial sedimentation. Additionally, the elevation of the Ganges and lower Meghna levees and of the tidal and young estuarine floodplains would further impede drainage from interior floodplain areas, thus tending to expand the extent of perennially wet land in and around the

Gopalganj-Khulna Bils (9), Arial Bil (10), and in the south of the Old Meghna Estuarine Floodplain (12c). Eventually, the effects of impeded drainage might extend up the Middle Meghna Floodplain (11a) to the very low-lying Sylhet Basin (13b).

Elevation of river levels in the lower Ganges and Meghna channels and the corresponding raising of the adjoining levees would create an increasingly unstable situation such that the Ganges or the Meghna might break out into one or more new channels (probably by enlarging an existing channel) across the Low Ganges River Floodplain (7c) and the Gopalganj-Khulna Bils (9), across Arial Bil (10) or across the Old Meghna Estuarine Floodplain (12c). Monitoring of flow through channels connecting the Ganges and the Meghna rivers with these interior floodplain areas might give an early indication of such changes; and so might monitoring by remote sensing of any trends in bank erosion bringing the main river channel closer to old channels or depressions (*bils*) on the adjoining floodplains.

It must be repeated that, given the historical changes in river channels that have occurred in the delta and the increased channel instability likely to result from increased flood flows and sediment loads consequent upon increasing degradation of vegetation and soils in the Himalayas and Assam Hills, such channel shifts could possibly occur within the next half-century anyway, regardless of any change in sea level. The most that could be claimed in this respect might be that a rise in sea level would increase the risk of such major channel shifts occurring and so might advance their time of occurrence.

Socio-economic consequences

The conjectured geomorphic and hydrological changes outlined above would have serious implications for people occupying the delta.

The increasing tidal and flood levels which are expected to occur would require rural settlement mounds to be raised. This probably would be done by the house-owners themselves. Institutionally, therefore, this would present less of a problem, and be less costly, than raising land levels in urban areas – e.g., in Barisal and Khulna – or protecting such towns from flooding by providing embankments and artificial drainage. Road and railway embankments would need to be raised by responsible authorities (including local councils for rural roads), as would the height of flood protection embankments (e.g., around the Chandpur Irrigation Project). This could probably be done as a part of periodic routine maintenance work. Drainage capacity would also need to be increased in the Chandpur Irrigation Project area, low-lying parts of which might eventually be below the sea level. Although increased provision for drainage might also be required for coastal embankment project polders, an easier solution there might be for project (or local) authorities to allow tidal sedimentation (warping) to occur occasionally so as to raise land levels within the polders concurrently with the rising external low tide levels.

The most serious effects of the conjectured changes would be on the agricultural economy. In this respect, the least affected areas might be on the Young Meghna Estuarine Floodplain and the Ganges Tidal Floodplain where land levels relative to sea level are expected to be unchanged – except possibly in mismanaged polders – thus allowing present agricultural practices to continue. Most affected in these regions, eventually, might be tree crops grown around homesteads which might need replacement by plants established at a higher elevation when house-holders raise their settlement mounds. No significant changes in land use would be expected to occur on the Active Ganges Floodplain, either: conditions will remain equally unstable as seasonal floods, bank erosion and neo-formation continue as at present.

If drainage capacity is increased appropriately, there need also be no change in land use within the Chandpur Irrigation Project area. However, the cost of drainage would increase because of the reduced effectiveness of tidal drainage as sea level rises and the higher heads against which drainage water would need to be pumped. Costs would increase still further if, as seems probable, a regulator must be built across the Chandpur outlet of the Dakatia river (in unit 11b) so as to prevent saltwater from the Meghna estuary entering in the dry season. In the latter situation, it would be necessary to provide a canal to bring irrigation water from the middle Meghna river (in Unit 11a) in the dry season.

It is in the interior where the most serious effects on agriculture may occur. The seasonal depth of flooding over interior floodplain land would gradually increase due to the rising levels of the Ganges and Lower Meghna levees and of the estuarine and tidal floodplains near the coast, and larger areas of basin land would remain wet or submerged in the dry season. The first areas likely to be affected are the southern parts of the Low Ganges River Floodplain (7c) and the Old Meghna Estuarine Floodplain (12c), together with the Gopalganj-Khulna Bils (9) and adjoining northern parts of the Ganges Tidal Floodplain (8a,b). Subsequently, areas further north might be affected, including Arial Bil (10) and adjoining parts of the Jamuna Floodplain; the Middle Meghna Floodplain (11a), together with adjoining parts of the Old Brahmaputra Floodplain (6c) and the Old Meghna Estuarine Floodplain (12c); valleys in the south of the Madhupur Tract (18); and eventually, perhaps, the southern part of the Sylhet Basin (13b).

The net result on agriculture is likely to be negative (unless counter-measures are taken, as discussed later). The main effect of increasing flood levels in the interior regions would be on the aus and deep-water aman paddy crops. The land area suitable for cultivating aus paddy (and jute) would be reduced and farmers would need to change to lower yielding aman paddy varieties which are tolerant of deeper flooding. Even deep-water aman could possibly be eliminated in some low-lying basins which become perennially wet or subject to flooding too early in the rainy season for reliable crop growth.

In the dry season, the slower drainage of basin land implies that there will be increasing areas of land which will be too wet to allow the sowing (or sowing in time) of dryland crops, e.g., wheat, pulses and mustard. Boro paddy may gradually be eliminated from some perennially wet land where traditional varieties are grown, due to the greater depth of water remaining in the basin centres, but concurrently this crop may be extended onto adjoining higher margins which presently drain too early for it to be grown without irrigation. The cultivation of modern varieties of boro paddy with irrigation on low-lying land may be jeopardized by the greater depth of water on the land at the time of planting and/or by the increased risk of early flooding before the crop can be harvested.

Since cultivators may be slow to modify their existing practices in response to the very slowly changing environment, increasing numbers will probably be marginalized economically by the increasing frequency of crop damage or loss, or by the reduced cropping opportunities. Given the existing unequal social relations and the likelihood of their continuation into the foreseeable future, especially in remote areas, it seems inevitable that poverty will increase and that big land holdings will continue to prevail in areas where they are found today. Where small holdings predominate, as on the Old Meghna Estuarine Floodplain, increasing pauperization resulting from a deteriorating agricultural environment may be expected to push increasing numbers of people off the land. The conjectured changes noted above would merely reinforce the trends that are already predictable as population pressure on the land throughout the delta mounts during the next few decades.

Salinity

One further environmental change needs to be considered: salinity. If, as predicted above, the tidal and young estuarine floodplains grow in height with increasing sea level, there seems no reason to expect that a rise in sea level, by itself, would push the saline limit further inland, except perhaps in the extreme west. Changes in flow through the estuary due to environmental changes and man-made interventions inland seem likely to be the major factor determining the freshwater–saltwater balance in the estuary and in the connecting tidal channels, both in the monsoon season and in the dry season.

In the west, however, where both the monsoon-season and dry-season river flows have already been reduced or interrupted by natural and man-made changes upstream, a rise in sea level might allow saltwater to penetrate further inland than at present, perhaps extending into the lower courses of rivers on the High Ganges River Floodplain (7b), and perhaps also extending the limit of saline groundwater northward in this region. This northward extension of the saline zone would probably have little effect in the monsoon season, during which rainfall would still desalinize soils sufficiently for cultivation (or shrimp farming) to continue as at present. There would be a greater effect on dry-season cropping, especially where present irrigation sources might become saline.

Considerable ecological changes could occur in the Sunderbans forest (8c). Mangrove species are very sensitive to changes in salinity. Changes in species vigour or composition due to changes in salinity in this region could have a considerable economic impact. An adverse impact, manifested in dying trees and reduced timber production, could lead to public (and political) demands to release part of this area for agricultural settlement and production.

Possible interventions

It must be assumed that the Government of Bangladesh and international aid agencies will not sit idly by and, Canute-like, let the rising water levels overwhelm the country. The superficially obvious response to the increasing flood hazards described above is flood-control, i.e., the erection of embankments and the installation of pump drainage. That was the response in the 1950s to a similar, but more rapidly aggravating, situation in which flood heights and associated damage to crops and property increased following the Assam earthquake (United Nations, 1957). The other more effective response was to emphasize dry-season crop production, especially by investment in irrigation and by promoting such crops as boro paddy, wheat, potato and mustard.

The 1950s flood experience provides a number of salutary lessons. One is that, in such situations, things continue to get worse for some time (even a long time) before they get better. This is because of the lags involved in recognising the problem, deciding upon a response, mustering financial support, creating or expanding appropriate institutions and executing the infrastructural works required. For example, the Brahmaputra Right-Bank Embankment, the need for which was identified in the 1950s (IECo, 1964), was not completed until 1968 (by when the severity of Brahmaputra floods had diminished); and the lower section of the Ganges River, where flooding (e.g., in the old Faridpur District) is even more frequent and serious than along the Brahmaputra, has remained without flood protection to the present.

Another lesson is that flood protection works are costly to build, operate and maintain, with the cost of pump drainage being especially high. Embankments along the major rivers are subject to erosion by shifting river channels, as has happened in the case of the Brahmaputra Right-Bank Embankment and the embankment of the Chandpur Irrigation Project. The cost of acquiring land to retire a section of embankment threatened by bank erosion is also high; and litigation over land acquisition may prevent a new embankment from being built before the earlier one is breached (as happened in several places along the Brahmaputra Right-Bank Embankment in 1984). To date, no satisfactory way has been found of recovering the capital and operational cost of such projects from the beneficiaries.

Also, embankments by themselves do not provide total flood protection to all the land they supposedly command since most floodplain land is flooded by the ponding of runoff from local rainfall and the raised groundwater-table when rivers are at peak levels. Depression sites in polder areas provided with pump drainage (e.g., in the Chandpur Irrigation Project area) continue to be flooded after heavy monsoon rainfall to such an extent that short-stem modern paddy varieties cannot be grown reliably in such sites, which may amount to 10–30% of the total command area (Brammer, 1982).

Where pump drainage is not provided, either tidal sluices are needed (as in the coastal embankment polders), or local runoff must be allowed to escape through channels behind the embankment (as behind the embankments along both sides of the Jamuna). The latter solutions may not be available, however, on the Old Meghna Estuarine Floodplain and the Low Ganges River Floodplain where gradients are less than on the Jamuna Floodplain and where the lowest parts lie close to the present mean sea level. It is on these floodplains, it may be recalled, where the most serious effects of deeper flooding and impeded drainage are predicted to occur in consequence of a rising sea level.

Even pump drainage does not offer a simple solution to the areas most likely to be adversely affected by rising sea levels. The active channels of Bangladesh's major rivers provide few sites where headworks for pumping stations can be built. This is because of the shifting river channels and the rapid silting up of outlet channels that may occur (as in the case of the intake channel for the Ganges–Kobadak Project near Kushtia).

The practicality of pump-draining areas as extensive as the Old Meghna Estuarine Floodplain and the Low Ganges River Floodplain must be questioned. Rainfall in these regions is high (mean annual > 2,000 mm) and short-period intensities can be very high (> 250 mm in a single day). If short-stem paddy is to be protected, it must not remain submerged for more than 2–3 days. That implies the need not only for very high pumping capacities but also for large channels to evacuate the expelled water (and that at a time, too, when such channels may already be brimful with water draining from upstream). The effect of such increased drainage flows on downstream areas of the Ganges Tidal Floodplain and the Old and Young Meghna Estuarine Floodplains needs also to be considered.

Double-embanking along major rivers such as the Brahmaputra, Ganges and Meghna – or even along their major distributaries – is still a controversial subject. The orthodox engineering view is that the construction of embankments along both sides of such mighty rivers, by concentrating their flow, increases their erosive capacity (thus exposing the embankments to a greater risk of breaching) and increases the flood hazard in downstream areas. As along the Mississippi River, it may also raise the river-bed level by sedimentation after flood peaks pass until the river eventually lies higher than land outside the embankments, thus creating a hazard of catastrophic overland floods occurring if an embankment is breached.

The implication of the above is that embankments and drainage do not provide a simple solution to the problems created by increased flood levels that might follow a rise in sea level. The hydrological problems involved are complex and will require much study. The solutions proposed may also be expensive to provide and to operate. The alternative – of doing nothing and learning to live with the new situation – may also be expensive, of course, whether measured in terms of increased human poverty and distress, in terms of public (and international) relief to alleviate such distress, or in terms of the cost of providing alternative livelihoods for the populations affected.

Conclusions

The complexity and dynamism of the Bangladesh delta present immense problems to researchers seeking to study the impact on Bangladesh of a slowly rising sea level. Regional and local diversity of geomorphology, hydrology, salinity, settlement, communications, land use and land tenure imply that great care will need to be taken in the selection of sampling sites, that a relatively large number of sites may need to be monitored and that such studies will need to be multi-disciplinary. Because of the close relationship between cropping patterns and land levels in relation to seasonal flooding, transects across neighbouring ridges and basins should be used for studying land use changes in response to changing flooding depth and duration.

The fact that year-to-year variability in seasonal flood levels, bank erosion, new land formation and land use are on a much bigger scale than the annual changes that might result from the predicted slow rise in sea level provides a particularly difficult problem. This problem is aggravated by the dynamic changes taking place in river flow, sediment load and channel instability, and in population density, land use and human interventions to regulate flooding. Any monitoring undertaken in an attempt to identify the specific effects of a rising sea level must take such dynamic changes into account.

The costs of conducting such comprehensive and complex long-term studies might be very great, possibly beyond the resources of any single national or international research institution. Such studies, therefore, – if considered feasible at all – might have to form a part of national, regional or project studies of, for example, estuary morphology, river hydrology or land use.

The possible consequences of a rising sea level on land use, urban settlements, roads, railways, flood embankments and drainage systems must be taken into account by both national and local government planners and by international donor agencies, especially those concerned with strategic economic planning. Ultimately, it is the responsibility of such planners and donors to ensure that appropriate institutional arrangements are made (and funded) to monitor and assess the impact of a rising sea level, including the relationship between survey benchmarks and reference tidal gauges. For such studies to be meaningful, it is particularly important that national institutions responsible for gathering agricultural, hydrological and other relevant statistics be strengthened so as to ensure the provision of reliable data for environmental impact assessments.

Postscript

Following severe floods in Bangladesh in 1987 and 1988, the Government of Bangladesh, with donor assistance, formulated a Flood Action Plan comprising 26 projects, studies and pilot projects which aim to increase the country's resilience to recurrent floods (see Brammer, 1990a, b).

References

Ascoli, F. D. (1910). The rivers of the delta. *J. As. Soc., Bengal: New Series*, **VI**, 543–51

Boyce, J. K. (1987). *Agrarian Impasse in Bengal*. Oxford University Press, Oxford, UK.

Brammer, H. (1990a). Floods in Bangladesh. Part I. Geograpical background to the 1987 and 1988 floods. *Geogr. J.*, **156**, No. 1, 12–22.

Brammer, H. (1990b). Floods in Bangladesh. Part II. Flood mitigation and environmental aspects. *Geogr. J.*, **156**, No. 2, 158–65.

Brammer, H. (1982). Agriculture and food production in polder areas. *International Symposium on Polders of the World, Lelystad, Netherlands*.

Coleman, J. M. (1969), Brahmaputra River: channel processes and sedimentation. *Sed. Geol.*, **III**, 129–239.

FAO (1988). *Land Resources Appraisal of Bangladesh for Agricultural Development. Report 2: Agroecological Regions of Bangladesh*. FAO, Rome.

Fergusson, J. (1863). On changes in the delta of the Ganges. *Q.J. Geol. Soc., London*, **19**, 321–54.

IECo (International Engineering Co Inc) (1964). *East Pakistan Water and Power Development Authority Master Plan*. 2 vols. IECO, San Francisco, USA.

Kingdon-Ward, F. (1955). Aftermath of the great Assam earthquake of 1950. *Geog. J.*, **121**: 290–303.

Land Reclamation Project (1987). *Feasibility Study of the Sandwip Crossdam Development Scheme: Final Report*. Dhaka: Bangladesh Water Development Board.

Montgomery, R. (1985). The Bangladesh floods of 1984 in historical context. *Disasters*, **9**, 163–72.

Morgan, J. P. & W. G. McIntire (1959). Quaternary geology of the Bengal Basin, East Pakistan and India. *Bul. Geol. Soc. Am.*, **70**, 319–42.

Pray, C. (1980). *An Assessment of the Accuracy of the Official Agricultural Statistics*. Bang. Dev. Stud., Univ. Dhaka, Bangladesh.

United Nations (1957). *Water and Power Development in East Pakistan*. UN Tech. Assist. Prog. Rpt, TAA/PAK/15. (Popularly known as the *Krug Mission Report*). UN, New York.

16

Possible impacts of, and adjustments to, sea level rise: the cases of Bangladesh and Egypt

J.M. Broadus

Abstract

The potential economic implications of two relative sea level rise scenarios for the years 2050 and 2100 are examined for Egypt and Bangladesh. The methods used to derive rough estimates of the potential scale of economic effects are very simplistic. Nonetheless, the results suggest that relatively large economic values are at stake in both cases. As a first-order approximation of these economic stakes, estimates are made of the *current* attributes and activities in the areas affected by *future* sea level rise. A 1 m rise, for example, would cover areas currently accounting for about 7% of habitable land, 5% of population and 5% of Gross Domestic Product (GDP) in Bangladesh; 12%, 14% and 14% of arable land, population and GDP, respectively, are in the area that would be affected in Egypt.

In order to demonstrate how the methods can be extended to estimate potential economic loss in the future, the case of Bangladesh is further developed. The analysis suggests that, without mitigating adjustments, a sea level rise of 1 m by 2050 could result in a cumulative loss (in present value terms) equivalent to 1–2% of Bangladesh's current GDP. This should be considered a 'worst-case' scenario, since human adjustments would tend to reduce future losses. Potential adjustments include: defensive measures, recombinations of productive factors and technological adaptation. In the near-term, institutional adaptations such as insurance, share markets, contracts and futures markets, mergers, and governmental risk-sharing schemes might be considered.

Introduction

This paper attempts to examine the potential economic implications for Egypt and Bangladesh of scenarios of relative sea level rise for the years 2050 and 2100. Obviously, such long-range projections are highly speculative. The best that can be achieved is a very coarse approximation of the economic scale of the potential problem.

There are several reasons why an economist might approach this topic with a strong sense of scepticism. First, there is a large uncertainty about whether sea level will rise in these regions, and, if so, by how much. Second, a sea level rise will be gradual compared to the human capacity to respond – to move and to adjust the allocation of resources under the changed conditions. Third, the bulk of potential costs would occur decades into the future, a time horizon over which even large economic values are apt to reduce

Table 16.1. *Summary of relative sea
level rise scenarios and economic
activities/assets in affected areas*

	1 m (2050)	3 m (2100)
Bangladesh		
Total sea level rise	0.83 m	3.4 m
(Global)	(0.13 m)	(2.2 m)
(Local subsidence)	(0.70 m)	(1.2 m)
Loss habitable land	7%	26%
Population	5%	27%
GDP	5%	20%
Egypt		
Total sea level rise	0.78 m	3.4 m
(Global)	(0.13 m)	(2.2 m)
(Local subsidence)	(0.65 m)	(1.2 m)
Loss habitable land	12%	20%
Population	14%	21%
GDP	14%	20%

Source: Source of inundation scenarios:
Milliman (1987).

to relatively trivial sums in present discounted value. Finally, the pace of socio-technological change promises to alter the economy substantially over the time period in question. Nonetheless, the *potential* scale of economic impacts suggested by the present analysis appears large enough to warrant further investigations with more refined techniques.

The general approach (Broadus *et al.*, 1986; Milliman *et al.*, 1989) employed here is as follows. First, along with scenarios of global mean sea level rise, geological information was used to construct scenarios of transgression of sea level beyond existing coastlines over the next 100 years. These scenarios were then used to define the geographic areas within Egypt and Bangladesh that would be affected by landward transgression. Next, demographic and economic data were used to examine the scale of economic activities within each nation that would currently be affected. (It is important to emphasize that estimates pertain only to current levels of economic activity within the areas potentially affected over the next 100 years.) Finally, using strong (but reasonable) parametric assumptions, demographic and economic projections were made for even cruder estimates of the present value of potential economic loss in the future, as illustrated with one of the Bangladesh scenarios below. The current economic activities or assets located or originating in the potentially affected areas of Egypt and Bangladesh are summarized in Table 16.1.

Egypt

Of Egypt's one million square kilometre area, only about 35,000 km², or 3.5% of total land area, is cultivated and settled (*Quarterly Economic Review of Egypt – Annual*, 1985). This results in an estimated population density in 1987 of about 1,500 people for every square kilometre of arable land in the country. The population is densely clustered about the banks of the Nile River and throughout the river's delta from the port city of Alexandria eastward to the northern entrance of the Suez Canal. The modern city of Alexandria, exposed since antiquity to the forces of the sea and historically reached by a causeway extending to its walled enclosure, contains nearly 4 million inhabitants. Port Said, at the eastern extremity of the delta, is the home of over 500,000 residents. Most of the delta between the two cities is densely inhabited and is devoted to intensive, multi-crop agriculture and to urban and industrial uses.

Fig. 16.1 illustrates the landward transgression that might result from a 1 m and a 3 m rise in relative sea level. These scenarios were based on estimates of both global mean sea level rise and vertical land movement, roughly rounded to the nearest metre. Thus the 1 m scenario corresponds approximately to a 13 cm global sea level rise and severe local subsidence of 65 cm by 2050, while the 3 m case represents just over 2 m of global sea level rise and 1 m of subsidence by 2100 (Milliman, 1987). The IPCC (1990) projected only a 30 cm global rise by 2050, so the 3 m case must be seen as very unlikely. The major existing natural defences against such rises in sea level are the series of sand dunes along the delta's coast and the increasingly brackish lakes that lie behind them. These lakes are a major source of about 100,000 tons of fish caught in Egypt annually, 80% of which is freshwater fish. The area affected by the 1 m relative sea level rise represents approximately 12–15% of the nation's arable land and (on the assumption that population is evenly distributed within each governorate) contains some 14% of the estimated 53 million population, or almost 7.5 million people. The area affected by the 3 m rise in relative sea level represents about 20% of the nation's arable land and is inhabited by over 10 million people, about 21% of Egypt's current population.

Assuming that agricultural output in the delta is distributed evenly within the arable area, and assuming that all other sectors of economic activity are spatially distributed with population, we estimate that approximately 14% of Egypt's current GDP originates in the area likely to be affected by a 1 m increase in relative sea level. If, as in historical times, Alexandria's elevation were to spare it from an encroaching sea, then the 1 m figures should be adjusted downward (El Sayed, 1991). Specifically, less than 7% of the nation's current population and about 6% of current GDP are then in the potentially affected area. Approximately 20% of the nation's current GDP originates within the area likely to be affected by a 3 m increase in relative sea level.

Bangladesh

The exposure to sea level rise is more serious in Bangladesh (Fig. 16.2). This densely populated nation of approximately 110 million people covers an area of 144,000 km². 80% of this area is made up of the complex Bengal delta system created by the Ganges, Brahmaputra and Meghna Rivers. Together, the Ganges and the Brahmaputra currently deliver approximately 1.6 billion tons of sediment annually to the face of the delta (Milliman and Meade, 1983). This sediment replenishment appears just to offset

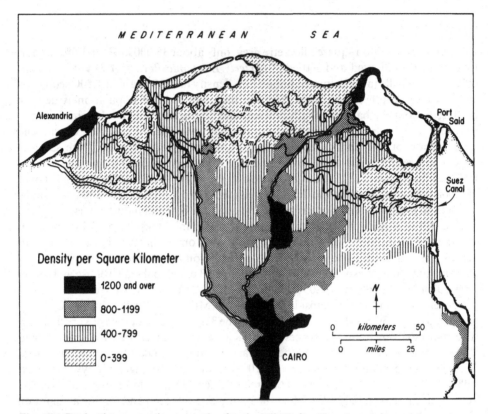

Fig. 16.1 Sea level transgression scenarios for the Nile Delta, Egypt, superimposed on population density. Adapted from Milliman *et al.* (1989).

the natural compaction and subsidence of the delta, keeping its size relatively stable (Milliman, 1987). There is some evidence, however, that the delta may be (or should be) experiencing net accretion as a result of increased upstream erosion (Brammer, this volume).

The country's population is widely distributed, with heavy concentrations in the major city of Dhaka (population over 4 million), in the southwestern city of Khulna (population over 800,000), and in the eastern port city of Chittagong (population about 1.5 million). Most of the remaining population is rurally dispersed and dependent on subsistence agriculture. Much of the population lives at the very edge of subsistence. Per capita Gross National Product (GNP) is approximately $150, compared to that of Egypt which is $700 and that of the United States which is over $18,000. Population increase is in the range of 2.5–3.0% per annum.

Population and economic activities at risk

For Bangladesh, the 1 m scenario applies roughly to the projected 13 cm global sea level rise and local subsidence of 70 cm by 2050. The 3 m scenario is comprised of 2 m of global sea level and 1 m of local subsidence by 2100 (Milliman, 1987). About 7% of the country's habitable land lies within the area covered by a 1 m rise and 26% within the area of the 3 m scenario. [These values are smaller than those reported earlier (Broadus

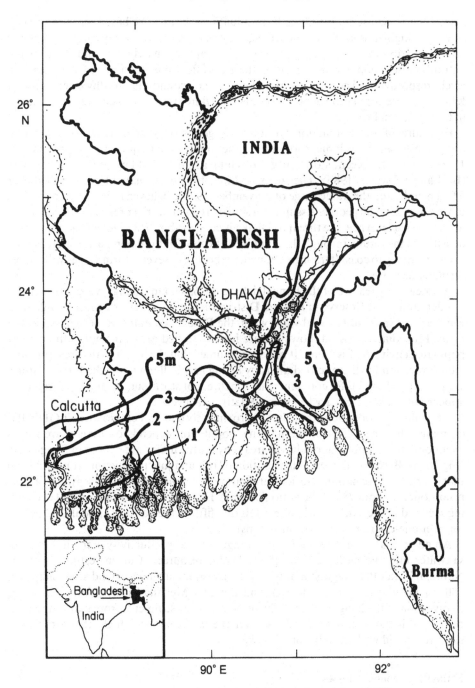

Fig. 16.2 Sea level transgression scenarios for Bangladesh. Adapted from Milliman *et al.* (1989).

et al., 1986) because they exclude large aquatic areas and are based on more precise computer digitization.] When the transgressions of sea level are superimposed on the spatial distribution of population (Fig. 16.3), approximately 5% and 27% of the nation's present population live within the areas affected by the 1 m and 3 m rise in sea level, respectively. The densely populated area around the southwestern city of Khulna, where population density exceeds 2,900/km², is situated between the 1 m and 3 m transgression lines.

Exposure of the population to storm surges is an extremely grave danger in Bangladesh. Severe cyclonic storms attack the country at a frequency of approximately 1.5 per year, with recent storm surges reaching as far as 160 km up the river courses (Bird and Schwartz, 1985). The storm of May, 1985 is estimated to have killed over 5,000 people, and the tragic surge of November, 1970 is believed to have taken the lives of well over 250,000 people. Assuming a similar 160 km up-river reach in the case of a 3 m relative sea level rise, the threat to Dhaka is conceivable. With increasing population densities in the future, through the combined effects of natural population increase and loss of territory to coastal transgression, the exposure to severe storm surges will almost certainly increase.

A noteworthy feature of the Bengal delta is the 6,000 km² mangrove and nepa palm Sunderban Forest Reserve in Khulna District on the southwestern coast (blank area, Fig. 16.3). A maze of forested waterways, the Reserve contains no permanent settlement. Immediately behind, however, are heavily settled areas, including the densely populated environs of Khulna. It may be presumed that the Sunderban forest provides vital protection for these settled areas by acting as a buffer against the force of storm surges. Loss of this buffer could increase the threat of storm surges, and the forest does appear vulnerable to even a 1 m rise in relative sea level.

Agricultural production comprises approximately one-half of the nation's GDP. It is estimated that over 85% of the nation's population depends on agriculture for its livelihood. Net cropped/agricultural area is by far the major use of land in Bangladesh (Fig. 16.4). Based on the areal distribution of rice and jute production, it is estimated that over 5% of the nation's agricultural output originates in the area seaward of the 1 m contour (or about 7% of the nation's crops). Estimated agricultural output originating seaward of the 3 m scenario line totals one-fifth of the nation's agricultural output (or over one-quarter of the nation's crops).

An additional, complicating factor for agricultural productivity is the intrusion of saltwater into the nation's fresh groundwater resources. Current estimates in the National Water Plan suggest that saltwater intrusion may now extend seasonally over 150 km inland in the western districts and along the Meghna River. If future damming or diversion of the Bengal rivers effectively cut off new sediments from maintaining the delta, and if the rate of sea level rise over the next century is high, the intrusion of saltwater could well reach beyond Dhaka.

Estimating economic losses

The approach reported above provides a rough approximation of the current population and economy potentially at risk as sea level rises (Mahtab, 1989). It does not take account of future mitigation measures and adaptive responses. Moreover, the approach does not allow for future changes in the growth and distribution of economic

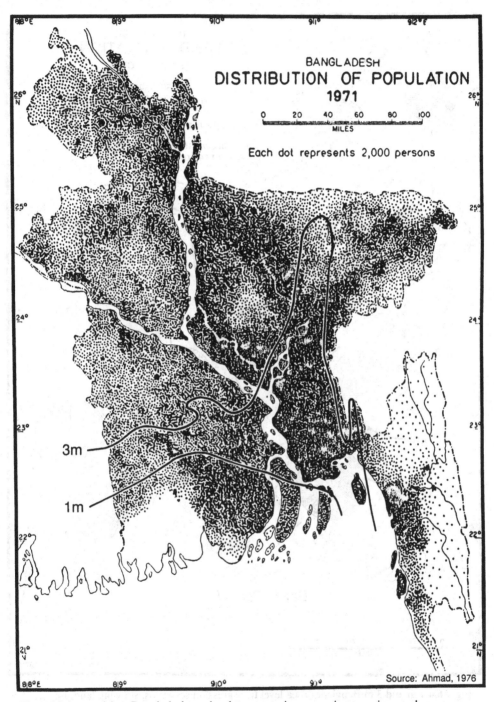

Fig. 16.3 1 m and 3 m Bangladesh sea level transgression scenarios superimposed on illustrative population density. Ahmad (1976) and Broadus *et al.* (1986).

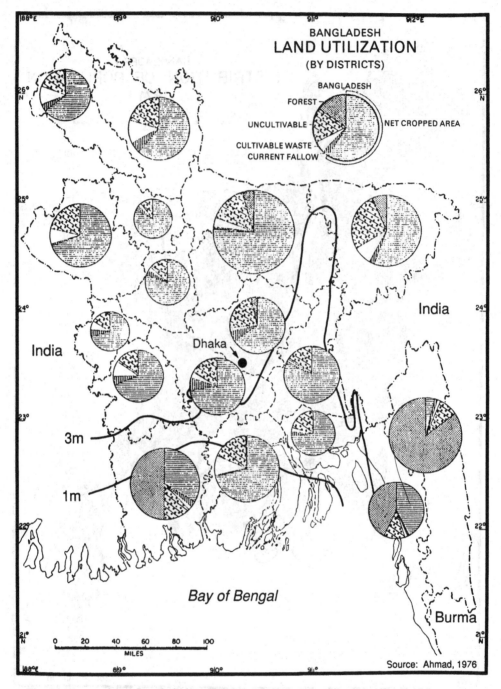

Fig. 16.4 1 m and 3 m Bangladesh sea level transgression scenarios, showing land use patterns by district. Ahmad (1976) and Broadus *et al.* (1986).

activities and population; in this sense, the results are probably conservative (see Barth and Titus, 1984, and Yohe, 1990, for discussion of alternative techniques for estimating economic losses).

By making further strong assumptions, however, it is possible to extend the analysis to make crude estimates of the potential economic losses associated with the inundation scenarios. The following assumptions are made: that human response is largely passive in the face of rising local sea levels; that the geographic distribution of future economic activities closely resembles the current distribution; that land loss is the only measurable loss (i.e., the approach abstracts from lost durable capital, increased storm surge exposure, saltwater intrusion, and the costs of passive adjustment); that the value of all lost land is in agricultural rents; that technological change does not materially affect the case; and that the social rate of discount is equal to the economic growth rate. The method provides a rough, 'order-of-magnitude' estimate of economic losses expressed in terms of net present value.

The method can be illustrated using the scenario of a 1 m rise in sea level by 2050. Let us begin with the current gross domestic product (GDP_o) which is known (use 418 billion Takas). With a time horizon (T) of 61 years (1989 to 2050) and a constant economic growth rate (r), the cumulative GDP in the absence of sea level rise can be estimated by the following expression

$$GDP_o \sum_{t=0}^{T} (1+r)^t$$

A share (a) of this cumulative value represents the share of GDP earned by agricultural land rents

$$aGDP_o \sum_{t=0}^{T} (1+r)^t$$

Empirically, it is known that agriculture accounts for about one-half of GDP in Bangladesh. It is reasonably assumed, further, that 50% is rent earned by the productive qualities of the land itself (with the other half accounted for by returns to capital and risk-bearing). Hence, for our purposes, a is set at 0.25 (=0.5 × 0.5). Examination of reported land-use practice suggests that the overwhelming bulk of land that is in the 1 m affected area and that is not uncultivable, waste, or forest reserve, is cropped (Fig. 16.4). Therefore, the area is treated simply as if all rent-earning land were agricultural. (Sensitivity tests adjusting a upward to account for reasonable proportions of other rent-earning land uses did not substantially affect the final loss estimate reported below.)

All the inundation will not occur at once, of course, and the last parcel lost will produce its full share of output right up until the last year of the time horizon. The annual rent value of the first parcels lost, on the other hand, accrues as an additional economic loss every year over the 61–year span. Although the actual rate of land loss would most likely be non-linear, the simplest case is a rent loss that is a linear function of time, $\lambda(t)$ giving a cumulative loss of land rent over the time horizon

$$aGDP_o \sum_{t=0}^{T} \lambda(t)(1+r)^t$$

Assuming linearity and an annual rate of loss totalling 7% at the end of 61 years, λ is 0.00115. The assumption of linearity can be relaxed later for comparison to other functional forms of loss rate.

Since all of this rent loss from sea level rise would occur in the future, much of it over a half-century from now, it is helpful to express the full stream of losses in terms of discounted present value. (A discussion of the rationale for, and effects of, economic discounting will be omitted since an upcoming assumption offsets the apparent effect of discounting in this example anyway.) With social rate of discount, ∂, the net present value of the cumulative loss (of agricultural land rent) becomes

$$NPVL = \alpha GDP_0 \sum_{t=0}^{T} \lambda(t)(1+r)^t(1+\partial)^{-t}$$

However, it can be assumed, realistically, that the economic growth rate of the economy just equals the social discount rate, $r = \partial$. This is clearly convenient because the 'back of the envelope' expression for estimated loss then simplifies to

$$NPVL = \alpha GDP_0 \sum_{t=0}^{T} \lambda(t)$$

Inserting values for $\alpha(=0.25)$, $GDP_0(=418)$, $T(=61)$, and $\lambda(=0.00115)$, generates an estimate of

$$NPVL \sim 7.5 \text{ billion Takas,}$$

or approximately 1.75% of current GDP.

If inundation approaches 7% of total land area exponentially, rather than linearly, over the time period, then the estimated value of NPVL would be smaller. This is because a larger proportion of the eventual land loss would occur later in the period, preserving the output of most land for longer periods. Such a non-linear rate of rise could generate a loss less than one-half the size of the linear case, or about 0.8% of current GDP. Therefore, subject to all the caveats and assumptions already described, it might be expected that the present discounted value of the economic loss in Bangladesh by 2050 of an unchecked 1 m rise in relative sea level would be of the order of 1% to 2% of current gross domestic product.

Responses

Unless the rate of inundation is very gradual, it is ridiculous to imagine that people would passively accept such losses. More likely, they would respond by doing something to cut the losses. Economists have expressed at least two different views of how effective responses emerge. One view stresses that people have a strong talent for adjusting to change, and that their adjustments are most effective when crafted little-by-little to incremental changes spread over time (Schelling, 1983). This view often also holds that decision-makers closest to the facts of each choice are likely to devise the most appropriate incremental responses (McFadden, 1984). In a prescriptive sense, this view tends to be non-interventionist and favours letting things sort themselves out as the facts emerge.

In contrast to such atomistic incrementalism, another view of human responses stresses the value of collective actions. Whether for reasons of political ideology or the failure of free exchange to account for effects that are spread across large social groupings, this view highlights the value of co-operative study and planning. It holds that it is the responsibility of governments and other collective institutions to anticipate and help engineer effective responses.

Both views share the assumption that economic decision makers act rationally. The reasonable assumption of rational decision making leads to a fairly strong expectation about human responses: that decisions will be made to reduce the expected costs and to maximize the expected net benefits of predicted change. Of course, mistakes may be made, but, in general, human responses can be expected to lower the net cost of sea level rise. If they did not, they would more likely not be undertaken. Nor is it necessary that responses will be confined to the areas most affected by the physical changes. If a nation loses its low-lying coastal rice production to sea level rise, for example, this will alter the allocation of resources in the economy in ways that extend beyond the loss of rice production. The resources formerly applied to rice (including materials, mobile capital and labour) will tend to be redirected into other activities that eventually will help replace at least part of the economic benefits lost with the rice fields. How quickly and how completely this economic adaptation will occur clearly depends on the suddenness of the initial loss, on the organization of the economy and on the availability of other opportunities for productive employment of the displaced resources.

Possible responses in Bangladesh

It would be difficult to replace croplands lost to sea level transgression in Bangladesh because the countryside is already so extensively cultivated. Of the 24.5 million acres in the nation estimated to be cultivable, more than 90% are already in cultivation (Zaman, 1983). Currently, fallow areas constitute a negligible percentage of land use (Fig. 16.4). Further prospects for increasing agricultural usage in northern districts (the area least likely to be affected by salt intrusion) are also relatively limited. Increasing the intensity of agricultural land use through multiple cropping strategies may show some promise. Of the nation's net cropped area, over one-half is currently used for only one crop per year, while 39% is used for two crops annually and about 7% is used fully in the production of three crops (*Quarterly Economic Review of Bangladesh – Annual*, 1985). However, much of the relatively high land in the delta that is least likely to be affected by coastal transgression consists of old delta sediments of distinctly poor fertility.

The nation annually produces about 700,000 tons of fish, 80% of which are freshwater species. Fish products are the nation's fourth leading export commodity after jute products, jute, and leather, exceeding even exports of tea. It seems likely that fishing centres are subject to considerable relocation and the fishing industry, in part due to its largely artisanal nature, may be in a position to respond with great flexibility over the coming century to changing configurations and conditions in the distribution of the nation's aquatic resources.

Many large industries in Bangladesh are associated with jute products, and as with the case of jute production, only a relatively minor proportion is found within the area affected by a 1 m relative sea level rise. In general, the relative sparsity of major industrial and urban development in the potentially affected area could be a decided advantage in responding to sea level rise. While the nation may be strapped for resources to mount a costly strategy of structural defence, neither has it sunk a commitment in the region into a durable infrastructure that demands defending.

Because the timing, magnitude and physical effects of sea level rise are still highly uncertain, potential near-term responses are more in the nature of risk management. This will tend to put a premium on research efforts that increase information and on institutional arrangements that permit diversification and pooling of risks. In the near term, responses to the increased risk of sea level rise might thus include, in some form, institutional adaptations such as simple insurance arrangements, share markets, contracts and futures markets, mergers, and governmental risk-sharing schemes. High risk is nothing new to the Bengal delta, and a number of existing practices in property management and redistribution of risk can be ascribed to this (Brammer, this volume). As physical changes call attention to the increased risk from sea level rise, medium-term responses are also likely to include increased investments in embankments, refuge mounds, and perhaps other defensive engineering measures (Brammer, this volume).

Recall that any significant reduction in the massive delivery of the sediments to the delta could disrupt its balance and expose it to net erosion and subsidence, thereby increasing the effects of sea level rise. Careful attention to this possibility seems warranted in the planning and design of upstream water management projects such as dams and barrages.

Assuring such careful attention, however, is greatly complicated by a sensitive international facet of the problem. Bangladesh shares the Ganges Basin with two of its neighbouring states, India and Nepal, and the Brahmaputra Basin with India, China, and Bhutan. Of the total drainage area of the Ganges–Brahmaputra–Meghna River system, only 7.5% lies within Bangladesh (Zaman, 1983). A serious dispute between Bangladesh and India concerns India's construction and use of the barrage on the Ganges at Farakka. This dispute led in 1972 to the creation of the Indo–Bangladesh Joint Rivers Commission and in 1977 to the Ganges Waters Agreement. Nonetheless, tensions continue over this issue, and it seems very likely that future difficulties can be expected in allocation and management of the aquatic and sediment resources of this complex international river system.

Summary

In Egypt, about 12–15% of the nation's current arable area, 14% of its population (7.5 million people) and 14% of its GDP fall within the region affected by the scenario of a 1 m rise in sea level. With the 3 m rise, the percentages of arable land, population and GDP in the affected region increase to about 20, 21 and 20%, respectively. In Bangladesh, approximately 7% of the land area, 5% of the population (5.5 million people) and 5% of GDP fall within the area affected by the 1 m rise; these percentages increase to 26, 27 and 20%, respectively, for the 3 m rise.

Rough estimates of economic loss were made for Bangladesh. The analysis suggests that, in the absence of mitigating adjustments, a relative sea level rise of 1 m by 2050 could result in a cumulative loss (in present value terms) equivalent to 1–2% of the nation's current GDP. However, human adjustments such as defensive measures, recombinations of productive factors, and institutional and technological adaptation would tend to reduce this loss. So it is best seen as a 'worst-case' estimate. Even so, the potential scale of economic impact in both countries suggests that further detailed investigations are warranted.

Acknowledgements

The author is indebted to John Milliman who conceived the project for which this work was developed and whose original conjecture is reflected throughout. Financial support from The Pew Charitable Trusts,, the US Environmental Protection Agency, and the Climatic Research Unit of the University of East Anglia is gratefully acknowledged. Thanks, too, to Mahmood Alam, Dave Aubrey, Hugh Brammer, Steve Edwards, Sarah Raper, Mike Schlesinger, Andy Solow, Richard Warrick and Tom Wigley for encouragement, patience and constructive feedback. Several useful suggestions were offered by an anonymous referee. Research assistance was provided by Frank Gable. Woods Hole Oceanographic Institution (WHOI) Contribution Number 7147.

References

Ahmad, Nafis, (1976). *A New Economic Geography of Bangladesh*. Vikas Publishing House PVT Limited, Bombay.

Barth, M.D. & Titus, J.G. (1984). *Greenhouse Effect and Sea Level Rise*. Van Nostrand Reinhold Company, New York.

Bird, Eric C.F. & Schwartz, M.L. (1985). *The World's Coastline*. Van Nostrand Reinhold Company, Stroudsburg.

Broadus, J.M., Milliman, J., Edwards, S., Aubrey, D. & Gable, F. (1986). Rising sea level and damming of rivers: possible effects in Egypt and Bangladesh. In *Effects of Changes in Stratospheric Ozone and Global Climate, Vol. 4: Sea Level Rise*, J.G. Titus (ed.). UNEP/EPA, pp. 165–89.

El Sayed, M.Kh. (1991). Implications of relative sea level rise on Alexandria. In *Impact of Sea Level Rise on Cities and Regions*, R. Frassetto (ed.). Marsilio Editoria S.p.a., Venice.

IPCC (Intergovernmental Panel on Climate Change) (1990). *Climate Change: The IPCC Scientific Assessment*. Final report of Working Group I.

Mahtab, F.U. (1989). *Effect of Climate Change and Sea-level Rise on Bangladesh*. Expert Group on Climate Change and Sea-level Rise. Commonwealth Secretariat, London.

McFadden, D. (1984). Welfare analysis of incomplete adjustment to climatic change. In *Advances in Applied Micro-Economics, Vol. 3*, V.K. Smith and A.D. Witte (eds.). pp. 133–49.

Milliman, J.D. (1987). Personal communication. Senior Scientist, Geology & Geophysics Department, Woods Hole Oceanographic Institution.

Milliman, J.D., Broadus, J.M. and Gable, F. (1989). Environmental and economic impacts of rising sea level and subsiding deltas: the Nile and Bengal examples. *Ambio*, **18**(6), 340–5.

Milliman, J.D. & Meade, R. (1983). World-wide delivery of river sediment to the oceans. *J. Geol.*, **91**, 1–21.

Schelling, T.C. (1983). Climatic change: implications for welfare and policy. In *Changing Climate: Report of the Carbon Dioxide Assessment Committee*. National Academy Press, Washington, DC, pp. 449–82.

Yohe, G. (1990). The cost of not holding back the Sea: toward a national sample of economic vulnerability. *Coastal Management*, **18**, pp. 403–431.

Zaman, Munir, (ed.). (1983). River basin development. In *Proceedings of the National Symposium on River Basin Development*. Tycooly International, Dublin.

17

Impacts of sea level rise on coastal systems with special emphasis on the Mississippi River deltaic plain

J.W. Day, W.H. Conner, R. Costanza, G.P. Kemp and I.A. Mendelssohn

Abstract

Numerous reports have recently emphasized the potential impact of global warming on future sea level rise in coastal areas. It is feared that many low-lying coastal areas, especially those dominated by wetlands, will be flooded as sea level rises over the next century. Thus, there is a need to consider critically the impacts of sea level rise and possible policy responses to these impacts. One area that can provide valuable information on both impacts and responses is the Mississippi delta. This area has been experiencing an 'apparent' sea level rise of about a metre per century for a very long time as a result of regional subsidence. This paper discusses the effect of this rise on coastal ecosystems in Louisiana and the response of public and private groups to the problem.

Introduction

There has been considerable speculation concerning the potential impacts of sea level rise on coastal ecosystems. One way to gain a better understanding of such impacts is by studying the response of coastal systems in areas where there has been an apparent sea level rise (ASLR) due to factors such as regional subsidence, subsurface fluid withdrawal or peat oxidation, as well as eustatic sea level rise. The Mississippi River deltaic region is a particularly good example of such an area, and provides a case study focus for this chapter in examining both natural and human responses to sea level rise.

As discussed in this chapter, sediment input is vital to maintaining the rate of vertical accretion in the Mississippi delta and other wetland areas. Areas with substantial sediment input can keep pace with a considerable rise in sea level and still maintain coastal wetlands. When sediment input to wetlands is lowered, either naturally or because of human actions, the rate of vertical accretion may not match the rate of sea level rise, resulting in submergence and salt intrusion.

Vegetation can also play a critical role in maintaining vertical accretion rates. Biomass production (roots and above-ground material) often contributes more to vertical accretion than mineral sediment input. For example, studies in The Netherlands have shown that vertical accretion increases by a factor of two to three when vegetation becomes established (Bouwsema *et al.*, 1986). For this reason, the effects of waterlogging and salt stress on plants are also discussed in this chapter.

It is recommended that a world-wide network should be established for monitoring impacts. Since the rate of sea level rise is low in many areas of the world, there may be a lag of decades before the effects of an acceleration in rise become apparent. Long-term monitoring is thus required for understanding and anticipating the changes in natural systems. Locations where monitoring could be initiated include the eastern shore of the Chesapeake Bay, the Gulf of Mexico (the Mississippi River delta, the Grijalva–Usumacinta delta, and the Everglades), the Mediterranean (the Ebro, Camargue, Po and Nile deltas), the Black Sea area, the Danube delta, and the Ganges–Brahmaputra delta.

From the standpoint of institutional response to sea level rise, the following recommendations for coastal management should be considered:

1. *Integrate information on long-term changes into the decision making process.* Because the rate of eustatic sea level rise is relatively low at present, the need for such planning may not be immediately obvious.
2. *Avoid short-term adjustments that aggravate the long-term problem.* Since responses of natural systems will be slow initially, approaches to rectify the short-term problems may, in the long term, make the situation worse. An example of this is impoundment and the use of gravity drainage to remove water: with rising water levels, gravity drainage has a finite lifetime.
3. *Avoid flood protection schemes that eliminate freshwater and sediment input to coastal areas.*
4. *Anticipate problems arising from land ownership which may complicate efforts to combat sea level rise.*
5. *Consider alternative agricultural practices in low-lying areas.* Dykes, spoil banks and drainage exacerbate the problem in the long term. A rotating system of crops and flooding might maintain accretion and help fertilize fields.
6. *Make maximum use of natural energies such as winds, river currents, and tides.*
7. *In general, encourage coordinated, long-term planning.*

Impacts of sea level rise

In most coastal areas, submergence and salinity increase are the two major direct impacts of sea level rise on coastal systems. For low-lying coastal areas, submergence will bring about poorer drainage and increased waterlogging of soils leading to lowered plant productivity and vegetation death. Submergence will generally be followed by an increase in salinity as saltwater intrudes into formerly fresher areas. These changes occur slowly and the effects of both submergence and salinity increase are incremental. These impacts will be most pronounced in large, near sea level areas such as deltas and lagoon-like systems. Because these areas are very important for fisheries and agriculture, the impact on local and possibly regional economies may be pronounced. Human actions, such as dyke and canal construction, can increase the rates of submergence and salinization.

Coastal wetland loss

Probably the dominant variable controlling long-term water tables, and thus vegetation patterns close to the coast, is sea level change (Olsvig *et al.*, 1979). In the United States, studies of sea level rise and the potential impact on wetland areas have been

Table 17.1. *Summary of mean rates (mm/yr) reported for accretion and apparent sea level rise (ASLR) (1940–80) for tidal marshes in the United States (adapted from Stevenson et al., 1986)*

Location	Accretion	ASLR
Barnstaple, MA	5.5	0.9
Prudence Island, RI	4.3	1.9
Farm River, CN	5.0	1.9
Fresh Pond, NY	4.3	2.2
Flax Pond, NY	5.5	2.2
Lewes, DE	4.7	2.0
Lewes, DE	> 10.0	2.0
Nanticoke, MD	6.1	3.2
Blackwater, MD	2.6	3.9
North River, NC	3.0	1.9
North Inlet, SC	2.5	2.2
Savannah River, GA	11.0	2.5
Sapelo Island, GA	4.0	2.5
Barataria, LA	7.2	9.5
Fourleague Bay, LA	6.6	8.5
Lake Calcasieu, LA	7.8	9.5

conducted in New York (Clark, 1986), Maryland (Stevenson *et al.*, 1985, 1986), North Carolina (Hackney and Cleary, 1987), South Carolina (Kana *et al.*, 1986), and Louisiana (DeLaune *et al.*, 1983; Baumann *et al.*, 1984; Conner *et al.*, 1986; Salinas *et al.*, 1986). (See Table 17.1 for various estimates of sea level rise and accretion rates.).

Coastal habitats respond to water level rise in various ways. In New York, for example, coastal forest stands have been able to maintain themselves in the face of a 3 mm/yr rise by colonizing new land as it becomes suitable for tree establishment (Hicks, 1972; Clark, 1986). In comparison, along the Atlantic coast many coastal marshes are not capable of surviving rising water levels (Kearney and Stevenson, 1985; Hackney and Cleary, 1987). The consequence of increasing marsh submergence is a decline in marsh vigour leading to the eventual formation of interior ponds that enlarge through time. In the Mississippi deltaic plain where ASLR is the highest in the US, wetland loss rates of as high as 100 km²/yr have been reported (Templet and Meyer-Arendt, 1988).

Waterlogging

When soils become waterlogged, the internal pore spaces, which are normally occupied with air, fill with water. This results in various stresses (oxygen-deficit stress, ion stress, and water-deficit stress) to plants (Levitt, 1980). Since diffusion of oxygen in water is about four orders of magnitude slower than in air (Gambrell and Patrick, 1978) continued oxygen demand of soil organisms and plant roots can rapidly deplete the oxygen content of a waterlogged soil (Turner and Patrick, 1968). Further activity of facultative and obligate anaerobic bacteria can cause the soil to become increasingly reduced.

Anoxic soil conditions can induce root oxygen deficiencies which may directly reduce plant growth (Drew, 1983; Crawford, 1978; Webb and Armstrong, 1983). The roots of many flood-tolerant plants can survive root hypoxia when flooded because they possess a network of cortical gas spaces (aerenchyma tissue) which allows the movement of oxygen from the atmosphere through the aerial tissue into the roots (Conway, 1977; Dacey, 1980; Gleason and Zieman, 1981). This observation has led to the conclusion that aerenchyma formation is a primary adaptation of plants to flooding (Armstrong, 1979). However, aerobic respiration even in aerenchymatous plants may become inhibited by a root oxygen deficiency when the rate of downward oxygen flux lags behind its consumption by respiration and by diffusive loss to the rhizosphere (Vartapetian, 1978; Saglio *et al.*, 1984). Armstrong and Gaynard (1976) speculate that the onset of respiratory inhibition in rice roots is associated with the appearance of anaerobic centres in the low porosity tissue of the apical meristem and stele.

Some flood-tolerant plants adapt physiologically and compensate for the absence of available oxygen and the resultant decrease in adenine triphosphate (ATP) production by accelerating the rate of alcoholic fermentation (John and Greenway, 1976; Crawford, 1978). Although this mechanism enables roots to maintain high rates of ATP production (Hochachka and Somero, 1973), it may lead to carbon starvation because much more glucose is used than in aerobic respiration. This can inhibit growth by affecting nitrogen metabolism (Givan, 1979) and cause irreversible damage to mitochondria (Vartapetian, 1978). Ethanol, a product of alcoholic fermentation, may be phytotoxic in some species (Crawford, 1976). Diffusion of ethanol to the rhizosphere may lessen this threat by preventing accumulation of ethanol in root tissue (Bertani *et al.*, 1980), but loss of ethanol may exacerbate the carbon deficit (John and Greenway, 1976).

Environmentally induced anaerobiosis is not uncommon in root tissues of salt-marsh plants (Gleason and Zieman, 1981). Accelerated rates of alcoholic fermentation and ATP production in the roots of *Spartina alterniflora* were found for plants located in highly reduced substrates of the salt-marsh (Mendelssohn *et al.*, 1981). The relatively low adenylate energy charge (AEC) ratio, a measure of plant energy status, of *Spartina alterniflora* leaves at these sites (Mendelssohn and McKee, 1981) suggests that the species experiences greater stress there than in better-aerated sites, possibly as a result of internal oxygen deprivation (Mendelssohn *et al.*, 1981) as well as toxicity from sulphide which accumulates in highly anaerobic saline marsh soils (Mendelssohn and McKee, 1988).

In forested wetlands, waterlogging can be deleterious to individual tree species. Unfortunately, physiological studies related to flood response in wetland species are limited in number. Among the known effects of flooding are early stomatal closure and the reduction in photosynthesis, which in turn lead to inhibition of growth (Kramer and Kozlowski, 1979; Kozlowski and Pallardy, 1979; Zaerr, 1983; Pezeshki and Chambers, 1985a and b). There is also some evidence for stomatal adaptation to flooding in some species, with stomatal reopening after a critical period of flooding (Regehr *et al.*, 1975; Pereira and Kozlowski, 1977; Kozlowski and Pallardy, 1979). The extent to which stomatal functioning and photosynthesis are related to flooding, and how they vary with the depth of flood waters and the developmental stage of the tree, are important in determining flood tolerance, seasonal growth responses and tree survival in flood prone environments.

The effect of flooding on wetland forests depends mainly on the season flooded, the depth of flooding, and the duration of the flood event (Teskey and Hinckley, 1977; Whitlow and Harris, 1979; McKnight et al., 1981; Kozlowski, 1982). Flooding during the dormant season seems to have no effect on tree growth (Broadfoot, 1967; Broadfoot and Williston, 1973), while seedlings of several species are very susceptible to flood damage after leaf emergence (Broadfoot and Williston, 1973). Short-term flooding events during the growing season seem to have very little effect on mature trees. Mitsch and Rust (1984) found very little correlation between growth of water-tolerant trees and flooding. However, they did find that years with a high percentage of flooding during the growing season were years of low tree growth. Johnson and Bell (1976), in a similar study, found no relationship between frequency of flooding and growth of trees with diameters exceeding 4 cm. Flooding extending into the growing season or for extended periods can have serious effects on the survival of bottomland trees (Bell and Johnson, 1974). Even though flooding over a single growing year may have no visible effect, only a few species have been reported to survive three years of continuous flooding (Green, 1947) and most species cannot survive two years of continuous flooding (Broadfoot and Williston, 1973). Hall and Smith (1955) reported that in Tennessee none of the 39 common deciduous tree species could survive flooding if the root system was covered for more than 54% of the growing season during an eight-year period.

Unfortunately, most of the investigations involving flooding impacts have come from studies of forests inundated after the construction of reservoirs or other flood-control structures (Hall and Smith, 1955; Bell and Johnson, 1974; Hall et al., 1946; Silker, 1948; Broadfoot and Williston, 1973; Green, 1947; Harms et al., 1980) in which existing stands of trees have been subjected to a sudden shock of continuous inundation. According to McKnight et al. (1981) this may not always indicate how a species will perform if it has to start as a seed and progress through the various stages of its life cycle under the influence of intermittent flooding. Even mature cypress and tupelo which do well under flooded conditions (Dickson et al., 1972; Kennedy, 1970) suffer if the mean depth of flooding exceeds 60 cm (Brown and Lugo, 1982). In Florida, Harms et al. (1980) found that in water from 20 cm to 100 cm deep, 0–16% of the cypress trees died in seven years. In water over 120 cm deep, 50% of the cypress died after four years. In Louisiana, a long-term study of cypress survival was conducted near Lake Chicot (Penfound, 1949; Eggler and Moore, 1961). After four years of flooding with water 60–300 cm deep, 97% of the cypress were still alive. Eighteen years after flooding, 50% of the cypress were still alive. However, most of the living trees in the deep water had dead tops (Eggler and Moore, 1961). From the available data on flooding stress and cypress survival, it appears that cypress can adapt to shallow (< 60 cm), permanent flooding, and even in deep water (> 60 cm), death and decline is a gradual process (Harms et al., 1980: Eggler and Moore, 1961).

Salt stress

Plants are susceptible to many different kinds of salt stress injury. The three major factors affecting plant growth are water stress, ion toxicity, and induced nutrient deficiency (Jones, 1981). There is, however, no consensus as to the relative importance of each of these stresses, and it is difficult to compare reported limits of stress because there is no standard method of measuring salt stress (Levitt, 1980). Plants range from highly tolerant to extremely sensitive to saline conditions, but species tolerance will vary depending upon environmental conditions and plant developmental stage (Levitt, 1980).

In coastal areas, sea level rise and the construction of canals allows the penetration of saline waters into brackish and freshwater areas that normally are not subjected to such salinity variations. Brinson *et al.* (1985) reported that salt intrusions into forested wetland areas of North Carolina created acute water stress in saline-intolerant tree species. In marsh areas, it is often difficult to place the blame for marsh deterioration on salinity increases alone. Mendelssohn *et al.* (1981) and Mendelssohn and McKee (1988) have studied areas of *S. alterniflora* dieback and found that root oxygen deficiencies rather than salinity was the cause of the problem. However, the combination of waterlogging and salinity increase in lower salinity wetlands can lead to lowered productivity (Mendelssohn and Burdick, 1988, Pezeshki *et al.*, 1989).

Vertical accretion potential of coastal habitats

The information in the preceding section indicates that rising water level is leading to loss of coastal wetlands in a number of areas. However, it is also known that certain elements of coastal landscapes such as barrier islands and wetlands have persisted for thousands of years in the face of rising sea level. Some coastal marshes can accrete quite rapidly. Even in parts of Louisiana where ASLR exceeds 1 cm/yr, some stream-side marshes can maintain their elevation; this is due to both higher mineral sediment deposition and marsh productivity than in inland marshes (DeLaune *et al.*, 1983), suggesting that mineral sediment input is a key process in the maintenance of marshes. Mangrove systems can also maintain their elevation in the face of rather high apparent water level rises, as demonstrated by the mangroves in the state of Tabasco, Mexico (Thom, 1969; Psuty, 1966).

The range of potential vertical accretion rates is suggested by Table 17.1 which compares observed accretion and ASLR rates for tidal marshes at various locations along the Atlantic and Gulf coasts of the United States. At Lewes, Delaware and Savannah River, Georgia vertical accretion exceeds 10 mm/yr, far outstripping the rate of ASLR. In contrast, the accretion rate for North Inlet, South Carolina is 2.5 mm/yr and is barely able to keep up with the local ASLR. The accretion rates for the three Louisiana tidal marshes are high, but face ASLR rates that are about 2 mm/yr higher.

What are the factors that affect these rates and thus cause gain or loss of coastal habitats? What are the lessons for coastal management? The case of the Mississippi delta can provide some insight to these questions.

The Mississippi delta as a case study

The Mississippi delta is an ideal area for considering the effects of sea level rise. The large delta consists of lakes, bays, near sea level wetlands and low-lying uplands. Water levels have increased substantially; this has resulted primarily from regional land subsidence rather than eustatic sea level rise (the impacts on natural and social systems are similar regardless of the cause). The particular characteristics which recommend the Mississippi delta as a case study are (Day and Templet, 1989): (1) the area has experienced a high ASLR over thousands of years; (2) in spite of this high ASLR, a large deltaic plain of several million hectares has formed over the last several thousand years; and (3) in recent decades, the historical trend of net land gain has been reversed and there is now a large net loss of land.

Apparent sea level rise (ASLR)

Apparent sea level rise currently averages greater than 1.0 cm/yr in the Mississippi deltaic plain (Swanson and Thurlow, 1973; Baumann and DeLaune, 1982; Hatton *et al.*, 1983; Salinas *et al.*, 1986). Most of this ASLR is due to regional subsidence (Gosselink, 1984). Deep layers of sedimentary rocks thousands of metres thick indicate that this process has been going on for millions of years. This is a regional phenomenon due to such factors as crustal downwarping, sediment compaction and dewatering. During this century the rate of ASLR has increased due to withdrawals of water, oil and gas, but still regional subsidence remains the dominant cause of ASLR. Eustatic sea level rise accounts for only 10–15% of the ASLR (Baumann, 1987).

Long-term growth of the delta

Global eustatic sea level stabilized around its present level about 5,000–7,000 years ago after being about 100 m lower during the last glaciation. Subsequently, there was a rapid net growth of the delta due to sedimentation from the Mississippi River, despite a continuing regional ASLR of about one metre per century. The average growth rate over the past 5,000 years was somewhat greater than 4 km²/yr. Land was formed by a series of shifting delta lobes (Kolb and van Lopik, 1966; Coleman and Gagliano, 1964; Frazier, 1967). There were high growth rates in active delta lobes and deterioration in abandoned lobes, but the overall net growth was positive.

Today, the large deltaic plain consists of about 5×10^6 ha (50,000 km²) of which over 2×10^6 ha (20,000 km²) are wetlands (Gould, 1970). The large majority of these wetlands are within one metre of sea level. The rest of the deltaic plain is made up of shallow aquatic systems and low (up to 5–6 m) upland ridges.

Recent high land loss rates and causes of wetland deterioration in the Mississippi delta

Over the past several decades, the long-term net gain of land has been reversed and there is now an accelerating land loss. Rates of wetland loss as high as 100 km²/yr have been reported (Gagliano *et al.*, 1981; Gosselink *et al.*, 1979; Wicker, 1980). This dramatic reversal is the result of a number of factors, all of which relate to the change in vertical accretion rates in the face of rising water levels. This phenomenon has been studied in detail over the past decade (Boesch *et al.*, 1983) and is not unique to the Mississippi delta. It is happening to a lesser degree on the eastern shore of Chesapeake Bay in Maryland and other locations (see Table 17.1, adapted from Stevenson *et al.*, 1986).

One of the main causes of the wetland deterioration is the lack of sediment input. There are two sources of allochthonous sediments which accrete on the surface of wetlands in the Mississippi delta. One source is direct input of riverine sediments during the annual spring flood. This source has been eliminated for most of the coastal zone as a result of levee construction along the Mississippi River which extends almost to its mouth (Day and Templet, 1989; Louisiana Wetland Protection Panel, 1987). In addition, a number of distributaries which carried lesser amounts of sediments to coastal wetlands have been dammed. In contrast, Atchafalaya Bay is one area of the Louisiana coast where there is still significant sediment input (van Heerden and Roberts, 1980a,b; Roberts *et al.*, 1980; van Heerden *et al.*, 1983; Wells and Kemp, 1981; Adams *et al.*, 1982) and the loss of existing wetlands has been slowed or reversed (Baumann and Adams, 1981; Johnson *et al.*, 1985; Rejmanek *et al.*, 1987).

A second source of sediments is the deposition of resuspended bay-bottom sediments (Baumann *et al.*, 1984). These sediments are resuspended by winds and deposited on the surface of wetlands. These resuspended materials are the major source of new sediments for most of the coastal zone. Resuspended sediments alone do not appear sufficient to maintain surface elevation against ASLR, but they greatly slow wetland loss. Along the Wadden Sea, the Dutch have used brush fencing baffles to encourage settling of resuspended sediments and to prevent resuspension in order to create wetlands (Bouwsema *et al.*, 1986).

In the Mississippi delta, the input of resuspended sediments has been greatly reduced by canal construction. During the past 30–50 years, more than 15,000 km of canals have been constructed for navigation, drainage, and oil and gas exploration and production (Scaife *et al.*, 1983). This network of canals has changed the regional hydrology of the coastal zone. Canals are generally straight, and along one or both sides there is a low spoil bank resulting from the deposition of excavated material. Since spoil banks are generally higher than mean high water, flow across the wetland surface is greatly altered and the input of resuspended sediments is reduced. Studies have shown that canals and associated spoil banks alter hydrology, reduce productivity of wetlands, cause saltwater intrusion and promote the deterioration of natural channels. A number of studies have correlated canal density to wetlands loss (Craig *et al.*, 1980; Swenson and Turner, 1987; Scaife *et al.*, 1983; Deegan *et al.*, 1984).

The dykes along the Mississippi River have stopped most of the direct freshwater input into the coastal zone. This has resulted in more rapid movement of saltwater into fresher areas. The intrusion of saltwater into abandoned delta lobes is a natural process, but dyking and canalization have greatly accelerated it. As described above, saltwater can kill or reduce the productivity of freshwater vegetation directly, causing loss of wetlands (also see Pezeshki *et al.*, 1987; McKee and Mendelssohn, 1989).

The institutional response in Louisiana

Coastal and wetland management in Louisiana is very complex. Federal, state, and local public agencies, scientists, and private groups (landowners, sportsmen, commercial fishermen and trappers, conservationists) are involved in the process. In addition, management interests vary widely (marsh management, aquaculture, waterfowl enhancement, fur mammal management, timber production) and only recently has management been focused on the problem of land loss and ASLR (Tripp, 1979; Houlk, 1983).

Based on past experience, it appears that many management actions are ill-suited for preventing, and may even exacerbate, the long-term adverse effects of sea level rise. The conventional approaches to management, however, are popular because they produce at least some positive results and/or are not immediately obviously harmful in the short term (10–20 years). Thus, the negative effects of inappropriate management practices (in terms of combating the long-term sea level rise problem) are cumulative.

Flood control

Flood control in the Mississippi delta area must contend with flooding from the river and from the sea (during tropical storms). To protect developed areas, the Mississippi River has been almost completely 'walled in' by dykes. Early flood control along the river was accomplished by ring (encircling) dykes; the city of New Orleans is still protected in this way. Despite this, the city occasionally experiences severe flooding due to extremely heavy short-term rainfall which occurs along the Louisiana coast (the record 24-hour rainfall in South Louisiana is of the order of 625 mm and 24-hour rains of about 250 mm occur almost annually). Dykes prevent sediments and freshwater from entering into much of the coastal zone, resulting in the problems discussed earlier. Such flood control measures have also stimulated agricultural and urban development in low-lying areas. What is obviously needed now is the reintroduction of freshwater and sediments into the coastal areas. However, development makes this difficult. Moreover, the widespread network of canals (with attendant spoil banks) retards the easy movement of sediment-laden water over wetlands. In the context of sea level rise, a clear lesson from the Mississippi delta is that flood protection should avoid practices that inhibit the introduction of freshwater and sediments to coastal areas. Similar problems of flood control limiting sediment input to deltas have been shown for the Nile (Sestini, 1991b) and Po deltas (Sestini, 1991a).

Because of continuing subsidence and ASLR, many low-lying areas are experiencing increased flooding. Gravity drainage systems are still being proposed that involve the construction of many more kilometres of canals, which in the long term will compound the problems of land loss (and water quality). For example, the State of Louisiana recently approved a gravity drainage system despite the fact that evidence was presented that water levels in the area could rise by more than a metre (due to the combined effects of subsidence and eustatic sea level rise) during the 50-year life of the project. Institutional inertia and overreliance on approaches used in the past discourage adoption of alternative practices.

Impoundment and semi-impoundment

Wetland impoundment (encirclement by dykes for some kinds of water management) has been carried out for over 100 years in the Mississippi delta area (Day *et al.*, 1990). During the second half of the 19th century and the early part of this century, extensive areas of wetland were reclaimed for agriculture. Most of these failed due to a combination of subsidence (due to regional sinking and oxidation of peat) and heavy rains. Most are visible today as large rectangular ponds in the coastal marshes (Okey, 1918; Turner and Neill, 1983).

More recently, semi-impoundment has become common as a technique for wetland management (Cowan *et al.*, 1986; Day *et al.*, 1989; Cahoon and Groat, 1990). In this approach, wetland areas are surrounded by low dykes and water movement is controlled by structures such as weirs, sluices and gates. Semi-impoundment has been practised since the 1950s to encourage the growth of grasses which are beneficial for waterfowl. Recently, semi-impoundment has become much more widespread as a form of marsh management designed specifically to address the problem of land loss. One of the main objectives is to control saltwater intrusion, although semi-impoundments have also been proposed in areas of salt-marsh vegetation where salinity is not clearly a problem. One concern with this type of marsh management is that the introduction of suspended sediments is reduced. Any future plans to introduce riverine sediments (which surely must come) will be hampered by the dykes surrounding these management areas. Even though serious questions have been raised about the effectiveness of this kind of marsh management, there are plans for semi-impoundment projects involving perhaps one-third to one-half of Louisiana's coastal wetlands. Recently a study of marsh management has been completed which provides information on the functioning of these systems (Cahoon and Groat, 1990). Sediment flux between the impounded areas and the surrounding estuary was significantly reduced compared to the control area and vertical accretion was 5–10 times lower in the impounded marshes. Access of migratory fish species was also reduced by impoundment. Vegetation productivity was higher than the control in one area and lower than the control in a second area. These data suggest that the use of various types of impoundment schemes for offsetting wetland loss due to rising water levels should be very carefully considered.

Land ownership

Land ownership may complicate efforts to deal with the effects of sea level rise (Day and Templet, 1989). Ecosystems, especially wetlands, are most effectively managed by natural landscape units such as the drainage basin. Politically and legally, however, it is much easier to implement management plans based on landscape units defined by ownership than by natural boundaries. Different landowners will likely have different objectives in the management of their lands. Private landowners often give preference to resources (such as furbearers or waterfowl) which maximize economic return rather than those which contribute to the elusive public good. All of this points to the need to include the land ownership issue explicitly in comprehensive planning. In areas where there is private ownership, landowners must be involved in comprehensive planning.

Impoundment as a response to sea level rise

It is likely that faced with the prospect of rising water level, impoundment schemes will be proposed to save wetlands. To determine the effectiveness of impoundments in dealing with rising sea levels, a survey of impoundments at 27 locations around the world was recently conducted (Day *et al.*, 1989). A number of general observations and conclusions emerged from this survey, as follows.

1. Comprehensive region-wide planning and management are necessary to ensure project success. For example, unplanned and uncontrolled wetland impoundment activities to exclude saltwater from one area have led to the removal of freshwater catchments causing higher salinities elsewhere, as reported in the

Northern Friesian Islands in Germany, the Philippines and South Florida. Cause and effect may be spatially and temporally separated, as in Indonesia where the uncontrolled spread of coastal fish ponds and rice culture has led to flooding, erosion and severe water quality problems both in inland and coastal waters. Similarly, in Louisiana, accidental impoundments created by uncontrolled canal construction have accelerated land loss, water quality deterioration, and salt-water intrusion.

2. Private costs generally end up as public costs. Impoundment activities often start out small, at private expense. However, as the overall scale of activities increases, maintenance and pumping costs tend to escalate beyond the capacities of private entrepreneurs. Problems arising from ill-defined property rights affect most wetland impoundment projects world-wide and often lead to inefficient resource management, as in Thailand, Bangladesh, and Louisiana. Abandonment is one common outcome as conditions change over time (wetland agriculture in Louisiana, Surinam and South Carolina). But unlike economic costs, environmental costs do not cease at this point. Furthermore, environmental impacts (which tend to be public costs) often materialize at a later time and require public funds. Some of the most striking examples of this classic evolutionary pattern are to be found in The Netherlands, in the Fenlands of England, and in the Sacramento–San Joaquin delta and Everglades regions of the US.

3. Environmental damages associated with wetland impoundment tend to increase over time, while benefits remain stable or decrease. Loss of habitat, decline in fisheries and water quality, and increased flooding are frequent adverse consequences associated with wetland modification by impoundment. This has led to a growing realization that wetlands are valuable and that their destruction or impairment leads not only to environmental deterioration, but also to economic costs. The erosion of coastlines in the Malay Archipelago as a consequence of the destruction of mangroves for aquaculture is one example. Many environmental impacts are cumulative and secondary, particularly in wetland systems whose extraordinary productivity is dependent upon maintaining hydrologic connections between the land and the sea.

4. Subsidence, whether due to oxidation of peat or to other causes, directly affects maintenance costs and therefore the potential for success in all impoundment projects. It is also clear that impoundment systems require high levels of maintenance. Subsidence tends to make impoundments permanent landscape features which become increasingly difficult to sustain as gravity drainage becomes less and less feasible. This has been observed in the agricultural and urban impoundments of Louisiana, the San Joaquin delta, Bangladesh and numerous other places.

5. Systems which are 'ecologically engineered' to use natural 'free' energies are less expensive to maintain and have a higher potential for success than those with a built-in dependence on fossil fuel. These natural energies include those associated with winds, tides and river flow. Much effort has been expended to gain acceptance of this idea, particularly in the Third World (e.g., Bangladesh and Surinam). Tidal and wind-driven currents and turbulence are being harnessed to suspend and move sediments which can be captured for beneficial use.

6. Intensity of use must balance drainage intensity. Impoundment projects that

require a sophisticated forced drainage system must provide high benefits to a large number of people if they are to be economically feasible. Where subsidence is a complicating factor, as it usually is, planners must anticipate and plan for an escalation in drainage intensity.

7. The surface of the land can be built up rapidly using riverine or resuspended sediments. Typically, these sediments have a low organic content which makes them less prone to compaction and oxidation, as shown in Louisiana, Bangladesh, The Wash in England, and Friesland in The Netherlands.

8. Reclamation and marsh management involving semi-impoundment often employ the same basic technologies, as is the case in the conversion of rice fields to waterfowl refuges in South Carolina. Once an impoundment has been constructed, priorities guiding the water level control programme implemented by the wetland manager determine the uses and habitats which will be enhanced within any given impoundment. Creation of an impoundment therefore places a new responsibility on regulatory authorities to determine which management approaches will be permitted, and to make sure that regional guidelines are followed. It was determined in South Carolina and Florida that restriction of new impoundments was the best way to avoid this difficult regulatory dilemma.

Management implications for dealing with sea level rise

The information reviewed above suggests a number of implications for dealing with problems arising from sea level rise. Although there are a number of reports that wetlands are disappearing due to sea level rise, it is also clear that wetlands can act as a buffer against rising water levels, as argued above. The case of the Mississippi delta clearly demonstrates that human activity can have very deleterious impacts by interrupting water and sediment movements, causing a high loss of wetlands. If a number of diversion outlets had been included in the lower Mississippi flood-control system, then the wetland loss problem would be much less severe. Moreover, rising water levels will likely lead to more, not less, intensive flood-control efforts (including dykes and impoundments) in coastal areas, which, unless corrective measures are taken, will compound the problems of sediment loss. This suggests that proper sediment management needs to be a central part of any plan to counteract the effects of rising sea level on coastal wetlands.

There are several other considerations which bear on the problem of management for sea level rise. The level of technology is one and should be carefully analysed. For example, the Dutch have developed an advanced technology for protecting their country from the sea. It is, however, very expensive and a significant portion of the national income is used for this purpose. For other countries, careful consideration should be given to lower technology approaches and the principle of 'ecological engineering' discussed earlier. If sea level is rising, then an alternative to protecting the area with dykes to keep out rising water levels is to build up the surface of the land. The use of riverine and tidal currents to move the necessary sediments is a low energy approach. In some cases, it may be feasible to mobilize sediments trapped behind reservoirs for use at the coast. The Nile delta is a place where the surface of the land needs to be built up as present rates of subsidence indicate that over one-half of the delta will likely disappear within the next 100 years (Stanley, 1988). However, the

amount of suspended sediment carried by the Nile has been reduced by over 90%
(Sestini, 1991b) and the large distance to the Aswan dam probably precludes sediment
remobilization. The Nile delta is an example of how human impacts can interact with
rising levels to produce potentially disastrous results. In other areas, it is probably still
possible to use riverine sediments. Since sea level rise will be gradual, a deposition
management plan could take place over many years.

Finally, it should be emphasized that dealing with the problem of sea level rise
demands an active approach to management rather than a passive one. Traditionally,
protection of natural habitats has often been a passive activity; that is, parks and
reserves have been created and human intervention has been severely limited. In the
case of sea level rise in which base conditions are changing, an active management
strategy is imperative. Action must be taken or many coastal ecosystems will disappear.
Anticipating the response of ecosystems to sea level rise is therefore an important
element in developing management strategies, a subject to which we now turn.

One approach to determining the response of ecosystems to sea level rise

The interaction among species and habitat level responses will lead to changes in
ecosystems. System-wide changes in hydrology, including such components as salinity
and turbidity, can be predicted in the short term using hydrodynamic models. But to
look at the integrated long-term effects on coastal ecosystems of sea level rise, requires
spatial ecosystem modelling (Sklar *et al.*, 1985, Costanza *et al.*, 1986, 1990). This
approach can simulate past behaviour and predict future conditions in coastal areas as
a function of various management alternatives and changes in forcing functions (such
as increased rates of sea level rise), both individually and in any combination. Models
can simulate both the dynamic and spatial behaviour of the system, and keep track of
the important variables in the system, such as habitat type, water level and flow,
sediment levels and sedimentation, subsidence, salinity, primary production, nutrient
levels, and elevation. Ecosystem models of this type are large and expensive to
implement, but once implemented can be used to quickly and inexpensively evaluate
various alternatives.

A considerable effort has been made to design and implement this type of model for a
coastal region in Louisiana. This model has been used to evaluate several current
management alternatives for this area. The Louisiana model consists of 2,479 intercon-
nected cells, each representing 1 km^2. Each cell contains a dynamic simulation model,
and each cell is connected to each adjacent cell by the exchange of water and suspended
materials. The volume of water crossing from cell to cell is controlled by habitat type,
drainage density, waterway orientation and levee heights. The build-up of land or the
development of open water in a cell depends on the balance between net inputs of
sediments and organic peats and outputs due to erosion and subsidence. The balance of
inputs and outputs is critical, and is important for predicting how marsh succession and
productivity is affected by natural and human activities.

Forcing functions (inputs) are specified in the form of time series over the simulation
period. Weekly values of Atchafalaya and Mississippi River discharges, Gulf of
Mexico salinity, river sediments and nutrients, rainfall, sea level, runoff, temperature
and winds are supplied to the simulation with each iteration. The location and

characteristics of the major waterways and levees are also supplied as input to the simulation. Water can exchange with adjacent cells via canals, natural bayous and overland flow or it may be prevented from exchanging with adjacent cells by the presence of levees. The overall water flow connectivity parameter is adjusted during the model run to reflect the presence and size of waterways or levees at the cell boundaries. If a waterway is present at a cell boundary a large value is used, increasing with the size of the waterway. If a levee is present, a value of 0.0 is used until water level exceeds the height of the levee. The model's canal and levee network is updated each year during a simulation run. Dredged canals and levees are added to the model's hydrologic structure at the beginning of the year they were built.

Each cell in the model is potentially connected to each adjacent cell by the exchange of water and suspended materials. Before this exchange takes place however, the ecological and physical dynamics within a cell are calculated. The volume of water crossing from one cell to another carries a specified sediment load. This sediment is deposited, resuspended, lost due to subsidence, and carried to the next cell. The amount of sediment in each 'tank' is a function of the habitat type. Included in the model is the fact that plants and nutrients within each cell will also influence these exchanges and flows.

Habitat succession occurs in the model (after a time lag) when the state variables within a cell become more like another habitat type due to the changing conditions. The biotic components in a cell (primary production) respond to the abiotic changes such as changing flooding regimes, nutrient levels, turbidity, and elevation.

The model can produce a high amount of output, the most useful of which is a contour map for each state variable as well as habitat maps for each week of the simulation. To produce these maps the model must solve over 17,000 simultaneous difference equations and generate over one million simulated data points for each year of simulation. With present-day computer models such large numbers of computations are feasible, and as computers continue to improve in speed and convenience, this type of modelling should become more practicable.

Model-predicted changes in habitat due to past and future sea level rise are presented in Fig. 17.1. The model predicts the gradual intrusion of salt into the system from the southeastern part of the study area with the concurrent freshening in the northwestern sector. It also illustrates a loss of elevation in the north and an increase in elevation in the south. Both of these trends are indicative of river water and sediments moving further south in recent times plus a lack of connectivity with the more northern fresh marsh areas. Predicted water volume and suspended sediments behaved in a similar way and are generally consistent with what is known about the historical behaviour of the system. These physical changes, in turn, have an impact upon the biology of the area. The relationship between plants and elevation of the marsh results in a feedback loop that enhances the rate of land loss as suspended sediments are diverted from an area of marsh. The model accurately predicted changes in salinity zones and generalized water flow patterns.

Overall, the present model predicts landscape succession fairly accurately. To validate the model, the 1978 habitat map generated by the model (SIM) is compared to the actual (REAL) 1978 habitat map. The most straightforward method to compare the maps is to calculate the percent of corresponding cells in the two maps which have

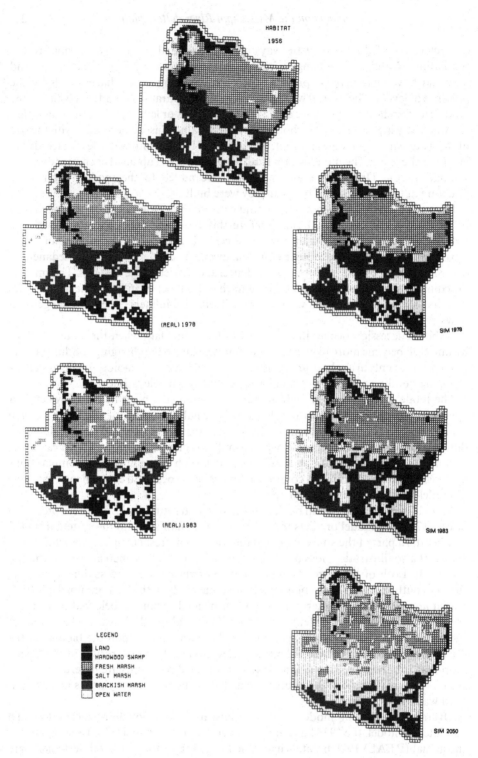

Fig. 17.1 Habitat maps for Louisiana coastal region: results of model runs for 1956, 1978, 1983 and 2050. REAL = actual; SIM = model-generated.

the same habitat type. In our current runs, a cell-by-cell comparison has resulted in a fit of 86% correct.

A co-ordinated monitoring plan

A co-ordinated monitoring plan should begin soon in a number of coastal areas of the world in order to gather data on the effects of sea level rise. In many cases, there are ongoing programmes which could be integrated and co-ordinated. Some possible areas include the following: River deltas including the Mississippi, Grijalva (Mexico), Ebro, Po, Nile, Camargue, Danube, and Ganges; and low-lying coastal areas of The Netherlands, and the Chesapeake Bay area. A number of factors should be measured including vertical accretion rates, rates of water level increase, vegetation response (productivity, tree ring growth, physiological stress indicators), habitat change over time (mapping), and shoreline retreat.

Such data are needed for the development of spatial models. As discussed above, spatial modelling is a powerful tool for predicting the response of low-lying areas to sea level rise. This type of model has great potential for considering the effects of ASLR in coastal areas and determining the impacts of different management scenarios. Spatial models could be developed for each area around the world as data are collected under a co-ordinated monitoring plan.

References

Adams, C.E., Jr., Wells, J.T. & Coleman, J.M. (1982). Sediment transport on the central Louisiana continental shelf: implications for the developing Atchafalaya River delta. *Contrib. Mar. Sci.*, **24**, 133–48.

Armstrong, W. (1979). Aeration in higher plants. *Adv. Bot. Res.*, **7**, 225–332.

Armstrong, W. & Gaynard, T.J. (1976). The critical oxygen pressures for respiration in intact plants. *Physiol. Plant.*, **37**, 200–6.

Baumann, R.H. (1987). Physical variables. In *The Ecology of Barataria Basin, Louisiana: an Estuarine Profile*, W.H. Conner and J.W. Day, Jr. (eds.). US Fish Wildl. Serv. Biol. Rep. 85(7.13), pp. 8–17.

Baumann, R.H. & Adams, R.D. (1981). The creation and restoration of wetlands by natural processes in the lower Atchafalaya River system: possible conflicts with navigation and flood control management. In *Proceedings of the Eighth Annual Conference on Wetlands Restoration and Creation*, Tampa, Florida, R.H. Stovall (ed.). pp. 1–25.

Baumann, R.H. & DeLaune, R.D. (1982). Sedimentation and apparent sea level rise as factors affecting land loss in coastal Louisiana. In *Proceedings of the Conference on Coastal Erosion and Wetland Modification in Louisiana: Causes, Consequences and Options*, D. Boesch (Ed.). US Fish Wildl. Serv., Office of Biological Services, FWS/OBS-82/59, pp. 2–13.

Baumann, R.H., Day, J.W., Jr. & Miller, C.A. (1984). Mississippi deltaic wetland survival: sedimentation versus coastal submergence. *Science*, **224**, 1093–5.

Bell, D.T. & Johnson, F.L. (1974). Flood-caused tree mortality around Illinois reservoirs. *Trans. Illinois State Academy of Sciences*, **67**, 28–37.

Bertani, A., Brambilla, I. & Menegus, F. (1980). Effect of anaerobiosis on rice seedlings: growth metabolic rate and fate of fermentation products. *J. Exp. Bot.*, **31**, 325–31.

Boesch, D.F., Levin, D., Nummedal, D. & Bowles, K. (1983). *Subsidence in Coastal Louisiana: Causes, Rates and Effects on Wetlands*. US Fish Wildl. Serv., Div. Biol. Serv., Washington, D.C. FWS/OBS-83/26, 30pp.

Bouwsema, P., Bossinade, J.H., Dijkema, K.S., van Meegen, J.W.Th.M., Reenders, R. & Vrieling, W. (1986). De ontwikkeling van de hoogte en van de omvang van de kwelders. In the *Landaanwinningswerken in Friesland en Groningen*. Rijksinstitut voor Natuurbeheer, Texel. RIN-rapport 86/3, 58pp.

Brinson, M.M., Bradshaw, H.D. & Jones, M.N. (1985). Transitions in forested wetlands along gradients of salinity and hydroperiod. *J. Elisha Mitchell Scient. Soc.*, **101**, 76–94.

Broadfoot, W.M. (1967). Shallow water impoundment increases soil moisture and growth of hardwoods. *Soil Sci. Soc. America Proc.*, **31**, 562–4.

Broadfoot, W.M. & Williston, H.L. (1973). Flooding effects on southern forests. *J. Forest.*, **71**, 584–7.

Brown, S. & Lugo, A.E. (1982). A comparison of structural and functional characteristics of saltwater and freshwater forested wetlands. In *Wetlands – Ecology and Management*, B. Gobal, R.E. Turner, R.G. Wetzel and D.F. Whigham (eds.). International Scientific Publications, Jaipur, India, pp. 109–30.

Cahoon, D. & Groat, C. (Eds.) (1990). *A Study of Marsh Management Practice in Coastal Louisiana, Volume III, Ecological Evaluation*. Final Report submitted to Minerals Management Service, New Orleans, Louisiana, Contract no. 14-12-0001-30410.

Clark, J.S. (1986). Coastal forest tree populations in a changing environment, Southeastern Long Island, New York. *Ecol. Monogr.*, **56**, 259–77.

Coleman, J.M. & Gagliano, S.M. (1964). Cyclic sedimentation in the Mississippi River deltaic plain. *Trans. Gulf Coast Assoc. Geol. Soc.*, **14**, 67–80.

Conner, W.H., Slater, W.R., McKee, K., Flynn, K., Mendelssohn, I.A. & Day, J.W., Jr. (1986). *Factors Controlling the Growth and Vigour of Commercial Wetland Forests Subject to Increased Flooding in the Lake Verret, Louisiana Watershed*. Final Report to Board of Regent Research and Development Programme, Baton Rouge, LA.

Conway, V.M. (1977). Studies on the autoecology of *Cladium mariscus* R.Br.III, The aeration of the subterranean parts of the plant. *New Phytol.*, **36**, 64–84.

Costanza, R., Sklar, F.H. & Day, J.W., Jr. (1986). Modeling spatial and temporal succession in the Atchafalaya/Terrebonne marsh/estuarine complex in south Louisiana. In *Estuarine Variability*, D. Wolfe (ed.). Academic Press, New York, pp. 387–404.

Costanza, R., Sklar, F. & White, M. (1990). Modelling coastal landscape dynamics. *Bioscience*, **40**, 91–107.

Cowan, J.H., Jr., Turner, R.E. & Cahoon, D.R. (1986). *A Preliminary Analysis of Marsh Management Plans in Coastal Louisiana*. Report to Lee Wilson and Associates, Inc., Santa Fe, New Mexico, for the US EPA, 30pp + Appendices.

Craig, N.J., Turner, R.E. & Day, J.W., Jr. (1980). Wetland losses and their consequences in coastal Louisiana. *Z. Geomorphol. Suppl.*, **34**, 225–41.

Crawford, R.M.M. (1978). Metabolic adaptations to anoxia. In *Plant Life in Anaerobic Environments*, D.D. Hook and R.M.M. Crawford (eds.). Ann Arbor Science Publ. Inc., Ann Arbor, Michigan.

Crawford, R.M.M. (1976). Tolerance of anoxia and regulation of glycolysis in tree roots. In *Tree Physiology and Yield Improvement*, M.G.R. Cannell and F.T. Last (eds.). Academic Press, New York.

Dacey, J.W.H. (1980). Internal winds in water lilies: An adaptation for life in anaerobic sediments. *Science*, **210**, 1017–19.

Day, J., Costanza, R., Kemp, G.P. & Teague, K. (1989). *Wetland Impoundments: a Global Survey with Emphasis on Louisiana Coastal Zone*. Centre for Wetland Resources, Louisiana State University, Baton Rouge, 140 pp.

Day, R., Holtz, R. & Day, J. (1990). An inventory of wetland impoundments in the coastal zone of Louisiana, USA: Historical trends. *Environ. Management*, **14**, 229–40.

Day, J. & Templet, P. (1989). Consequences of sea level rise: implications from the Mississippi delta. *Coastal Management*, **17**, 214–57.

Deegan, L.A., Kennedy, H.M. & Neill, C. (1984). Natural factors and human modifications contributing to marsh loss in Louisiana's Mississippi River deltaic plain. *Environ. Manage.*, 8, 519–28.

DeLaune, R.D., Baumann, R.H. & Gosselink, J.G. (1983). Relationships among vertical accretion, coastal submergence and erosion in a Louisiana gulf coast marsh. *J. Sed. Petrol.*, 53, 147–57.

Dickson, R.E., Broyer, T.C. & Johnson, C.M. (1972). Nutrient uptake by tupelogum and baldcypress from saturated or unsaturated soil. *Plant Soil*, 37, 297–308.

Drew, M.C. (1983). Plant injury and adaptation to oxygen deficiency in the root environment: a review. *Plant and Soil*, 75, 179–99.

Eggler, W.A. & Moore, W.G. (1961). The vegetation of Lake Chicot, Louisiana, after eighteen years of impoundment. *Southwest. Nat.*, 6, 175–83.

Frazier, D.E. (1967). Recent deltaic deposits of the Mississippi River, their development and chronology. *Trans. Gulf Coast Assoc. Geol. Soc.*, 17, 287–315.

Gagliano, S.M., Meyer-Arendt, K.J. & Wicker, K.M. (1981). Land loss in the Mississippi River deltaic plain. *Trans. Gulf Coast Assoc. Geol. Soc.*, 31, 295–300.

Gambrell, R.P. & Patrick, W.H., Jr. (1978). Chemical and microbiological properties of anaerobic soils and sediments. In *Plant Life in Anaerobic Environments*, D.D. Hook and R.M.M. Crawford (eds.). Ann Arbor Science Publishers, Inc., Ann Arbor, Michigan, pp. 375–423.

Givan, C.V. (1979). Review, metabolic detoxification of ammonia in tissues of higher plants. *Phytochem.*, 18, 375–82.

Gleason, M.L. & Zieman, J.C. (1981). Influence of tidal inundation on internal oxygen supply of *Spartina alterniflora* and *Spartina patens*. *Estuarine Coastal Shelf Sci.*, 13, 47–57.

Gosselink, J.G. (1984). *The Ecology of the Delta Marshes of Coastal Louisiana: A Community Profile*. US Fish Wildl. Serv. FWS/OBS-84/09, 134pp.

Gosselink, J.G., Cordes, C.L. & Parsons, J.W. (1979). *An Ecological Characterization of the Chenier Plain Coastal Ecosystem in Louisiana and Texas, Vol 1*. US Fish Wildl. Serv. FWS/BS-78/09, 302pp.

Gould, H.R. (1970). The Mississippi delta complex. In *Deltaic Sedimentation, Modern and Ancient*, J.P. Morgan (ed.). Soc. Econ. Paleontol. Mineral. Spec. Publ. 15, pp. 3–30.

Green, W.E. (1947). Effect of water impoundment on tree mortality and growth. *J. Forest.*, 45, 118–20.

Hackney, C.T. & Cleary, W.J. (1987). Saltmarsh loss in Southeastern North Carolina lagoons: importance of sea level rise and inlet dredging. *J. Coastal Res.*, 3, 93–7.

Hall, T.F. & Smith, G.E. (1955). Effects of flooding on woody plants, West Sandy Dewatering Project, Kentucky Reservoir. *J. Forest.*, 53, 281–5.

Hall, T.F., Penfound, W.T. & Hess, A.D. (1946). Water level relationships of plants in the Tennessee Valley with particular reference to malaria control. *J. Tenn. Acad. Sci.* 21, 18–59.

Harms, W.R., Schreuder, H.T., Hook, D.D. & Brown, C.L. (1980). The effects of flooding on the swamp forest in Lake Ocklawah, Florida. *ECOL.*, 61, 1412–21.

Hatton, R.S., DeLaune, R.D. & Patrick, W.H., Jr. (1983). Sedimentation, accretion and subsidence in marshes of Barataria basin, Louisiana. *Limnol. Oceanogr.*, 28, 494-502.

Hicks, S.D. (1972). On the classification and trends of long period sea level series. *Shore and Beach*, April, 20–3.

Hochachka, P.W. & Somero, G.N. (1973). *Strategies of Biochemical Adaptation*. Saunders, Philadelphia.

Houlk, O.A. (1983). Land loss in coastal Louisiana: causes, consequences and remedies. *Tulane Law Review*, 58, 3–168.

John, C.D. & Greenway, H. (1976). Alcoholic fermentation and activity of some enzymes in rice roots under anaerobiosis. *Aust. J. Plant. Physiol.*, 3, 325–36.

Johnson, W.B., Sasser, C.E. & Gosselink, J.G. (1985). Succession of vegetation in an evolving river delta, Atchafalaya Bay, Louisiana. *J. Ecol.*, **73**, 973–86.

Johnson, F.L. & Bell, D.T. (1976). Plant biomass and net primary productivity along a flood-frequency gradient in the streamside forest. *Castanea*, **41**(2), 156–65.

Jones, R.G. Wyn (1981). Salt tolerance. In *Physiological Processes Limiting Plant Productivity*, C.B. Johnson (ed.). Butterworths, London, pp. 271–92.

Kana, T.W., Baca, B.J. & Williams, M.L. (1986). *Potential Impacts of Sea Level Rise on Wetlands around Charleston, South Carolina*. US EPA 230–10–85–014, 65 pp.

Kearney, M.S. & Stevenson, J.C. (1985). Sea level rise and marsh vertical accretion rates in Chesapeake Bay. *Coastal Zone*, **85**(2), 1451–61.

Kennedy, H.E. (1970). Growth of newly planted water tupelo seedlings after flooding and siltation. *For. Sci.*, **16**, 250–6.

Kolb, C.R. & Van Lopik, J.R. (1966). Depositional environments of the Mississippi River deltaic plain – Southeastern Louisiana. In *Deltas and Their Geologic Framework*. Houston Geological Society, pp. 17–61.

Kozlowski, T.T. (1982). Water supply and tree growth, II, Flooding. *For. Abstr.*, **43**, 145–61.

Kozlowski, T.T. & Pallardy, S.G. (1979). Stomatal response of *Fraxinus pennsylvanica* seedlings during and after flooding. *Physiol. Plant.*, **46**, 155–8.

Kramer, P.J. & Kozlowski, T.T. (1979). *Physiology of Woody Plants*. Academic Press, New York.

Levitt, J. (1980). *Response of Plants to Environmental Stresses, Vol. 2: Water, Radiation, Salt and Other Stresses*. Academic Press, New York.

Louisiana Wetland Protection Panel (1987). *Saving Louisiana's Coastal Wetlands: the Need for a Long-term Plan of Action*. US Environmental Protection Agency and Louisiana Geological Survey, EPA-230-02-87-026, 102pp.

McKee, K. & Mendelssohn, I. (1989). Response of a freshwater marsh plant community to increased salinity and increased water level. *Aquat. Bot.*, **34**, 301–16.

McKnight, J.S., Hook, D.D., Langdon, O.G. & Johnson, R.L. (1981). Flood tolerance and related characteristics of trees of the bottomland forests of the Southern United States. In *Wetlands of Bottomland Hardwood Forests*, J.R. Clark and J. Benforado (eds.). Elsevier Scientific Publishing Company, Amsterdam, pp. 26–69.

Mendelssohn, I.A. & Burdick, D.M. (1988). The relationship of soil parameters and root metabolism to primary production in periodically inundated soils. In *The Ecology and Management of Wetlands*, D. Hook *et al.* (eds.). Timber Press, Portland, Oregon, pp. 398–428.

Mendelssohn, I.A. & McKee, K.L. (1988). *Spartina alterniflora* die-back in Louisiana: time-course investigation of soil waterlogging effects. *J. Ecol.*, **76**, 509–21.

Mendelssohn, I.A. & McKee, K.L. (1981). Determination of adenine nucleotide levels and adenylate energy charge ratio in two *Spartina* species. *Aquat. Bot.*, **11**, 37–55.

Mendelssohn, I.A., McKee, K.L. & Patrick, W.H., Jr. (1981). Oxygen deficiency in *Spartina alterniflora* roots: metabolic adaptation to anoxia. *Science*, **214**, 439–41.

Mitsch, W.J. & Rust, W.G. (1984). Tree growth responses to flooding in a bottomland forest in Northwestern Illinois. *For. Sci.*, **30**, 499–510.

Okey, C.W. (1918). The subsidence of muck and peat soils in Southern Louisiana and Florida. *Trans. Am. Soc. Civil Eng.*, **82**, 396–422.

Olsvig, L.S., Cryan, J.F. & Whittaker, R.H. (1979). Vegetation gradients of the pine plains and barrens of Long Island, New York. In *Pine Barrens: Ecosystem and Landscape*, R.T.T. Forman (ed.). Academic Press, New York, pp. 265–82.

Penfound, W.T. (1949). Vegetation of Lake Chicot, Louisiana in relation to wildlife resources. *Proc. LA. Acad. Sci.*, **12**, 47–56.

Pereira, J.S. & Kozlowski, T.T. (1977). Variation among woody angiosperms in response to flooding. *Physiol. Plant.*, **41**, 184–92.

Pezeshki, S.R. & Chambers, J.L. (1985a). Stomatal and photosynthetic response of sweetgum,

Liquidambar styraciflua to flooding. *Can. J. Forest Res.*, **15**, 371–5.

Pezeshki, S.R. & Chambers, J.L. (1985b). Responses of cherrybark oak (*Quercus falcata var. pagodaefolia Ell.*) seedlings to short-term flooding. *For. Sci.*, **31**, 760–71.

Pezeshki, S.R., DeLaune, R.D. & Patrick, W.H., Jr. (1989). Assessment of saltwater intrusion impact on gas exchange behaviour of Louisiana Gulf coast wetland species. *Wetlands Ecology and Management*, **1**, 21–30.

Pezeshki, S.R., DeLaune, R.D. & Patrick, W.H., Jr. (1987). Response of the freshwater marsh species, *Panicum hemitomon Schult.* to increased salinity. *Freshw. Biol.*, **17**, 195–200.

Psuty, P.N. (1966). Beach-ridge development in Tabasco, México. *Ann. Assoc. Am. Geographers*, **55**(1), 112–24.

Regehr, D.L., Bazzaz, F.A. & Boggess, W.R. (1975). Photosynthesis, transpiration and leaf conductance of *Populus deltoides* in relation to flooding and drought. *Photosynthe.*, **9**, 52–61.

Rejmanek, M., Sasser, C.E. & Gosselink, J.G. (1987). Modelling of vegetation dynamics in the Mississippi River deltaic plain. *Vegetatio*, **69**, 133–40.

Roberts, H.H., Adams, R.D. & Cunningham, R.H.W. (1980). Evolution of the sand-dominant subaerial phase, Atchafalaya delta, Louisiana. *Am. Assoc. Petrol. Geol. Bull.*, **64**, 264–79.

Saglio, P.H., Rancillac, M., Bruzau, F. & Pradet, A. (1984). Critical oxygen pressure for growth and respiration of excised and intact roots. *Plant. Physiol.*, **76**, 151–4.

Salinas, L.M., DeLaune, R.D. & Patrick, W.H., Jr. (1986). Changes occurring along a rapidly submerging coastal area: Louisiana, USA. *J. Coastal Res.*, **2**, 269–84.

Scaife, W., Turner, R.E. & Costanza, R. (1983). Coastal Louisiana recent land loss and canal impacts. *Environ. Manage.*, **7**, 433–42.

Sestini, G. (1991a). Implications of climatic changes for the Po delta and Venice lagoon. In *Impact of Climatic Change on the Mediterranean Region*, J. Milliman (ed.). (in press).

Sestini, G. (1991b). Implications of climatic changes for the Nile delta. In *Impact of Climatic Change on the Mediterranean Region*, J. Milliman (ed.). (in press).

Silker, T.H. (1948). Planting of water-tolerant trees along margins of fluctuation level reservoirs. *Iowa State Coll. J. Sci.*, **22**, 431–47.

Sklar, F.H., Costanza, R. & Day, J.W., Jr. (1985). Dynamic spatial simulation modeling of coastal wetland habitat succession. *Ecol. Modeling*, **29**, 261–81.

Stanley, D. (1988). Subsidence in the Northeastern Nile delta: rapid rates, possible causes and consequences. *Science*, **240**, 497–500.

Stevenson, J.C., Ward, L.G. & Kearney, M.S. (1986). Vertical accretion in marshes with varying rates of sea level rise. In *Estuarine Variability*, D.A. Wolfe (ed.). Academic Press, New York, pp. 241–59.

Stevenson, J.C., Ward, L.G., Kearney, M.S. & Jordan, T.E. (1985). Sedimentary processes and sea level rise in tidal marsh systems of Chesapeake Bay. In *Wetlands of the Chesapeake*, H.A. Groman *et al.* (eds.). Environmental Law Institute, Washington, DC, pp. 37–62.

Swanson, R.L. & Thurlow, C.I. (1973). Recent subsidence rates along the Texas and Louisiana coasts as determined from tide measurements. *J. Geophys. Res.*, **78**, 2665–71.

Swenson, E.M., & Turner, R.E. (1987). Spoil banks: effects on a coastal marsh water level. *Estuarine, Coastal Shelf Sci.*, **24**, 599–609.

Templet, P.H. & Meyer-Arendt, K.J. (1988). Louisiana wetland loss: a regional water management approach to the problem. *Environ. Manage.*, **12**, 181–92.

Teskey, R.O. & Hinckley, T.M. (1977). *Impact of Water Level Changes on Woody Riparian and Wetland Communities, Vol II: Southern Forest Region*. US Fish Wildl. Serv. FWS/OBS-77/59.

Thom, B.G. (1969). Problems of the development of Isla del Carmen, Campeche, México. *Z. Geomorphologie, Neue Folge Bd.*, *Heft* **13**(4), 406–13.

Tripp, J.T.B. (1979). The unique Louisiana coastal zone management problem: meandering groping for adequate institutional arrangements. In *Proc. Third Coastal Marsh and Estuary Symposium*, J.W. Day, Jr., D.D. Culley, Jr., R.E. Turner and A.J. Mumphrey, Jr. (eds.).

Louisiana State University Division of Continuing Education, Baton Rouge, LA, pp. 481–502.

Turner, F.T. & Patrick, W.H., Jr. (1968). Chemical changes in waterlogged soils as a result of oxygen depletion. *Transactions of the 9th International Congress of Scientists*, **4**, 53.

Turner, R.E. & Neill, C. (1983). Revisiting the marsh after 70 years of impoundment. *Water Quality and Wetland Management Conference Proceedings, New Orleans, LA*, pp. 309–32.

van Heerden, I.L. & Roberts, H.H. (1980a). The Atchafalaya Delta – Louisiana's new prograding coast. *Trans. Gulf Coast Assoc. Geol. Soc.*, **30**, 497–506.

van Heerden, I.L. & Roberts, H.H. (1980b). The Atchafalaya Delta – rapid progradation along a traditionally retreating coast (South Central Louisiana). *Z. Geomorph. NF*, **34**, 188–201.

van Heerden, I.L., Wells, J.T. & Roberts, H.H. (1983). River-dominated suspended sediment deposition in a new Mississippi subdelta. *Can. J. Fish Aquatic Sci.*, **40**, 60–71.

Vartapetian, B.B. (1978). In *Plant Life in Anaerobic Environments*, D.D. Hook and R.M.M. Crawford (eds.). Ann Arbor Science Publishers, Ann Arbor, Michigan, pp. 1–2.

Webb, Y. & Armstrong, W. (1983). The effects of anoxia and carbohydrates on the growth and viability of rice, pea and pumpkin roots. *J. Exp. Bot.*, **34**, 579–603.

Wells, J.T. & Kemp, G.P. (1981). Atchafalaya mud stream and recent mud flat progradation: Louisiana Chenier Plain. *Trans. Gulf Coast Assoc. Geol. Soc.*, **31**, 409-16.

Whitlow, T.H. & Harris, R.W. (1979). *Flood Tolerance in Plants: A State of the Art Review*. Tech. Rep. E-79-2. Chief, Corps of Engineers, Washington, DC, 169pp.

Wicker, K.M. (1980). *Mississippi Deltaic Plain Region Ecological Characterisation: a Habitat Mapping Study – a User's Guide to the Habitat Maps*. US Fish Wildl. Serv. Office Biol. Serv. FWS/OBS-79/07.

Zaerr, J.B. (1983). Short-term flooding and net photosynthesis in seedlings of three conifers. *For. Sci.*, **29**, 71–8.

18

Sea level rise: assessing the problems

H.G. Wind and E.B. Peerbolte

Abstract

This chapter describes the rationale for, and the development of, the first phases of an integrated, inter-disciplinary approach to assessing the impacts of sea level rise in The Netherlands, based on a workshop held by Delft Hydraulics in 1986. This work laid the foundation for subsequent, refined model developments which have since been applied to other parts of the world potentially vulnerable to sea level rise. This presentation of the reasoning behind the early development of the approach is intended to provide guidance to others in the field of impact assessment as the need for such integrated, comprehensive approaches to the sea level rise problem becomes increasingly evident.

Introduction

In 1986, a workshop was held on the *Impact of Sea Level Rise on Society* (ISOS) which focused on the consequences of sea level rise and possible counter-measures (Wind, 1987; Delft Hydraulics *et al.*, 1988). There were two major objectives of the workshop: to investigate the possibilities of designing a decision-supporting model by a group of specialists from five different relevant disciplines; and to collect relevant information from these disciplines to develop further the concept of such a model. The ISOS workshop succeeded in clarifying the disciplinary perspectives required for an integrated sea level rise impact model. It was recommended that further work should be initiated following the framework of analysis developed during the workshop.

The first phase of this work was reported in Delft Hydraulics *et al.* (1988), which described how the model would be developed for The Netherlands. The actual construction of the model was started at the end of 1988. In the period 1988–90 a detailed study was carried out on The Netherlands resulting in a computational model for the assessment of the various consequences of sea level rise and related counter-measures (Delft Hydraulics *et al.*, 1988; Rijkswaterstaat and Delft Hydraulics, 1991). In this period case studies on the Maldives (Delft Hydraulics, 1989a) and Egypt (Delft Hydraulics, 1991) were also carried out. On a global-scale, attempts were made to highlight regions which are vulnerable to an increase in sea level rise (Delft Hydraulics, 1989b, 1990; IPCC, 1990).

The purpose of the present paper is to summarize the set-up of the first provisional model and to present some typical computational results. (See also Baarse and Rijsberman, 1987.) The historical reflection on the rationale and steps behind the development of such a model serves as an example to those in the field of sea level rise impact assessment in which the need for integrated, inter-disciplinary approaches is becoming increasingly evident.

Impacts of sea level rise and responses: a schema

A schema of effects of global warming and counter-measures is shown in Fig. 18.1. Global warming due to increasing atmospheric concentrations of carbon dioxide and other 'greenhouse' gases will cause various marine changes, such as a rise in sea level. The consequent effects depend on the area under consideration (e.g., estuaries, cliff coasts). Some of the resulting effects will call for remedial actions, which in turn will cause new effects. With knowledge of such effects, strategies can be formulated. Such a strategy contains a set of measures aimed at an efficient reduction of impacts. These can be subdivided into source-oriented measures (for example, reduction of emissions) and impact-oriented measures (flood protection, change of land use, retreat from coastline, etc.). This paper is concerned with the assessment of consequences of sea level rise and possible measures; the evaluation of the best strategy and methods used for that purpose are not dealt with here.

Marine changes

For a given location in the nearshore region, an increase in mean water level will affect waves, currents and bottom pressure in the nearshore region. In general, an increase in mean water depth will be accompanied by an increase in mean wave height, resulting in more serious wave attack on the coast and greater wave-induced littoral drift. However, the sea bed may tend to adapt to the increasing water level, resulting in smaller (or no) changes in the water depth. In such a situation, the above increase in wave attack will be partially or completely neutralized. Furthermore the tidal amphidromic points will generally shift leading to changes in tidal conditions along the coast. In a note on mean sea level rise (Rijkswaterstaat, 1986) it is indicated that the tidal range will both increase and decrease, depending upon the coastal location. This will obviously affect the tidal flow. In addition, climatic changes will also lead to a change in wave climate, possibly storm frequency and related phenomena. However, current climate models are unable to specify such regional details of climate change with much certainty. In the present approach these effects have been left out, but could be included without complication as knowledge improves.

Impacts of sea level rise

The impact of a rise in mean sea level depends on the type of area and its associated land uses. Here, the following types of areas are considered: coastal lowlands, wetlands and estuaries; shores and beaches (possibly with coastal defence systems); rivers, inlets and outlets; shipping channels and ports.

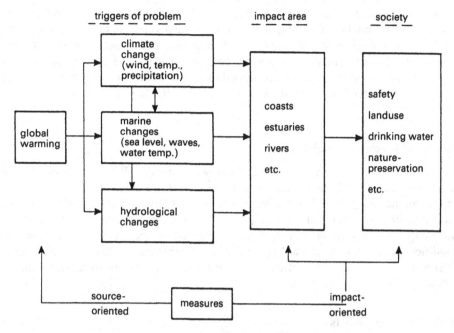

Fig. 18.1 Effects of sea level rise and possible counter-measures.

Coastal lowlands, wetlands and estuaries

Both the morphological and ecological systems of coastal lowlands, wetlands and estuaries will be affected by a rise in mean sea level.

In the case of a coastal area with a non-erodible sea bed and shore, sea level rise will result in a landward shift of the mean water line with consequent loss of land. However, in most cases sea level rise will affect the present coastal morphology because of the changed water and sediment motion. Approximately 70% of the world's sandy shorelines are presently eroding and it is generally expected that erosion processes will be aggravated by an increase in sea level.

However, relatively little is understood about wave-induced onshore transport processes and the consequent building-up of the foreshore and beach. Therefore, it might well be that under certain conditions sea level rise will stimulate onshore sediment movement. The present configuration of the Dutch coast is probably partly due to onshore transport processes, in spite of continued rise in mean sea level.

River deltas are often accreting due to material supply by contributing rivers. However, a number of major deltas (for example the Mississippi delta; see Day *et al.*, this volume) are also subsiding due to the increased sediment load and the settlement of lower subsoil layers. Increased global sea level rise may affect the balance of these two processes and, in turn, affect the rate of local relative sea level change, just as man-made structures (barrages) have disturbed the sediment budget in river deltas in many places of the world.

With respect to ecological systems, a rise in sea level is most likely to cause the system to shift shoreward in situations of a gentle (fixed) slope. However, in areas protected by dykes there may not be sufficient space to allow for such a shift. In that case, part or the whole of the ecological system may be lost. In accreting regions, the ecological system must follow the changing bottom topography in order to survive. Similar adaptations are being made under present-day conditions and hence no major changes are expected. Under conditions of erosion a breakdown of the ecological system may occur (see Day, 1987).

Shores and beaches (coastal defence)

As explained previously, an increase in mean water depth will lead to an increase in mean wave height and wave-related effects. For hard coastal defence structures this implies an increased risk of overtopping, damage and failure. The same holds for soft coastal defence structures like dunes. In addition, the foreshore of the coastal profile will adapt to a new dynamic equilibrium governed by the changed hydrodynamic conditions and sediment characteristics. For dunes this implies that a 1 m rise in sea level will induce a shoreline retreat of the order of one hundred to several hundred metres.

Rivers, inlets and outlets

The mean water level of a river may increase as a result of a rise in mean sea level because of the back-water effect. Such an increase may extend many tens of kilometres inland depending on local conditions, the river gradient and the rise in mean sea level. One of the effects of an increase in mean water level is a reduction in the sediment transport capacity of the river (if river discharges remain the same). In the case of sufficient supply of bed material this will lead to an increase in sedimentation. This morphological reaction aims at restoring the equilibrium depth of the river.

As a result of a sea level rise the tide will propagate farther inland. If the river bed does not build up (because of a lack of sediment supply, for example) salt intrusion will increase. However, if river bed aggradation keeps pace with the sea level rise (and if the tide at the seaward boundary is not changing) in the extreme case neither tidal propagation nor saltwater intrusion will change. In the case of The Netherlands the critical zone (from the point of freshwater users) in the northern part of the Rhine delta might shift 10–30 km landward for a sea level rise of 4.5 m, if the river bed does not adapt to the new hydraulic conditions and remains the same (Rijkswaterstaat, 1986).

River embankments will be subject to an increased risk of damage, overtopping and failure due to an increase in the mean water level as is the case with coastal defence structures.

Seepage under the embankments and the dunes increases with sea level rise. The ruling factors are the head difference between both sides of the embankment and the soil permeability. In areas affected by saltwater intrusion the seepage flow will be brackish or saline, resulting in damage to agriculture, horticulture and cattle farming.

Inlets and outlets are used for cooling water, agricultural and industrial use, drinking water and drainage (agricultural drainage, industrial and urban runoff). The effectiveness of inlets and outlets is reduced by the rise in mean water level and related effects such as saltwater intrusion and accretion of the river bottom. Measures in the realm of water resource management are required to counteract these negative effects.

Shipping channels and ports

Increase in mean water level in shipping channels and ports could have a positive effect on the manoeuvrability of ships which presently have a small keel clearance. However, increased penetration of low-frequency components of the wave field could (partly) counteract this positive effect, because vertical ship motion strongly depends on these components. Furthermore, there are possible negative effects related to the effectiveness of quays and the reduction in passage height below bridges. On the whole, the impact of sea level rise on the shipping and port system seems to be low relative to the other impacts.

Summary of the impacts

The impacts outlined above can broadly be combined into the following impact categories: safety (of coastal and river defence systems), land loss (and environment) and economy (water resources management system, ports and shipping system). Safety is related to the number of people at risk. Land loss can be expressed in economic terms as well as in km². Both damages to the water resources management (WRM) and to the ports and shipping system lead to economic losses.

A conceptual framework for modelling impacts and response

To counteract or prevent the impacts of sea level rise, individual adaptive measures or strategies (sets of measures) can be adopted. Initially, the effectiveness of the measure in counteracting the impact should be known; and, secondly, the effects induced by the measure itself should be estimated. To take such effects into account, the simple schemata presented earlier have been expanded, as shown in Fig. 18.2, based on the discussion presented above. This suggests the types of impact which can be expected. In order to quantify the impacts, a computational framework, following the scheme in Fig. 18.2, was developed and forms the basis of the ISOS model.

The interactions between the elements presented in Fig. 18.2 represent processes which range from very simple to extremely complex. For the development of a computational framework these interactions have to be specified. This often requires extensive studies. Here, a few simplified relations will be used for demonstration purposes. As a test-case, The Netherlands has been selected.

Test-case: The Netherlands

The Netherlands is situated in West Europe at the mouth of the rivers Rhine, Maas and Scheldt, and covers an area of approximately 37,000 km². With a population of 14.5 million, it is among the most densely populated countries in the world: 424 persons per km² of land. If dunes and dykes did not protect the country the most densely populated part of The Netherlands, including cities like Amsterdam, Rotterdam and The Hague,

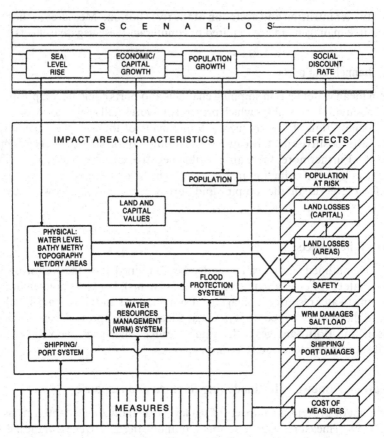

Fig. 18.2 Simplified ISOS system diagram (Baarse and Rijsberman, 1987. Reprinted with permission from A.A. Balkema, Rotterdam, The Netherlands).

would be inundated. More than one-half of the area of the country is below (MSL + 1) m (mean sea level plus 1 metre) and about 27% lies below mean sea level.

Geographic representation

The Netherlands has been subdivided into three sections: A, B, C (see Fig. 18.3). Section B, Holland, is largely protected by dunes, while sections A and C represent estuarine areas protected by dykes. This division is rather crude, but is suitable for the demonstration purposes. A finer grid would show more detail, but more time would be required to select and to represent the data. Furthermore, a finer grid would require an adaptation (refinement) of the impact area characteristics. Application of a GIS (Geographic Information System; see Shennan, this volume), which enables the overlay of administrative and geographical data bases, would then be required to handle the large quantities of data connected to the finer grid and the higher-resolution model.

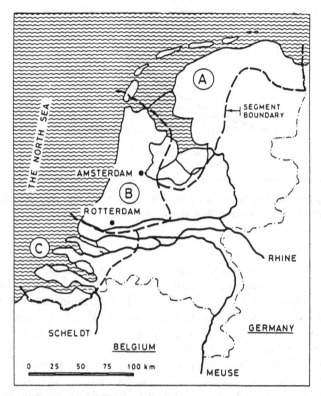

Fig. 18.3 Area and segments involved in The Netherlands case study (Baarse and Rijsberman, 1987. Reprinted with permission from A.A. Balkema, Rotterdam, The Netherlands).

Impact area characteristics

Loss of land is a primary impact. In the simplified version of the ISOS model morphological processes have not been modelled. The loss of land is simply defined as the submerged area using the present-day depth contours. In order to put economic values to the land, the following land types are distinguished: intertidal, urban/ industrial, agricultural and environmental. Where a flood protection system is present, it is assumed that only intertidal areas are lost due to sea level rise. Basically, land loss follows from the existing depth contours. The total value of the land lost is obtained by adding the separate losses pertaining to the relevant land types which are inundated in the section under consideration.

An obvious effect of sea level rise is a reduction in *safety*. Flooding frequencies will increase in the long run; the question is how to express the consequent impacts. For the purpose of the present demonstration model, a log–linear relationship is assumed between sea level and the overtopping frequency. Safety is defined as the reciprocal value of the flood frequency.

In the event of flooding (when the existing flood protection system is overtopped), the society within the protecting embankments, etc. will suffer damage. Klaassen

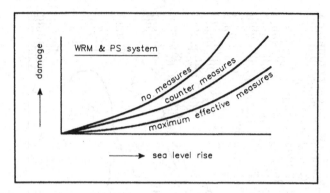

Fig. 18.4 Principle of damage and damage reduction. WRM = water resource management; PS = ports and shipping (Baarse and Rijsberman, 1987).

(1987) proposes a method to quantify the damage to both people and property from flooding. For the purpose of the present model, no attempt has been made to quantify the different parameters of floods (e.g., intensity and duration) in order to assess the damage to The Netherlands following his approach. As a first approximation, the number of people living in the endangered area and the estimated value of the endangered area are taken as impact indicators.

Sea level rise may cause increasing damage in the water resources management system (WRM) as a result of salt intrusion via surface waters and groundwater. In the past, extensive studies have been undertaken to investigate the mechanisms of *salt intrusion*. Salt intrusion via groundwater (saline seepage) is most difficult to model. The difference in height between sea level and the phreatic surface, and soil characteristics are probably important parameters governing the process of saline seepage. However, it has been concluded that, based on the literature, a general method to quantify saline seepage is not available.

Saltwater intrusion in estuaries and the lower branches of rivers is related to river discharge, the tidal range and the water depth, to name a few factors. As far as water depth is concerned, salt intrusion is approximately proportional to its square root.

In the present ISOS model the step from salt intrusion to monetary damage is skipped and an exponential relationship is adopted between sea level rise and salt damage. Different curves reflecting this relationship can be distinguished, depending on the investment counter-measures. The maximum curve represents the situation when no counter-measures are taken; the minimum curve shows the remaining damage at maximum (useful) investment in counter-measures, such as reflushing of polders with fresh water, adaptation or relocation of freshwater inlets, etc. The principle is shown in Fig. 18.4.

For the effects on *ports and shipping (PS) systems*, basically the same procedure is followed as above: a set of exponential relationships represent the dependency of the damage in this sector on sea level rise, taking account of the effects of counter-measures, such as adaptation works in harbours, harbour entrances, shipping locks, etc.

Table 18.1. *Investment scheme*

Year	WRM investments (US$ × 10⁶) Section			Raising dykes (m) Section		
	A	B	C	A	B	C
2000	30	30	30			
2010	30	30	30			
2015					0.5	
2020	30	30	30			
2030	30	30	30			
2045					0.5	
2050				0.5		0.5

Scenarios

The scenarios of sea level rise consists of two components. The first component is the present relative sea level rise, which for The Netherlands is estimated at 0.20 m per century. The second is the increase in sea level rise, due to the thermal expansion of sea water, melting of glaciers and ice caps etc, which is set at 0.80 m in 2085. The eustatic sea level rise follows a hyperbolic tangent. In the present example no additional variations in the sea level rise scenarios were made.

For the remaining scenario variables the following values were selected: population growth, 0.2%; economic/capital growth, 3.0%; and social discount rate, 0.0%.

Test-runs and measures

Two test-runs of the ISOS model were made for demonstration purposes. In the first test-run no remedial actions were assumed; in the second test-run an investment scheme is used involving investments in WRM measures and raising the height of dykes in order to increase the safety of the structures. The investment scheme shown in Table 18.1 has been applied in case 2. (For additional test-cases the reader is referred to Wind, 1987).

The impacts of sea level rise for base case 1 (no remedial actions) are shown in Figs. 18.5(*a*), (*c*) and (*e*). Sea level rise has been added to each figure for reference purposes. With respect to monetary losses (Fig. 18.5(*a*)), the WRM damages and PS losses of US$ 3.5×10^9 well exceed land capital losses of US$ 0.1×10^9. This is due to the low capital value which has been put on the intertidal area. As regards the return period of flooding (Fig. 18.5(*c*)), the value in 1985 has been set at 10^{-4} years; by 2085, the value is reduced to 4.65×10^{-3} years. The loss of land (Fig. 18.5(*e*)) consists mainly of the inter-tidal area in the estuaries. If morphological changes are added to the model, the loss of land may be considerably different. Salt load increases in the order of 50–60%.

In case 2 the remedial actions, as scheduled in Table 18.1, are carried out. The results are shown in Figs. 18.5(*b*) and (*d*). The effect of raising the height of the dykes can be discerned from a comparison between the Figs. 18.5(*c*) and (*d*). It is clear that raising

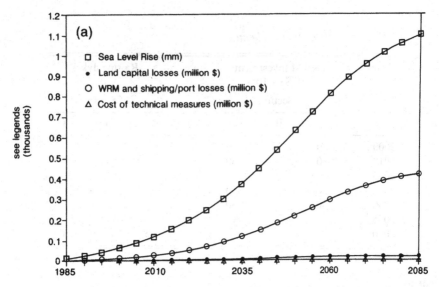

Fig. 18.5(*a*) Monetary losses – base case 1 (expressed in cumulative values over five-yearly intervals).

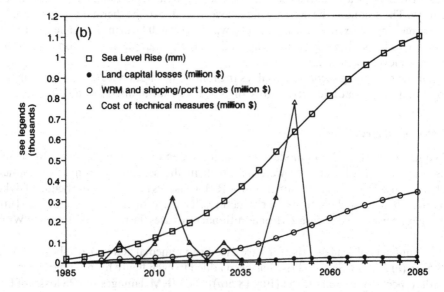

Fig. 18.5(*b*) Monetary losses, case 2 (expressed in cumulative values over five-yearly intervals).

the dyke by 0.5 m only partially and temporarily increases the safety of the dykes. Depending on political acceptability, further investments in raising the height of the dykes are required. A comparison of Fig. 18.5(*a*) and (*b*) shows that due to a total investment of US\$ 0.1×10^9 the WRM losses by 2085 are reduced from US\$ 2.0×10^9 to US\$ 1.5×10^9. Particularly during the first decades the rapid rise in WRM damages is reduced by the scheduled WRM investments.

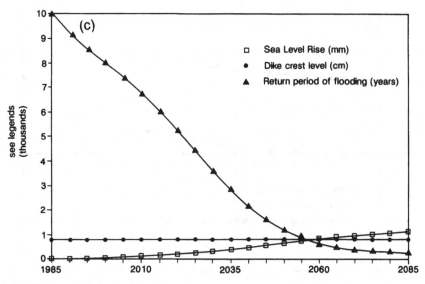

Fig. 18.5(*c*) Return periods of flooding – base case (segments A, B, C).

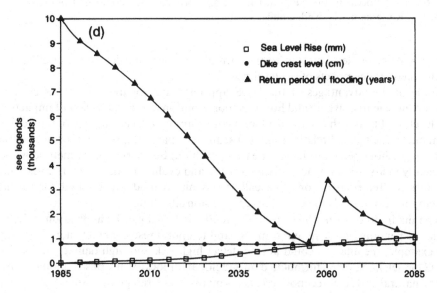

Fig. 18.5(*d*) Return periods of flooding for case 2, segment A.

Discussion, conclusions and recommendations

This paper summarized a method for assessing the Impact of Sea Level Rise on Society (ISOS), as was originally proposed by Baarse and Rijsberman (1987) at the occasion of the ISOS workshop in November 1986 in Delft, The Netherlands. The impacts were first discussed in qualitative terms; within the computational framework the impacts were then expressed quantitatively. The model attempts to provide an integrated

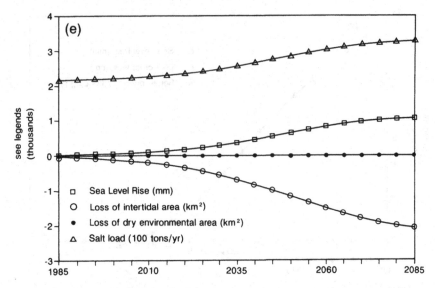

Fig. 18.5(e) Land losses and salt load – base case.
(Baarse and Rijsberman, 1987. Figs. (a)–(e) are all reprinted with permission from A.A.
Balkema, Rotterdam, The Netherlands).

analysis, from sea level rise up to the formulation of strategies aimed at efficiently
counteracting the impacts of sea level rise.

Some of the advantages of the ISOS approach are apparent. It makes clear to
supporting scientists where and how their contribution to the assessment of impacts of
sea level rise fits into the analysis. Once a contribution to a certain type of impact has
been made, the applied relations and consequent impacts reappear consistently in the
analysis, without being lost in new ideas and fashion. Furthermore, the method works
efficiently in assessing impacts of sea level rise and evaluating the effect of investment
schemes for remedial actions. Basically the same method can be used for further
development of models to be used in real decision situations.

The methodology on which the ISOS model is based can be applied on different
scales, depending on the required spatial and temporal resolution. On a global-scale,
for example, the analysis could aim to detect high-risk and high-damage areas. The
ISOS model, however, is formulated on a regional- or local-scale. Based upon analyses
of the natural and socio-economic systems involved, a strategic assessment of impacts
due to sea level rise is made; next, remedial actions are formulated and evaluated. The
whole process of natural development and the effects of remedial actions interacting
both with the natural system and the measures themselves, is integrated in the 'impact
area'. The resulting 'impacts on society' are expressed in monetary terms, in areas (of
lost lands) and in numbers of people at risk.

The future scenarios concerning the effect of greenhouse gases, sea level change and
societal development play an essential role in assessing the impacts of sea level rise.
Asking an expert is a common approach for obtaining values for scenario variables.
Because of its important role in projecting impacts, the methodology of scenario
development should be explored to arrive at realistic input scenarios for the model. For

instance, expressing the value of scenario variables in terms of statistical distributions could, in many cases, better represent the present knowledge of future developments.

Each of the impact relations represents a complex set of processes, the physical and mathematical description of which is based upon the best present knowledge. However, too often such knowledge relies on direct experience and does not include periods of rapid sea level rise; hence extrapolation of existing knowledge can be hazardous. Thus, existing data should be constantly updated and evaluated with respect to their applicability to projecting future conditions.

The positive results of the use of the ISOS approach call for other applications on a regional-scale. For such an application the geographical representation should be improved, the impact relations adapted to local-scale and the actual data used for verification. Then the resulting model can be used to assess impacts of sea level rise on society.

References

Baarse, G. & Rijsberman, F.R. (1987). Policy analysis. In *Impact of Sea Level Rise on Society*, H.G. Wind (ed.). A.A. Balkema, Rotterdam, Brookfield, pp. 21–75.

Day, J.W., Jr. (1987). Consequences of sea level rise: implications from the Mississippi delta. In *Impact of Sea Level Rise on Society*, H.G. Wind (ed.). A.A. Balkema, Rotterdam, Brookfield, pp. 146–52.

Delft Hydraulics (1991). *Implications for Sea Level Rise on the Development of the Lower Nile Delta, Egypt; Pilot Study for a Quantitative Approach*. Final report, H927, Delft Hydraulics, Delft (subject to EC approval).

Delft Hydraulics (1990). *Sea Level Rise: A World-wide Cost Estimate of Basic Coastal Defence Measures*. H1068, Delft Hydraulics, Delft.

Delft Hydraulics (1989a). *Republic of Maldives, Implications of Sea Level Rise; Report on Identification Mission*. H926, Delft Hydraulics, Delft.

Delft Hydraulics (1989b). *Criteria for Assessing Vulnerability to Sea Level Rise: A Global Inventory to High Risk Areas*. H838, Delft Hydraulics, Delft.

Delft Hydraulics *et al.*, December (1988). *Impact of Sea Level Rise on Society, A Case Study for The Netherlands: Report of Phase I*.

Intergovernmental Panel on Climate Change (IPCC) (1990). *Strategies for Adaption to Sea Level Rise*. Response Strategies Working Group, Ministry of Transport and Public Works, Rijkswaterstaat, Tidal Waters Division, The Hague.

Klaassen, L.H. (1987). Some societal consequences of the rising sea level. In *Impact of Sea Level Rise on Society*, H.G. Wind (ed.). A.A. Balkema, Rotterdam, Brookfield, pp. 189–91.

Rijkswaterstaat (Ministry of Transport and Public Works), May (1986). *Sea Level Rise* (in Dutch), Note on consequences of a sea level rise of 5 m.

Rijkswaterstaat and Delft Hydraulics (1991). *Impact of Sea Level Rise on Society, A Case Study for The Netherlands*, Final Report, H750, Delft Hydraulics, Delft.

Wind, H.G., Ed. (1987). *Impact of Sea Level Rise on Society*, Report of a project-planning session Delft Hydraulics, 27–9 August 1986, A.A. Balkema, Rotterdam, Brookfield.

19

Adjustment to greenhouse gas induced sea level rise on the Norfolk Coast – a case study

K.M. Clayton

Abstract

The Norfolk coast is developed on relatively weak rocks of Cretaceous and Quaternary age and has a history of relatively rapid natural change; accretion on the low-lying northern coast, and rapid cliff erosion ($\sim$ 1m/yr) between Sheringham and Happisburgh. The low coast between Happisburgh and Great Yarmouth is protected by beaches and dunes built of sand and derived from the cliffs of the north. With 88% of the coast already defended, the cost of attempting to maintain the status quo would be very high and increased selectivity and an increasing willingness to live with natural change are seen as keys to a long-term coastal strategy. The other needs are improved controls on land use in the coastal hazard zone and development of the flood warning system to improve its effectiveness when tidal surges occur in future. Future sea level rise will bring no new problems to coastal managers, but it will make informed and well-planned action even more important than it is today.

Description of the coastline

The Norfolk coast extends from just east of the Nene estuary in The Wash, to Corton cliffs between Great Yarmouth and Lowestoft. In sequence from west to east the physiographic units are as shown in Table 19.1.

With the exception of the town of King's Lynn on the Ouse, the Wash coast is backed by arable land. The small seaside resort of Hunstanton has chalk cliffs which form part of the western side of The Wash. Most of the North Norfolk coast is a low barrier island coast with sand dunes and marshes, largely managed as nature reserves. However, large parts of the former salt-marsh are defended by flood banks and used for grazing and some reclaimed areas are under arable cultivation, as at Holkham.

On the higher ground along the cliffed coast from Weybourne to Happisburgh, settlements have been built nearer the sea, notably the two resorts of Sheringham and Cromer. There are smaller villages along the coast road, but most of the rest of the area is farmland with only scattered houses. This cliffed coastline has a long record of erosion and villages mentioned in the Domesday Book of 1086 are now totally lost. Just north of a short break in the cliffs at Walcott, the Bacton gas terminal is built on the top of the cliffs.

Table 19.1. *Physiographic units of the Norfolk coast (clockwise)*

Region	Coastal (km)	length (%)
Reclaimed shore of The Wash	25	18
Chalk cliffs (Hunstanton)	2.5	2
Low barrier island coast	44	31
Cliffs in glacial sediments	33.5	24
Generally low coast with dunes	32	23
Cliffs in glacial sediments	3	2

Beyond the end of the cliffs at Cart Gap, just southeast of Happisburgh, farmland lies behind the long sea wall, built after the 1953 floods and fronting a narrow belt of sand dunes. Tourist development is limited to the houses and holiday homes of Sea Palling and some isolated groups of houses as at Cart Gap. However, from Winterton south to the northern edge of Great Yarmouth the littoral zone is developed by holiday camps, tourist villages and caravan sites. Great Yarmouth is a small port on the estuary of the River Yare, but the seaward side of the spit east of the Yare estuary is developed as one of Britain's larger holiday resorts. The tourist beach at Gorleston lies south of the mouth of the Yare. To the south the remaining few kilometres of the Norfolk coastline rise to low cliffs at Hopton on Sea and Corton before descending into Suffolk towards Lowestoft Ness, the easternmost point of England. Figure 19.1 illustrates the areas of Norfolk which are below high tide level.

The coastal sediment budget

In general, the cliffed coast supplies sediment to the beaches lying between Happisburgh and Lowestoft (Clayton *et al.*, 1983). It moves off-shore, particularly at the projecting nesses, passing into the off-shore bank system where it is moulded by tidal currents. Though a little sediment moves westwards from west of Sheringham, the bulk of the North Norfolk coast sediment as well as the mud entering The Wash probably comes from the eroding Holderness cliffs north of the Humber, about 100 km to the north.

Erosion and sedimentation over the last 100 years

A record exists of the position of the coastline and of High Water and Low Water Mark of Ordinary Tides (HWMOT and LWMOT respectively) on the Ordnance Survey (OS) 1:10,560 or 'Six Inch' (now 1:10,000) maps since about 1880. The changes are summarized in Table 19.2, for the sectors identified in Table 19.1, and for subsectors where these show consistent variation.

The consistent feature of this table is that the beach (HWM–LWM) profile has steepened in every case except one. This is probably due to a combination of the attempt to stabilize the coast by engineering structures and the tendency for beach

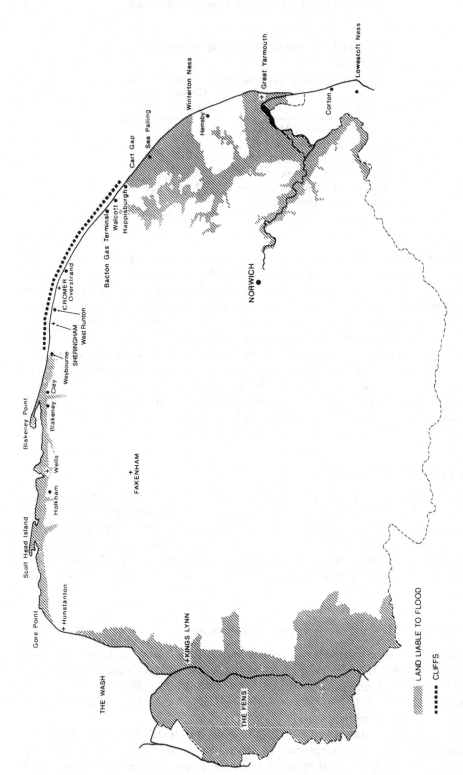

Fig. 19.1 The Norfolk coast showing the major towns and the areas liable to tidal flooding or coastal erosion.

Table 19.2. *Norfolk coast: mean rates of erosion (m/yr) over the last 100 years (for the coastline (CL), HWMOT and LWMOT as indicated on the OS 1:10,560 maps). Positive values indicate progradation*

Region	CL	HWM	LWM
Reclaimed shore of The Wash	+4.49	+5.79	−1.77
Chalk cliffs (Hunstanton)	+0.10	−0.00	−2.26
Gore Point	−0.10	+1.32	−1.22
Barrier coast of N Norfolk	+0.40	+0.85	−1.39
Undefended cliffs	+0.09	−0.13	−0.52
Sheringham – Happisburgh	−0.76	−0.47	−1.28
Sea Palling – Winterton	−0.38	−0.63	−0.61
Winterton – North Denes	+0.44	+0.33	−0.09
Great Yarmouth – Hopton	−0.02	+0.09	−0.13

Table 19.3. *Engineering structures along the Norfolk coast*

Region	Dominant structures	Coastal length (km)
The Wash	Embankments	25
Hunstanton	Revetment, wall and groynes	2.5
Gore Point	Undefended	7
North Norfolk	Embankments	39.5
Norfolk cliffs	Undefended	5.5
Norfolk cliffs	Groynes, revetments and breastwork	27
Sea Palling	Groynes and sea wall	13.5
Winterton	Undefended	7.5
Great Yarmouth	Walls and groynes	15

volumes to fall. This loss of beach volume may be due to slowly rising sea level and/or reduced sediment supply as a result of the defences now protecting part of the cliff system.

Coastal defences – sea defence and coastal protection

By far the greater part of the coast is affected by engineering structures as Table 19.3 shows.

Some of these structures are very old. For example, reclamation in The Wash dates back to the Roman occupation, 2,000 years ago, though the present shoreline was reclaimed within the last century; a wall was built at Cromer in 1845. However, the majority were added or rebuilt from earlier structures after the serious storm surge of 1953. Following that surge event in which considerable areas were inundated and over 300 people drowned, the Waverley Committee (1954) recommended that future defences should be constructed with a height of one foot (0.3 m) above the local surge level. Whilst most of the sea defence structures thus date from the decade after 1953, coastal protection structures continued to be built in front of the cliffs, the most recent being completed in 1988.

With only about 12% of the coast undefended, the Norfolk coastline involves considerable investment of public money. However, a larger proportion of the coast is in a reasonably natural state because most of the 40 km or so of the North Norfolk coast has the embankments landward of a natural barrier coast, often with a broad belt of salt-marsh behind the coastal dunes. Even the reclamation along The Wash has left a belt of salt-marsh in front of the latest embankments, giving a relatively natural edge to the sea. Furthermore, the revetments fronting some of the most active Norfolk cliffs are not particularly effective. They are able to reduce erosion but after about 15 years erosion recovers to about two-thirds its natural rate, keeping the cliffs relatively free of vegetation and with active landslides, mudflows and channelling streams in winter.

With future sea level rise the cost of maintaining and improving the defences along the Norfolk coast will be high and probably more than the coast defence authority can afford. In addition the steepening beaches are reducing the strength of existing defences and many are coming to end of their useful life. For these reasons research commissioned by the Anglian Water Authority and now the responsibility of the National Rivers Authority concluded that some of the coast should no longer be protected so that the remainder can be defended effectively (Anglian Water, 1988; Fleming, 1989). The remainder of this chapter discusses the basis on which decisions about abandonment or continued defence might be made. It then examines the alternative strategies for both the defended and the 'back to nature' coasts.

The future effect of rapid sea level rise – general review

The effect of sea level rise on beach volume/position has been modelled by Bruun (1962). When tested in wave tanks it is found to predict behaviour well. Essentially this is because sediment volume is conserved in a confined tank; on natural beaches the situation is usually complicated by the longshore transport of sediment, both into and out of the section under consideration.

This approach was one of the first attempts to relate erosion rate to rate of sea level rise in a quantitative fashion. This simple model is two-dimensional and does not give reliable estimates of erosion in situations where the third dimension of longshore movement of sediment is important. Thus, when longshore transport occurs, which is the situation in all but very confined bays between headlands, other approaches are required. Most of these imply, either directly or indirectly, the conservation of beach and nearshore slope over time, since these likely to be adjusted to local wave energy, tidal currents, and sediment size and availability. With modest changes in sea level these will not change appreciably so long as the beach face is free to migrate landwards or seawards in response to sea level changes. Thus as a first approximation, the perpetuation of existing slopes from the beach to the limit of wave action off-shore is an acceptable basis for predicting coastal changes in response to changes in sea level. There are subsidiary and complicating effects where changes in the rate of coastal erosion as a result of sea level rise produce changes in the supply of sediment to beaches and in the input to littoral drift systems. These will be considered after the direct effects of predicted sea level rise have been assessed.

If we consider a stable sector of coast, with no littoral drift, we have a situation where there must be dynamic equilibrium between the wave energy and the off-shore and beach slope. In such a situation, the Bruun approach, or some modification of it, is generally applicable. British beaches in this situation respond to seasonal changes in wave energy, such that beaches are generally high during summer and low during winter. As the sand removed to the nearshore zone in winter is placed back on the upper beach by spring and summer waves, the essential equilibrium between wave and tidal current energies and the nearshore geometry is clearly persisting over time.

In this situation, a rise of sea level will lead to an increase in water depths off the beach, resulting in an increase in wave energies at the beach face. Winter drawdown of beaches will be intensified and summer return of sand will be less effective, and if the beach is wholly natural, it will migrate inland a little way until equilibrium is regained. If sea level rise persists, the landward transgression of the beach will persist. The rate will be a function of the geometry of the system, and will be controlled by the underwater slope from the base of the beach to the limit of wave action on the sea bed. In practice the distance from the -2 m to the -10 m contour is a simple measure. If this distance is 4 km, then the off-shore slope is 1:500 and a 1 mm rise in sea level will result in the landward displacement of the coastline by 0.5 m. Pro rata, a 10 mm rise, whether in a single year or over perhaps ten years, will require a 5 m landward shift of the beach.

This approach suggests that more steeply sloping coasts will respond more conservatively to sea level rise and those with very low gradients off-shore will be most affected. For example, off Scolt Head Island, the extensive shallow water of Burnham Flats gives a ramp slope of about 1:3,000, so that we expect 1 mm of sea level rise to require a landward shift of 3 m. Off the Norfolk cliffs at Overstrand the slope down to -10 m is about 1:300, so we would only expect 0.3 m for each 1 mm rise in sea level. However, the off-shore slope is not uniform, and were we to take the distance to the -20 m contour in the case of Overstrand, we would get a slope of 1:550 and thus erosion of 0.5 m for each 1 mm rise in sea level. It should also be noted at this stage that for whatever reason, the coastline at Scolt Head Island is currently stable, whilst the cliffs near Overstrand are retreating at an average rate of about 1.65 m/yr. Thus since any effects of sea level change will be additional to the current situation, a rise of sea level of 5 mm/yr would give 15 m/yr retreat at Scolt Head Island, wholly related to the gentle off-shore slope. In contrast, the rate would increase to about 4 m/yr at Overstrand, 2.5 m/yr additional retreat from the steeper off-shore slope, on top of the current rate of about 1.5 m/yr.

We cannot assume that these changes in position would be the sole coastal response to sea level rise. The Norfolk coast is already dominated by longshore transport of sediment away from the eroding cliffs between Weybourne and Happisburgh, and particularly the sector east and southeast of Cromer. Cliff erosion can continue because the sediment reaching the beach is removed by wave energy; beaches downdrift are sustained by the sand delivered by longshore drift. Not all the sand is retained in the longshore drift system, and perhaps 25% of the littoral drift is lost off-shore from each sector (Simmonds, 1984).

The sand retained on the beach below the cliffs must, to some extent, buffer the erosion of the cliffs – the sediment from a 60 m high cliff represents an input of 60 m³/yr for each metre width, of which about two-thirds (i.e., 40 m³/yr) is sand and gravel (Cambers, 1973). However, Cambers (1973, 1976) showed that cliff height was positively correlated with long-term erosion rate for the Norfolk cliffs, i.e., even the highest cliffs do not produce enough sediment to inhibit wave attack. But we could not expect this to persist if sea level rise caused far higher rates of erosion, and there will be some upper limit to the rate at which the higher cliffs can erode. Unfortunately, we have no basis on which to estimate that limit.

More rapid erosion of the Norfolk cliffs would put more sand into the longshore drift system. Thus the beaches downdrift would be better able to resist erosion. The extent to which this would offset the tendency for erosion brought about by sea level rise is difficult to say: it may be that computer modelling of the entire sediment system and changes in water depth could produce a satisfactory simulation of the evolving system.

Changes on the North Norfolk coast are less easy to predict than elsewhere in Norfolk because the sediment supply comes from distant cliffs and input may be constrained by the capacity of the transport path rather than the volume supplied from the cliffs. Were more rapid erosion of the Holderness cliffs to result in greater sand transport to this coast, its current stability could persist even in the face of sea level rise. One line of evidence here is geological: this barrier and salt-marsh coast developed during the Flandrian transgression and has survived at least one major oscillation of sea level as shown by dated sediments at about 3,500 BP (Funnell and Pearson, 1984).

The future effect of rapid sea level rise – implications for coastal management

The Wash

There can be no question that the embankments along The Wash which keep the sea out of the Fens must continue to be effective. The total area of arable land and associated housing at risk is of the order of 5,000 km² and the farmland is some of the most productive in the UK. Raising the banks to provide a higher crest will involve widening them, and is thus rather expensive, but it will have to be done. The issue of whether they should be built entirely of clay, or of coarser materials with a higher shear strength (but with a clay core or face) deserves study. Dutch sea banks are already built to the height required for these Wash embankments in future, so experience exists.

However, it will also be necessary to consider how far the present position of the outermost bank can be retained. These banks were built on the assumption that the salt-marshes in front of them would continue to grow upward and so provide protection against wave attack, especially during surge events. With a more rapid rise in sea level it may be that these rather narrow strips of marshes do not adjust by accretion, in which case the banks will become endangered through the combination of loss of vegetation in front of them and higher sea level. Where this becomes serious there may be a case for abandoning the latest belt of reclamation and allowing (or encouraging by careful management) it to go back to salt-marsh, strengthening the older bank a kilometre or so inland as the new outer defence.

An alternative approach would be to reconsider the proposal for a barrage across the mouth of The Wash, providing the opportunity to reduce bank lengths within the barrier. However, it would not avoid the continuing threat of river flood, so that most river banks would require continued maintenance. The deep channels in the outer Wash and the lack of any great advantage in a road link along the barrage have made this a low priority in the past. Whether tidal power generation could in due course put a barrage back on the agenda remains to be seen.

Hunstanton

At Hunstanton itself, attempts will no doubt be made to retain the line of the sea wall, and to try and keep the cliff erosion close to its present level. Here, as at most coastal resorts of this type, the best solution will probably involve beach feeding. This may well prove successful in the relatively sheltered environment, at least in the first few decades of sea level rise.

North Norfolk

The largely natural North Norfolk coast has nevertheless an almost continuous flood bank for a distance of almost 40 km. Even the apparently natural shingle bank at Cley at the eastern end is maintained almost two metres above its natural crestline by bulldozing the shingle each winter, though the bank is permeable enough for flooding to occur during exceptional tides. Only at Holkham does this bank enclose any appreciable area of agricultural land, though at many sites the marsh has been modified with freshwater ponds to encourage birds.

Several of the small villages and the town of Wells along this coast will require some sort of protection from the sea. Providing flood protection will not really be feasible because the settlements lie at the head of creeks which are open to the sea. Where the settlement extends up onto rising ground there may only be a few buildings at risk, but often, as at Blakeney, such buildings have high amenity value. These are attractive and deserve protection if at all possible.

For the rest of the north Norfolk coast there are strong reasons for arguing that most areas should be allowed to revert to a more natural state, with most of the flood bank between the settlements removed. In general, such action would open up the reclaimed upper parts of the former salt-marshes; this should allow old creeks to reopen, and active creeks to carry larger flood and ebb volumes. The overall effect would be to encourage the movement of mud onto the marshes so that they can build up and so keep pace with rising sea level.

Norfolk Cliffs – Weybourne to Cromer

The cliffs from Weybourne to Cromer are partly defended by revetments, with walls at Sheringham and, for a short distance at West Runton, west of Old Woman Hythe. Analysis of the rate of erosion of this northward-facing sector (Clayton, 1989a) has shown that this is a swash-aligned coast in the sense of Davies (1980) and that littoral transport rates are relatively low. Most of the coast shows movement alongshore to the east towards Cromer. The combination of low longshore transport and the presence of chalk everywhere below low tide level gives a slowly eroding coastline, with an overall rate of erosion where there are no defences of 0.26 m/yr.

At Cromer and Sheringham substantial concrete walls have been in place for a century or more, with groynes in front of the walls. The beach at Sheringham is shingle, and loss of volume has been causing increasing problems in recent years. In the long run it will be necessary to plan for a strategic retreat to a new wall and the future line should be established so that sensible decisions may be made about the redevelopment of buildings in the zone to be lost.

Land values at Cromer are higher than at Sheringham, the beach is dominantly sandy, and the town has one of our few surviving Victorian piers (at the end of which is the lifeboat station). Although longshore drift values are higher than at Sheringham, they are still relatively low; sand placed on the beach will not be lost at an uneconomical rate if the groyne system is maintained. Holiday use of the sandy beach will reinforce the economic argument for beach nourishment, for the beach has become increasingly dominated by flints as its volume has fallen.

Norfolk Cliffs – Cromer to Cart Gap

The cliffs from east of Cromer to Cart Gap beyond Happisburgh are a drift-aligned system (Davies, 1980) with current rates of retreat exceeding 1 m/yr in the north and around 0.5 m/yr in the south. At current natural rates of retreat this 20 km of cliffs supplies about 600,000 m^3/yr of sediment, of which two-thirds is sand and gravel. Both the sand and gravel which moves through the beach system, and the mud which moves off-shore but in part finishes up on salt-marshes and in estuaries, are significant contributors to the sediment budget of this part of the North Sea littoral.

The one major investment along this coast is the gas terminal at Bacton. Based on earlier estimates of the life of the North Sea gas fields, this facility might have been running down by now, but it seems that further discoveries will keep it in use for some time to come. It may prove expensive to defend in the last few decades of its life.

With the single exception of the Bacton gas terminal, there is no long-term case for defending any of this cliffed coast. Its role in supplying sediment will increase in significance as sea level rises, and although theoretical rates of retreat could increase by three, four or even five times, this may well be moderated by the feedback of larger beach volumes at the cliff foot. Once the concept of natural erosion is established, the cliff-top area can be zoned and the life expectancy of each zone can be stated. Planning decisions, the compensation of house owners and the realignment of coastal roads could then all follow in an ordered and efficient way.

Cart Gap to Winterton Ness

Beyond the end of the cliffs at Cart Gap there is a length of coast past Sea Palling to Winterton Ness. This coast with its narrow lines of dunes was seriously breached in the 1953 surge, threatening a breakthrough into the low land of the Broads behind. A concrete wall was constructed in front of the restored dunes after 1953 and for two decades this gave little trouble, indeed at several points the wall became buried by a foredune. However, north of Sea Palling (but also at other points from time-to-time), the beach has been eroded and more of the wall has been exposed.

In the summer of 1989 a scheme for securing higher beach levels in front of this wall was proposed. This suggested building widely spaced and very massive groynes composed of large blocks of rock to be shipped from Sweden; concern has been expressed by the Countryside Commission about the visual impact of these huge structures on this coast and their effect on the southward littoral drift. It seems inevitable that this length of coast will be defended in the future as now. Hence, the issue is not the decision to defend, but the cost effectiveness and physical and visual impact of this particular solution.

Winterton Ness to Great Yarmouth

South of Winterton Ness and as far as the northern limits of Yarmouth at North Denes, the coast is often backed by higher land and erosion could not cause a breakthrough. Indeed, were it not for the promontory at Caister which is protected by a rebuilt wall, this coast would be free of serious erosion problems. The coast is eroding slowly at a rate which may increase as sea level rises. This will depend on the amount of sand delivered south of Winterton Ness. Locally there are dunes, usually badly battered by holiday makers and in serious need of proper management. Holiday homes on the dunes (as at Hemsby) need to be removed to reduce trampling and damage to the marram grass. Everywhere the upper beach should be managed to encourage the growth of foredunes. It should prove possible to manage this coast without large investment in defences by encouraging high beach levels and accepting moderate amounts of erosion from time-to-time.

At Great Yarmouth, the beach north of the river mouth is wide. If the beach suffers as a result of sea level rise, there will be no problem justifying a sand feed. The resort is one of the biggest in Britain and the tourist value of the beach is consequently very high. Even in 1973 it was estimated to be over £60,000/km/yr (Simmonds, 1976).

South of Great Yarmouth

Beach erosion is a serious problem south of the Yare entrance where Gorleston Beach has lost sand over the last 25 years. Here a sand feed is overdue; southward drift north of the harbour entrance may even justify a bypassing scheme to pump sand across the harbour mouth. Any sand reaching the beach would move away to the south and help to reduce erosion on the cliffs at Corton and on the northern side of Lowestoft Ness in Suffolk. However, renewing the defences south of the former borough boundary at Gorleston is unlikely to prove cost effective. In the long run these cliffs will have to be allowed to retreat at their natural rate. The benefits of such action will include improved sand supply to Lowestoft Ness which is protected by a sea wall and has industrial development at its southern end.

The storm surge warning and response system

After the 1953 surge, the Waverley Committee made a number of recommendations. As already noted, the sea defences were to be constructed to a level one foot (0.3 m) above the 1953 surge level. It was appreciated that this left the risk of overtopping by higher surges (the 1953 surge did not coincide with high water springs) and in any case, overtopping could occur should large waves reach the banks (a likely situation when the wind comes from east of north). For this reason, the Committee also proposed a warning and evacuation system which has been in place since 1954. It has been refined over time and, in terms of the warning given, is now sophisticated and efficient.

General warnings of the possibility of a surge are remarkably frequent – perhaps 30 in a winter. Only one or two of these pass the stage of alerting the appropriate officials and result in warnings being passed to the police. Actual public alerts, the sounding of sirens and radio/TV warnings are no more common than minor overtopping events, about once every 10–15 years.

Like many such hazard warning systems, this one is technically excellent up to the issuing of the information to the local authority and the police. Beyond that point it is virtually untested and all the evidence from the various relatively minor floods which have occurred suggest that it could never work in the way the Waverley Committee intended. Although we cannot be certain whether future climatic changes will alter the return period of catastrophic surges, the threat is real enough to be taken very seriously. This conclusion is reinforced by the considerable increase in the winter population within the areas at risk since 1953.

Conclusion – a time-scale for effective action

Some proposals for the management of the Norfolk coast made in this chapter involve acceptance of the coast as a zone of hazard. Parts of that zone should be left to natural evolution, and other parts should be covered by the surge warning system should coastal flooding threaten. Other considerable lengths of the coast should be defended whatever the costs, in places by wider beaches and dunes as much as by built structures.

Any examination of the costs and benefits of future coastal defence strategy will soon show that action needs to start now – it is no good delaying until the rate of sea level rise has increased to twice its present value – and that could be as little as a decade hence. The first stage is a careful appraisal of the coast along the lines attempted in outline here, drawing up maps of areas to be handled in the various different ways. The boundaries of land at risk from tidal surge must be established; the zone likely to be lost due to unimpeded cliff erosion must be outlined. Once these coastal hazard zones have been mapped, planning policies can develop in a rational way. If the land is lost within the next few decades, it is sheer waste to allow development, or even the extension of existing houses, especially if future policies allow full compensation to those who lose their homes and land in these circumstances. If land is at risk from surge floods, then the houses should be substantial and able to withstand immersion without collapse. Bungalows must be built with floored lofts, loft ladders and dormer windows – if they are to be permitted at all. Even in some of the developed coastal resorts where considerable expenditure on beach nourishment to retain the existing frontage is feasible, it may be wise to plan for a fall back position should sea level rise force judicious retreat.

The key to the rational management of the coastal zone depends on intelligent and early decisions about the use of land (Clayton, 1989b). By contrast, the improvement of coastal defences is relatively routine and simply involves existing techniques and policies. If these hazard zones are established early and then managed intelligently, the costs of adjustment to rising sea levels will be greatly decreased. It is human use of the coast which has caused the problems of coastal erosion and tidal flood. A rationalization of that use in the face of increasing hazard will keep those problems under control.

References

Anglian Water (1988). *Anglian Coastal Management Atlas*. Sir William Halcrow and Partners.
Bruun, P. (1962). Sea level rise as a cause of shore erosion. *Proceedings A.S.C.E., Journal of the Waterways and Harbours Division* **88**, 117–30.
Cambers, G. (1976). Temporal scales in coastal erosion systems. *Trans. Inst. Brit. Geogr.*, 1(2), 246–56.

Cambers, G. (1973). *The Retreat of Unconsolidated Quaternary Cliffs*. PhD thesis, University of East Anglia, Norwich.

Clayton, K.M. (1989a). Sediment input from the Norfolk cliffs, eastern England – a century of coast protection and its effect. *J. Coastal Res.*, **5**(3), 433–42.

Clayton, K.M. (1989b). The implications of climatic change. In *Coastal Management*. Thomas Telford, London, pp. 165–76.

Clayton, K.M., McCave, I.N. & Vincent, C.E. (1983). The establishment of a sand budget for the East Anglian coast and its implications for coastal stability. In *Shoreline Protection*. Thomas Telford, London, pp. 91–6.

Davies, J.L. (1980). *Geographic Variation in Coastal Development. 2nd edition.* Longman.

Fleming, C.A. (1989). The Anglian sea defence management study. In *Coastal Management*. Thomas Telford, London, pp. 153–64.

Funnell, B.M. & Pearson, I. (1984). A guide to the Holocene geology of the North Norfolk coast. *Bulletin Geological Society Norfolk*, **34**, 123–40.

Simmonds, A.C. (1984). *The Evolution of the East Anglian Beaches 1974–1980*. PhD thesis, University of East Anglia, Norwich, 2 volumes.

Simmonds, A.C. (1976). *The Recreational Value of the Beaches. Report 4, East Anglian Coastal Research Programme*. University of East Anglia, Norwich.

Waverley Committee (1954). *Report of Departmental Committee on Coastal Flooding*. HMSO, London, Cmd. 9165.

20

What will happen to The Netherlands if sea level rise accelerates?

J.G. de Ronde

Abstract

The likely impacts of arbitrary increases in sea level of 1.0, 2.5 and 5.0 m over 100 years on the defence structures, morphology, ecosystem stability and water resources of The Netherlands are investigated. A two-dimensional computer model of the North Sea and part of the continental shelf was used to calculate the tidal system changes as a result of sea level increases of 2.5 and 5.0 m. In general, greater tidal motion and increases in wave height resulted, assuming an unchanging sea bed. In order to determine the complex morphological response the coast of The Netherlands was considered in three different parts. Dune areas are likely to retreat unless strengthening occurs, and in the Wadden Sea and delta areas erosion will be governed by changes in the cross-section of the tidal inlet and changes in sedimentation rates. Impacts of sea level rise include greater salt intrusion, creating problems for agricultural and domestic water supplies; in order to counteract this, an increase in the number and strength of pumps may be necessary. Ecological impacts may include a decrease in dune and intertidal areas and a subsequent decrease in the number of summer breeding bird species as well as migratory species. In order to maintain and strengthen coastal and river defence systems to cope with a 1 m rise in sea level over 100 years the maintenance costs will approximately double.

Introduction

There is little doubt that due to global warming the sea level along the coast of The Netherlands will rise more rapidly in the coming century than in the recent past. During the last 100 years relative sea level rose by 15–20 cm (de Ronde, 1983). There are large uncertainties in predictions of future global mean sea level rise (Bolin *et al.*, 1986; van der Veen, 1986; Wigley and Raper, this volume). However, recent best estimates (see Wigley and Raper, this volume; Woodworth, this volume; Warrick and Oerlemans, 1990) imply that The Netherlands will experience a future rate of sea level rise that, at a minimum, is at least twice that of the recent past.

In this presentation arbitrary increases in sea level of 1.0, 2.5 and 5.0 m over the next 100 years, up to 2090, are chosen in order to investigate the possible impacts of future sea level rise on the Dutch lowlands. Whereas others (e.g., van der Kley, 1987; Hekstra, 1986) have discussed such impacts on The Netherlands in general, the present study focuses on the effects of sea level rise in more detail. Some of the results presented herein are based on a Dutch report entitled '*Zeespiegelrijzing, Worstelen met Wassend Water*' (J.G. de Ronde *et al.*, 1986).

A sea level rise of 1 m or more has serious implications for:

* *The safety of dykes*, and other defence structures along the coast and mouths of the main rivers. Sea level rise increases the risks of overtopping.
* *The morphology of dunes, the shore face, and estuarine systems*. A critical issue is whether the sea bed can rise as fast as the sea level.
* *The stability of ecosystems*. How will vegetation and animal life change with sea level rise? The Wadden Sea is of great importance for North Sea fish and migratory birds.
* *Water resources*. Increasing seepage of saltwater into inland polders and increasing salt intrusion via the rivers are of great concern. The present natural drainage of the IJssel Lake during low tide may have to be replaced by a pumping system. Storage of drinking water below the dunes may diminish.
* *Other structures*. Harbours, bridges, locks etc. may require alteration.

Preliminary impact studies by Rijkswaterstaat were first conducted with an extreme scenario of a 5 m sea level rise in order to gain a better understanding of the major mechanisms involved; alternative scenarios of 2.5 and 1.0 m sea level rise were later included. The results obtained with the various scenarios will be discussed in the following sequence: impact on the tidal system in the North Sea, on wave propagation, on coastal morphology, on salt intrusion, on water management and on ecology. Finally, the costs involved will be discussed.

Impacts on the tidal system in the North Sea

A two-dimensional hydraulic computer model (WAQUA) of the North Sea and part of the Continental Shelf (Verboom *et al.*, 1991; see Fig. 20.1) was used to calculate tidal system changes compared to the present situation for sea level scenarios of 2.5 and 5 m higher, and 2.5 m lower, than today (de Ronde and de Ruijter, 1986; Bavelaar, 1987).

The main effect was that at higher sea levels and greater water depths the tidal wave propagated faster and tidal amplitudes (water levels and currents) increased. A second effect was the change in the positions of the amphidromic points (equivalent to nearly zero amplitude; Fig. 20.2). Thus, a location from which the amphidromic point retreats will experience a higher amplitude, and a location approached by the amphidromic point will experience a lower amplitude. The combination of these two effects resulted in an increase of the tidal amplitude along most North Sea coastlines (Fig. 20.3). Only locations near the Cromer–Den Helder line experienced a decrease as the southern amphidromic point drew nearer. Here the decrease in tidal amplitude due to the approaching amphidromic point is stronger than the increase due to the greater water depth.

For the North Sea in general (except near the Cromer–Den Helder line) a higher sea level caused a greater tidal motion. Water levels not only rose because of the sea level rise but also because of the increased amplitude, although the latter impact was much smaller. On the English coast, for instance (Bavelaar, 1987), a sea level rise of 2.5 m gave a rise of the high water levels of between 2.4 and 2.7 m; the smallest rise was found in Cromer, the greatest rise in Southend. Along The Netherlands coast the maximum

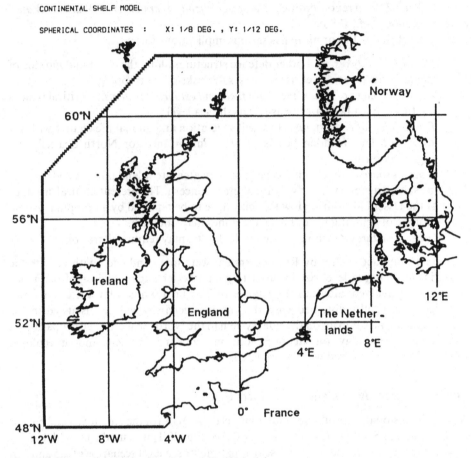

Fig. 20.1 The outline of the Continental Shelf Model.

increase was at Hoek van Holland and the minimum at Den Helder. A rise of 1 m would thus cause a rise of mean high water between 0.95 and 1.1 m.

In the case of storm surges, model results showed (de Ronde and de Ruijter, 1986) that a 5 m increase in sea level caused a change in the storm surge heights of between − 20 cm and + 10 cm while wind velocities remained unchanged. A sea level rise of 1 m would cause changes of less than 5 cm (also see Flather and Khandker, this volume). The effect of sea level rise on storm surges is thus very small and the possible increase or decrease in the number or severity of storms will be much more important for changes in storm surge heights. On the global-scale the temperature gradient between the equator and poles is likely to decrease with greenhouse warming, perhaps resulting in generally less and minor depressions, but the effects on regional storms in general, and on the North Sea in particular, cannot yet be predicted (Mitchell *et al.*, 1990).

Another feature which may be important is a change in the residual transports, i.e., the surplus of water going in (or out) of the tidal inlet averaged over an astronomical tidal cycle. This will have impacts on the morphology and the marine ecosystem, especially in the Wadden Sea. Model results (Elorche and Plieger, 1988) show a change

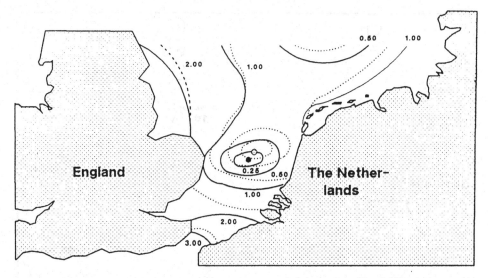

(a) Iso-amplitude lines in m

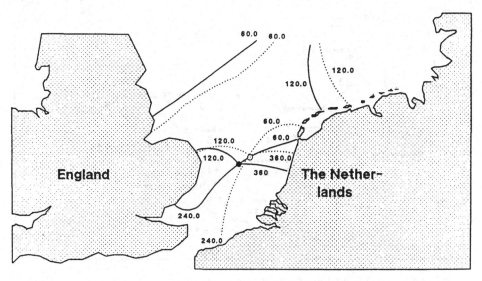

(b) Iso-amplitude lines in degrees

————— present situation

············· 5m plus situation

Fig. 20.2 The change in the semidiurnal amphidrome in the Southern North Sea, if sea level rises 5 m.

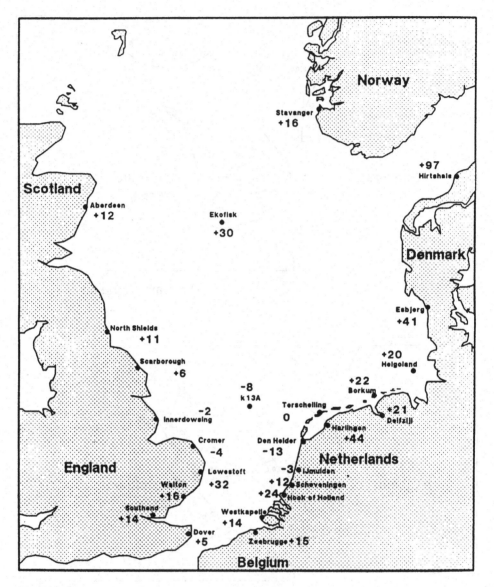

Fig. 20.3 Change (%) in tidal amplitude, if sea level rises 5 m.

in the residual transports when the sea level rises by 1 m (Fig. 20.4), but results are still too inaccurate for quantitative use.

Impact on waves and swell

A higher sea level and a greater water depth cause less dissipation of waves and swell. Wave heights increase, especially in shallow water. Many of the smaller waves, which break and dissipate on shallow sandbanks in the present situation, will continue unbroken in the case of increased water depth. For instance, calculations of the wave

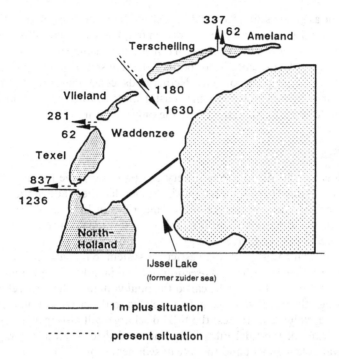

Fig. 20.4 The change in the residual transports in the Wadden Sea, if sea level rises 1 m.

height with the HISWA wave model (Holthuijsen *et al.*, 1988) near the Haringvliet sluices, a location behind a shallow sandridge called the Hinderplaat, showed an increase in wave height from 90 cm to 210 cm when the sea level rose 5 m over an unchanged sea bed. However, if the sea bed were to rise as fast as the sea level, the wave height would hardly change. In reality the outcome will depend on the change in the morphology.

Impact on the morphology

The Netherlands coast can be roughly divided into three different parts: the Delta area in the south with the Eastern and Western Scheldt; the 'closed coast' between Hoek van Holland and Den Helder without important inlets; and the Wadden Sea area with the inlets between the islands. The easiest part to study is the closed coast.

The closed coast

The coastal defences of the closed coast consist mainly of dunes. The present situation is close to stable but a sea level rise threatens to destroy this delicate balance and will cause the coastline to retreat. The present beach slope is about 1:60, so a rise of 1 m may result in a retreat of 60 m. A complicating factor is that the dunes are composed of finer grain sand than that found on the beach and shore face. The expected slope of a beach consisting of this finer sand will be less than 1:60 and consequently the retreat of the coastline may be as great as 80–150 m. After such a retreat the dunes may not be strong enough to withstand a design storm surge (design frequency 10^{-4}) resulting in the need to strengthen parts of the dunes. A retreat of dunes and coastline would be undesirable in many cases due to the presence of harbours, other constructions and valuable dune areas. Hence, for large parts of The Netherlands coast measures such as sand replenishment may be necessary. For the coastal stretch protected by dykes (like the Hondsbossche Zeewering) strengthening is the only solution.

The Wadden Sea area

The morphology of the Wadden Sea is very complex. Even the present situation is difficult to model. Added to this uncertainty is the fact that the Wadden Sea has still not attained equilibrium since the closure of the Zuider Sea in 1932 (now IJssel Lake). In this section some hypotheses are put forward in order to make a rough prediction of what will happen to the Wadden Sea.

The morphological adjustment of an estuarine system to a rising sea level is mainly determined by the change in the cross-section of the tidal inlet. (The volume of water going in and out during a tidal cycle, called the tidal volume, is linearly related to the discharge through the inlets which is, in turn, related to the cross-sectional area of the inlet). If the tidal volume increases, the tidal discharge will also increase as will the cross-sectional area of the tidal inlet. The time needed to reach a new equilibrium depends on available sediment and the rate of sedimentation.

In the case of a sea level rise three situations may occur:

a. The tidal volume may increase slightly. This will happen if there are few intertidal flats between the high- and low-water line which are flooded at every tidal cycle. The cross-section of the inlet will increase and as a result the tidal velocities will decrease. Smaller velocities result in sedimentation which reduces the tidal volume until a new balance has been reached with the cross-sectional area of the tidal inlet.

b. The tidal volume may slowly increase, but proportionally slower than the cross-sectional increase of the tidal inlet. The resulting sedimentation will be less rapid.

c. The proportional increase of the tidal volume is more rapid than that of the cross-section of the tidal inlet. This is the situation when the intertidal area is relatively large. In this case erosion will be strong especially in the tidal gullies.

Which of the above will occur in the Wadden Sea? There is presently a transport of sand from the North Sea into the Wadden Sea of 30×10^6 m^3/yr (standard deviation 15×10^6). If sea level rises 1 m in 100 years, an addition of 28×10^6 m^3/yr sand is needed in order for the sea bed to keep pace. For the area at large, bottom rise can probably keep up with this rate of sea level rise.

For more detail, the Wadden Sea was divided into 10 smaller tidal systems, each with its own inlet which can be characterized by the situations (a), (b) and (c) described above. Situation (a) is applicable for only one of the tidal systems (inlet South of Texel with 25% of the total Wadden area) where sedimentation will occur. Situation (c) is also applicable for only one tidal system (inlet west of Ameland with 10% of the total area) where erosion will occur. For the other 8 tidal systems, situation (b), with sedimentation, will be applicable, although sedimentation will probably not be enough in two or three cases for the bed to keep pace with the sea level rise.

It should be stressed once more that the morphology of the Wadden Sea is very complex and that the sedimentation theory used here is very simple. For example, only 25% of the present sedimentation (30×10^6 m³/yr) can be traced to the coast of North Holland, the coasts of Texel and Vlieland and the adjacent North Sea, whereas the origin of the remainder is more diffuse.

The Delta area

In the Eastern and Western Scheldt, the proportional increase in tidal volume will be relatively larger than that of the cross-sectional area of the tidal inlets (situation (c)). Hence erosion will occur and the intertidal area will probably decrease. But here, too, it has to be stressed that the processes involved are very complex.

Salt intrusion

Besides pollution of the Rhine by the French salt mines, there are two major ways in which saltwater can penetrate into The Netherlands. The most direct way is from the sea into the river mouths, for which the Rotterdam Waterway is the most notorious and the major problem. With the increase in the size of shipping coming to Rotterdam, the depth of the Waterway had to be increased. This caused the salt intrusion to become more severe and penetrate farther inland, creating problems for agricultural and domestic water supplies at water inlets. Since the new harbour area, Europoort, was completed a decade ago further dredging of the Rotterdam Waterway is no longer necessary, and the depth could even be decreased.

A rise in sea level, however, could cause new problems of salt intrusion. Figs. 20.5 and 20.6 show the results of model calculations with a sea level rise of 5 m over the present bottom topography. The Haringvliet–Hollandsch Diep area, important for freshwater supply, was contaminated with saltwater. Saltwater built up in the deeper parts of the Haringvliet where it would be extremely difficult to remove. However, if the bottom of the river system (notably the Waterway) were to rise as fast as the sea level, the salt intrusion of course would stay as it is today.

The second major route of saltwater intrusion is via groundwater flow through the subsoil (seepage). Large areas of The Netherlands are below sea level. This causes brackish groundwater to penetrate (although very slowly) parts of the groundwater system which are deeper and further inland. In many polders the seepage rate is more than 0.25 mm/day which, without counter-measures, would change the pastures into salt-marshes. The solution is the flushing and rinsing of polders and canals with freshwater. However, in very dry summers not enough freshwater is available and crop damage is unavoidable. The amount of seepage is linearly related to the difference in sea and polder water levels, to the distance from the sea and to the permeability of the subsoil. Thus a higher sea level would cause a larger gradient and hence an increase in seepage.

In certain areas where little or no seepage presently occurs, a sea level rise would cause seepage to start. Fig. 20.7(a) shows the present seepage rate and Fig. 20.7(b) the rate if sea level were 5 m higher. Simple calculations showed an increase in the total seepage of up to 300% when sea level increased by 5 m. The amount of freshwater needed for flushing of the polders would consequently increase by about 50%. In the case of a 1 m rise this increase will be about 10%. This extra amount is already

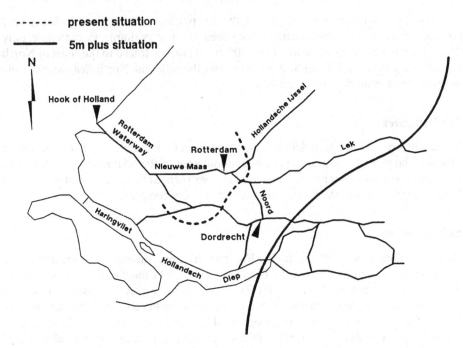

Fig. 20.5 The change in the salt intrusion in the Rhine estuary, if sea level rises 5 m.

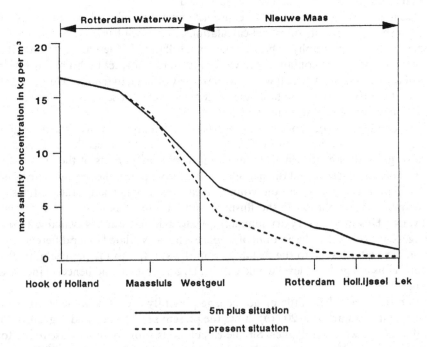

Fig. 20.6 The change in the maximum salinity concentration during a tidal cycle in the Rhine estuary, if sea level rises 5 m (see Fig. 20.5 for the locations).

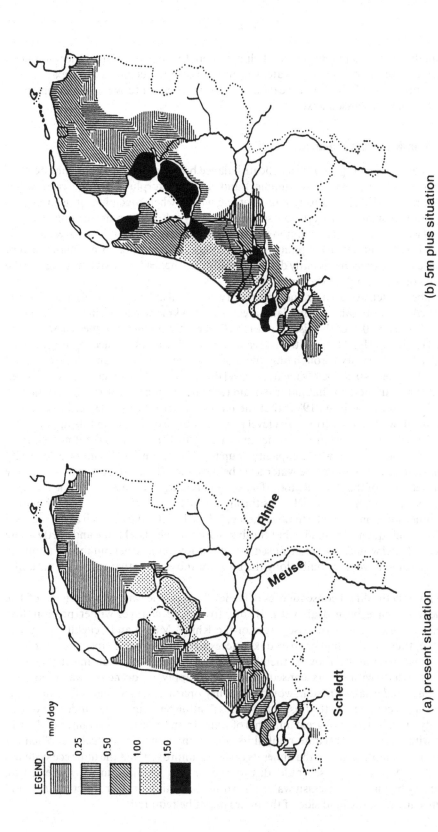

LEGEND

mm/day

0

0.25

0.50

1.00

1.50

Rhine

Meuse

Scheldt

(a) present situation

(b) 5m plus situation

Fig. 20.7 The change in the seepage rate in the Netherlands, if sea level rises 5 m.

unavailable in dry summers and so, if climate were to change towards drier summers, stringent conservation of winter water surpluses to counteract increased seepage might become necessary. This would create a conflict with regard to water management at large, as will be discussed next.

Impact on water management

Water management is presently a delicate balance between counteracting the effects of saltwater intrusion (mainly in summer) and draining surplus rain and river water (Rhine, Meuse, Scheldt) during and after the winter when precipitation is large and evaporation is small. In areas below sea level the only solution is pump drainage. In the event of a higher sea level all pumping stations would have to be rebuilt to overcome the greater height, and stronger pumps would be needed, in order to maintain current standards. Areas which presently drain naturally into the sea by gravity may require the building of new pumping stations.

The most extensive area drained by gravity, which also has a large discharge, is the IJssel Lake and the surrounding land area of 15,000 km² which drains into the IJssel Lake. Moreover 10% of the discharge of the River Rhine comes into the Lake from the River IJssel. The IJssel Lake with an area of only 1,200 km², is drained directly into the Wadden Sea by gravity through only two sluices in the enclosing dam. The water level of the lake is kept 40 cm below mean sea level during winter and 20 cm below mean sea level during summer, so that natural drainage is permanently possible at low tide. A study has shown (de Jong, 1985) that the natural drainage of the IJssel Lake can be maintained, when considering a sea level rise of about 50 cm, without raising the water level of the lake. A substantial further increase up to 1 m might make it necessary to build a pumping station with a capacity of up to 120×10^6 m³/day. This large capacity will be needed to keep the lake water level below certain limits to cope with extremely high discharges of the River Rhine. (The design discharge of the River Rhine is 16,500 m³/s, about 1,560 m³/s of which goes through the River IJssel into the IJssel Lake).

Another solution would be to raise the level of the IJssel Lake as much as the sea level rise. The consequence would be that the dykes around the IJssel Lake and partly along the River IJssel would have to be raised as well. Moreover a larger number of pumping stations would have to be built along the Lake for those areas that still drain naturally into it.

So far only the direct consequences of sea level rise on the water management of The Netherlands have been dealt with. If, due to climate change, winter precipitation increases, the extreme river discharges of the Rhine, Meuse and Scheldt might also increase, thus making higher river dykes necessary. If summer evaporation becomes higher and precipitation lower, then the shortage of freshwater during the summer could increase, with all its consequences. The supply of domestic water, already mentioned under salt intrusion, would be put in jeopardy. A major part of the domestic water supply comes from the freshwater lenses floating on saline water under the dunes. Already, the extraction of great amounts of water from these lenses requires replenishment with river water because rainfall itself over the dunes is insufficient. When the river water is infiltrated into the dunes its quality is improved, but the dune ecosystem is degraded. A sea level rise could distort the present equilibrium of these lenses, essentially by pushing brackish water up. More infiltration of river water and/or sand replenishment on the land side of the dunes might be required.

Ecological impact

For a sea level rise of 1 m within 100 years, we briefly speculate on the ecological impacts on the following three coastal areas and the lowland rivers.

The closed coast

The total dune area would diminish, with a strip of width 80–150 m disappearing (as discussed above). Assuming a strengthening of the dunes (where needed) with sand nourishment (not by dykes) the ecological consequences along the closed coast would be small. It is assumed that the beach and coastal slope can be maintained, naturally or artificially.

The Wadden Sea area

The total extent of the intertidal area is of great ecological importance. This is the only place where summer breeding bird species can forage. A decrease in the intertidal area would certainly reduce breeding populations and the number of breeding species. However, it is the migratory species from Scandinavia and the Arctic that either over-winter here or migrate through, which will suffer the most from a reduced intertidal area. In the case of the Wadden Sea the assumed sea level rise would probably not significantly affect the extent of the intertidal area, so that the ecological impact will be small. As mentioned previously the residual transports in the Wadden Sea may change. This can affect the underwater ecosystem, although this is still difficult to predict or model.

The Delta area

In the Eastern and Western Scheldt the intertidal area is likely to decrease. This may have the same negative ecological consequences as mentioned above on the migratory birds and, to a lesser extent, on the breeding birds.

The River area

In a great part of the lowland river system in The Netherlands mean and extreme water levels would increase with a sea level rise. Floodplains would be inundated more frequently during the winter (or most of the year) and this would greatly change their ecosystems.

Economic implications

With a 1 m sea level rise over a century, present maintenance costs for coastal and river defence systems would approximately double. The present cost of coastal maintenance amounts to about 60×10^6 guilders per year. The strengthening of dykes, dunes, beaches and shore faces over the next century would cost about 14×10^9 guilders (7×10^9 US dollars). A storm surge barrier in the Rotterdam Waterway is currently under construction as an alternative to the very high cost of strengthening the dykes along the lower river branches. The prospect of an increasing rate of sea level rise makes this alternative all the more attractive.

Although the structural adjustments required would be more complex than for the coastal defence, the amount of money involved is less. An investment of 1×10^9 guilders would be needed for pumps and changes in the infrastructure. The annual input costs (fuel) will increase by about 10×10^6 guilders per year. Furthermore, harbours, locks, bridges, etc. would have to be adapted to the new situation; these costs would be roughly 3×10^9 guilders (de Ronde and de Vrees, 1991).

The total costs of adaptation to a 1 m sea level rise over one century can be estimated at about 20×10^9 guilders (10×10^9 dollars). For comparison, the costs of the Delta Works (1958–88) were about 14×10^9 guilders.

Conclusions

A future acceleration of sea level rise will have great consequences for a low lying country like The Netherlands. As a result of a 1 m sea level rise:

* the tidal motion in the North Sea and along the Dutch coast will hardly change. Mean high water levels will rise 0.95–1.1 m.
* the storm surge heights will change by less than 5 cm. The effect due to changes in storm frequency or severity of storms will be much more important, although at present it is unknown if frequency and severity will increase or decrease.
* the wave heights will change. To predict the changes in wave height it is necessary to know the morphological changes first.
* the dune coastline will retreat by 80–150 m.
* the sedimentation in the Wadden Sea will increase and probably keep pace with this sea level rise, so mean water depth and intertidal area will probably remain constant.
* the sedimentation in the Eastern and Western Scheldt will hardly increase, so mean water depth will increase and intertidal area decrease.
* the salt intrusion via rivers and subsoil will increase. During summer the amount of freshwater needed for flushing of the polders will increase by about 10%.
* the water management of The Netherlands will have to be adapted. Large areas that presently drain by gravity will be unable to do so in future and so large pumps will have to be installed.
* the ecological systems will be affected. Intertidal area in the Eastern and Western Scheldt and valuable dune area will diminish. The ecological value of these areas will be lost.
* the extra costs to raise the defences against high water and to adapt the water management system will be about 20×10^9 guilders (10×10^9 US dollars).

Acknowledgements

This report is based on the results of the work of many people. I would specially mention C.H. de Jong for his work on the consequences of sea level rise for the IJssel Lake and the authors of the Dutch report 'Zeespiegelrijzing-worstelen met wassend water' (GWAO-86.002 of Rijkswaterstaat, Tidal Waters Division): J.P. Boon, B.J.E. ten Brink, A. van der Giessen, D.J. de Jong, V.N. de Jonge, L.H.M. Kohsiek, D.J.

Kylstra, R. Misdorp, M. Pluijm, J.H. de Reus, J.G. de Ronde, W.P.N. de Ruijter, W. Verbakel, J.A. Vogel, A. van der Wekken, J. Wiersma.

References

Bavelaar, A. (1987). *Verandering van de getijbeweging in de Noordzee t.g.v. zeespiegelrijzing.* GWAO-87.205, Rijkswaterstaat, Tidal Waters Division (in Dutch).

Bolin, B., Döös, B.R., Jäger, J. & Warrick, R.A. (eds.) (1986). *The Greenhouse Effect, Climatic Change and Ecosystems, SCOPE 29.* John Wiley and Sons, Chichester, UK.

de Jong, C.H. (1985). *De veiligheid van Nederland bij grote zeespiegelrijzing.* Rijkswaterstaat, Dir. Waterhuishouding en Waterbeweging (in Dutch).

de Ronde, J.G. (1983). Changes of relative mean sea level and of mean tidal amplitude along the Dutch coast. In *Proceedings NATO Advanced Research Workshop Utrecht, Seismicity and Seismic Risk in the Off-shore North Sea Area, June 1–4, 1982,* A.R. Ritsema and A. Gürpinar (eds.).

de Ronde, J.G. & de Ruijter, W.P.M. (eds.) (1986). *Zeespiegelrijzing, worstelen met wassend water.* GWAO-86.002, Rijkswaterstaat, Tidal Waters Division (in Dutch).

de Ronde, J.G. & de Vrees, L.P.M. (1991). *Rising Waters; Impacts of the Greenhouse Effect for the Netherlands.* GWAO-90.026 Rijkswaterstaat, Tidal Waters Division.

Elorche, M. & Plieger, R. (1988). *Verandering van de waterbeweging in de Westelijke Waddenzee t.g.v. een zeespiegelrijzing van 1 meter.* GWAO-88.404 Rijkswaterstaat, Tidal Waters Division (in Dutch).

Hekstra, G.P. (1986). Will climatic change flood The Netherlands? Effects on agriculture, land-use and well-being. *Ambio,* 15(6), 316–26.

Holthuijsen, L.H., Booij, N. & Herbers, T.H.C. (1988). A prediction model for stationary, short-crested waves in shallow water with ambient currents. *Coastal Eng.,* 13, 23–54.

Mitchell, J.F.B., Manabe, S., Tokioka, T. & Meleshko, V. (1990). Equilibrium climate change. In *Climate Change: The IPCC Scientific Assessment,* J.T. Houghton, G.J. Jenkins and J.J. Ephraums (eds.). Cambridge University Press, Cambridge.

van der Kley, W. (1987). Sea level rise, evaluation of the impacts of three scenarios of sea level rise on flood protection and water management of The Netherlands. In *Impact of Sea Level Rise on Society,* H.G. Wind (ed.). A.A. Balkema, Rotterdam, Brookfield.

van der Veen, C.J. (1986). *Ice sheets, Atmospheric CO$_2$ and Sea Level.* Dissertation University of Utrecht.

Verboom, G.K., de Ronde, J.G. & van Dijk, R.P. (1992). A fine grid tidal flow and storm surge model of the North Sea. *Continental Shelf Res,* 12 (2/3), 213–33.

Warrick, R.A. & Oerlemans, J. (1990). Sea level rise. In *Climate Change: The IPCC Scientific Assessment,* J.T. Houghton, G.J. Jenkins and J.J. Ephraums (eds.). Cambridge University Press, Cambridge.

21

The vulnerability of the east coast of South America to sea level rise and possible adjustment strategies

E.J. Schnack

Abstract

The east coast of South America is located on a 'passive' margin. There exists a range of coastal environments, from sandy shorelines to wetlands, from rock to poorly consolidated cliffs. Two major river systems, the Amazon and Parana-La Plata, drain extensive catchments.

The coastal plains, despite their climatically controlled diversity, exhibit a striking similarity in their Quaternary sea level history. Pleistocene and Holocene emerged shorelines have been recorded from French Guyana to Tierra del Fuego. Except for a few localities, the observational record does not provide sufficient evidence to identify a secular rise or fall in mean sea level for most of this region.

Whether or not sea level is rising, the coast is already experiencing serious problems. Shore erosion is severe, particularly in densely-populated areas of Argentina, Brazil and Uruguay. Beach erosion and cliff retreat are common in several localities, very often due to mismanagement of the coastal zone (beach sand mining, engineering structures). Dune fields are being urbanized, causing greater impermeability of the terrain and preventing the sediment exchange within the littoral zone. Much of the wetland areas are still undeveloped, but as in some places in Brazil, human occupation is causing their deterioration and disappearance.

A sea level rise as a result of the 'greenhouse' effect would only exacerbate the already existing processes of coastal degradation. Adjustment strategies should, therefore, take into account integrated approaches in which 'sea level' will have to be considered an additional factor. Both corrective and preventive strategies will be needed to avoid serious environmental and economic disturbances over the next decades. However, due to the pressing needs of economic development in these countries, it is difficult to foresee that the 'coastal issue' will be given high priority.

Regional background

From a geological standpoint, the east coast of South America (Argentina, Brazil and Uruguay) is located on a 'passive' margin. The structural features include different major geological units. As is shown in Fig. 21.1, the general basement is predominant in a large area of Brazil and Uruguay, while Paleozoic/Mesozoic and Cretaceous/Cenozic sedimentary basins are distributed along the margins of the whole region (Urien and Martins, 1979).

The climate varies from equatorial humid in the north at the Amazon region to sub-

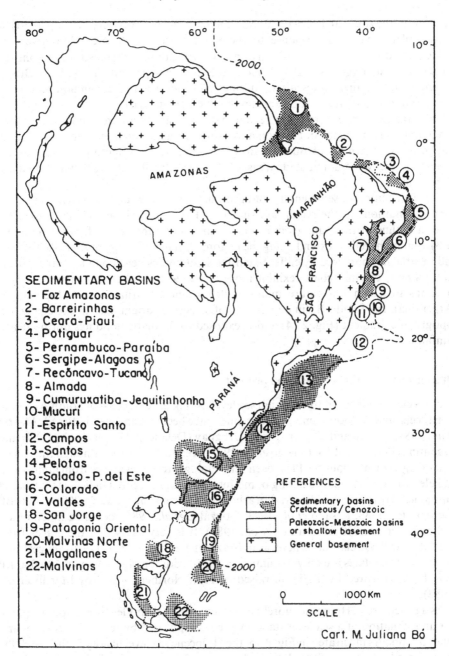

SEDIMENTARY BASINS
1- Foz Amazonas
2- Barreirinhas
3- Ceará-Piauí
4- Potiguar
5- Pernambuco-Paraíba
6- Sergipe-Alagoas
7- Recôncavo-Tucano
8- Almada
9- Cumuruxatiba-Jequitinhonha
10-Mucuri
11-Espirito Santo
12-Campos
13-Santos
14-Pelotas
15-Salado-P. del Este
16-Colorado
17-Valdes
18-San Jorge
19-Patagonia Oriental
20-Malvinas Norte
21-Magallanes
22-Malvinas

REFERENCES

Sedimentary basins
Cretaceous/Cenozoic

Paleozoic-Mesozoic basins
or shallow basement

General basement

0   1000 Km

SCALE

Cart. M. Juliana Bó

Fig. 21.1 Main sedimentary basins and basement distribution in east South America (after Urien and Martins, 1979, modified in Martins and Willwock, 1987).

antarctic in the south at Tierra del Fuego. Different ecosystems have developed on the coastal plains partly in response to these differences in climate. Mangroves are particularly dominant in the northern part of Brazil, although they also occur along the southern coast. Lagoons and salt-marshes are mostly present in Southern Brazil, Uruguay and northern Argentina, although marshes are also a common feature of meso- and macro-tidal environments on the Patagonian coast.

Tidal ranges vary according to region. Macro-tidal ranges are found in northern Brazil and in many areas in Patagonia. Meso-tidal ranges predominate in Patagonia and in the southern tip of Buenos Aires Province, whereas micro-tidal regimes are typical of the wave-dominated areas of Eastern Brazil, Uruguay and Northern Argentina.

The distribution of sedimentary and geomorphic features and the general environmental conditions of the region are shown in Fig. 21.2. Schnack (1985) presents a general outline of the main geomorphic features along the coast of Argentina (Fig. 21.3). Sandy shorelines extend along the region, both in low-lying, barrier-like settings and in stretches bordered by cliffs. Extensive sand bodies develop in low areas, while mountainous or cliffy coasts exhibit more restricted littoral environments. Coastal lowlands are a common feature, mainly at deltaic and estuarine areas. At the Pampas (Argentina), the Salado Basin depression, of structural origin, has an extremely low topographic gradient (Fig. 21.4) and is exposed to dramatic floodings during heavy rainfalls.

Quaternary sea levels – past and present

There is clear evidence of former high sea levels along the east coast of South America. Pleistocene and Holocene emerged shorelines have been documented in Brazil (Martin et al., 1987; Suguio and Tessler, 1984), Uruguay (Delaney, 1966; Jackson, 1984) and Argentina (Feruglio, 1950; Fidalgo et al., 1973; Frenguelli, 1950). Although there is no general agreement about the Pleistocene chronology, recent evidence from amino acid and electron spin resonance dating of mollusc shells in Patagonia indicates that raised shorelines, previously yielding radiocarbon ages within the range 25,000–40,000 BP (Codignotto, 1983; Fasano et al., 1983; González et al., 1986), could be much older, perhaps over 100,000 BP (Rutter et al., 1987). It is very likely that these ages from Argentina will correlate elsewhere, as in Brazil where Suguio et al., (1980) suggest that the ^{14}C derived dates are only 'minimum' ages. In addition, ages of about 120,000 BP have been measured by Th/U on raised coral in Northern Brazil by Martin et al., (1982).

As a general rule, Holocene shorelines are lower than their Pleistocene counterparts, but an estimation of a sea level stand may be complicated by tectonics, isostasy and other factors. As illustrated in Fig. 21.5, the Holocene sea level history in the last 8,000 years, based on radiocarbon dating, shows a maximum peak of about 5 m above present mean sea level around 5,000 BP in Brazil (Martin et al., 1987; Suguio and Tessler, 1984); sea level was 2.5 m higher at the southeastern coast of Buenos Aires Province (Mar del Plata area; Schnack et al., 1987 and submitted) and nearly 5 m higher in Tierra del Fuego (Porter et al., 1984) between 5,000 and 6,000 BP.

Recent sea level variations have been analysed by Pirazzoli (1986). Of the 10 stations corresponding to East South America (Fig. 21.6), only three show sufficiently long

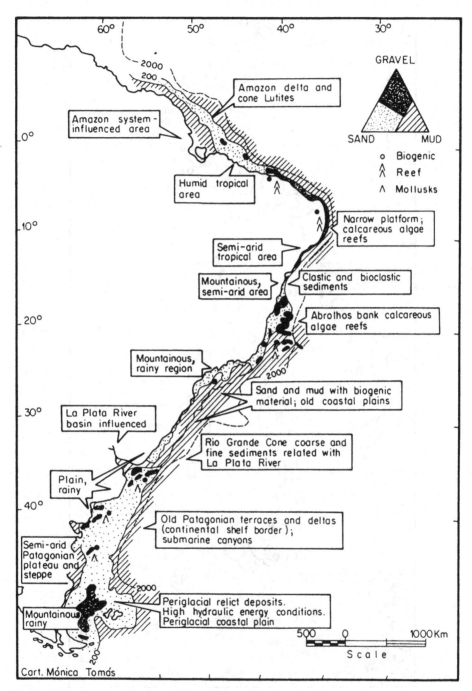

Fig. 21.2 Distribution of main surface sedimentary textures and general environmental settings along the east South America margin (after Urien and Martins, 1979, modified in Martins and Willwock, 1987).

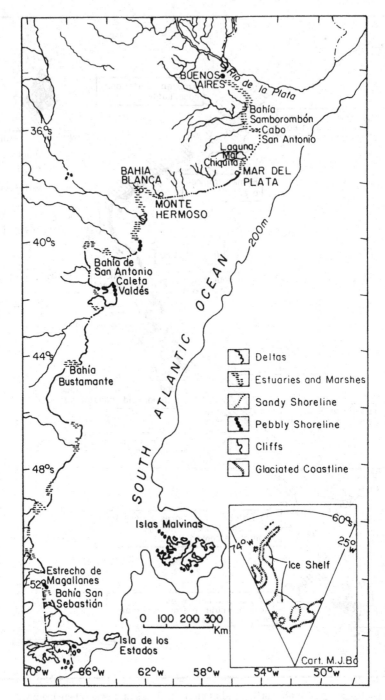

Fig. 21.3 Predominant geomorphic features along the Argentine coast (after Schnack, 1985).

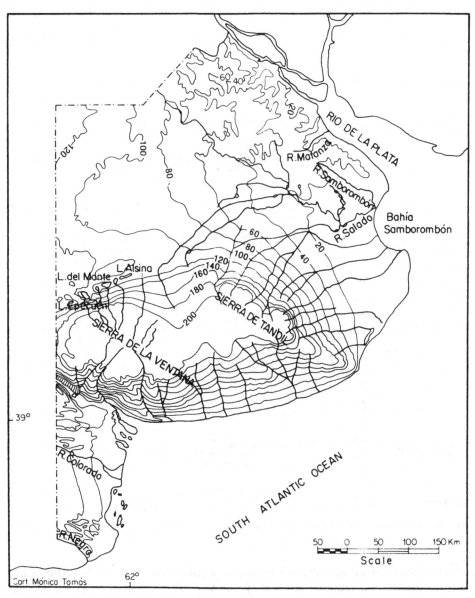

Fig. 21.4 General topography of the Province of Buenos Aires, to show the very low gradient in the Salado Basin region. Contours in metres.

records to allow for an interpretation of secular tendencies. The long-term trend seems to be quite stable in Mar del Plata, slightly rising (+ 0.7 mm/yr) in Buenos Aires where fluvial discharge and meteorological effects should be influential, and rapidly rising in Comodoro Rivadavia. Brandani *et al.*, (1985) suggest a slightly rising trend for Mar del Plata, but the records are only about 30 years long. In the case of Comodoro Rivadavia, the fact that oil and gas pumping have been occurring over many decades calls for a careful interpretation of this striking trend. Here the observed relative sea level rise

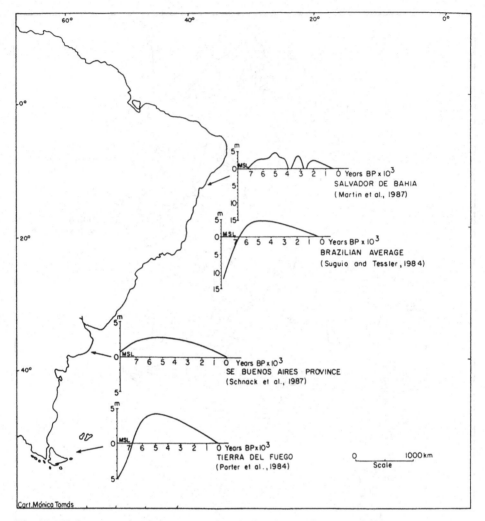

Fig. 21.5 Holocene sea level changes on the east South America coastal plains.

could be caused by surface lowering due to compaction of the reservoir. In addition, the records show an interruption which may lead to a misinterpretation. The remaining station records along the coasts of Uruguay and Brazil are too short to say anything meaningful about secular trends.

Effects of sea level rise

If global mean sea level rises as a consequence of a global warming due to CO_2 and other greenhouse gases, shorelines will experience changes according to the nature of the particular ecosystems, but mainly as an exacerbation of already existing phenomena.

In a region as varied as the east coast of South America several responses can be

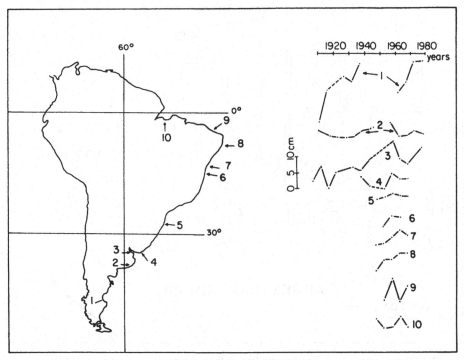

Fig. 21.6 Average 5 year relative sea level variation on the east coast of South America from tide-gauge records (after Pirazzoli, 1986): (1) Comodoro Rivadavia; (2) Mar del Plata; (3) Buenos Aires; (4) Montevideo; (5) Imbituba; (6) Canasvieiras; (7) Salvador; (8) Recife; (9) Fortaleza; (10) Belem.

expected, as indicated by Bird (1987) on a world-wide basis and Leatherman (1986) for the entire coast of South America. Some of the expected coastal responses will be:

a. Increase in beach erosion by the simple application of Bruun's Rule (Bruun, 1962);
b. Acceleration of coastal retreat where cliffs are exposed to wave action, mainly in areas where the substrate is poorly consolidated and beaches are being eroded by (a);
c. Restriction of sandy barriers;
d. Wetlands migration inland and modifications in their vegetation pattern; eventual losses where migration is restricted by highlands;
e. Lowlands exposure to increasing floodings;
f. Salinization of coastal aquifers.

As pointed out by Leatherman (1986) beach erosion and cliff retreat are critical processes on the east coast of South America, particularly in wave-dominated areas where the coastline is exposed to severe storm damage. By coincidence, these areas are of critical economic significance, as they are densely occupied by occasional and permanent residents. Rio de Janeiro (Brazil), Punta del Este (Uruguay) and Mar del Plata (Argentina) are the most important beach resorts in the region. The Rio de

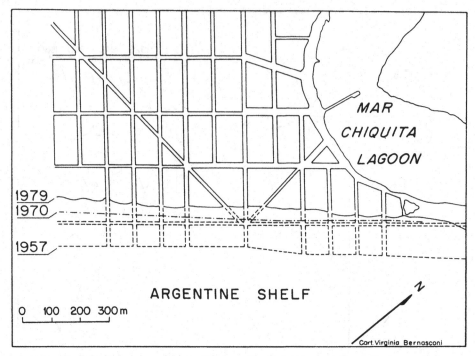

Fig. 21.7 Coastal retreat at Mar Chiquita beach, from air photographs and field surveys (after Schnack, 1985). For location see Figs. 21.3 and 21.4.

Janeiro beaches have been replenished artificially (Leatherman, 1986, from personal communication by D. Muehe).

An interesting case is that of Mar del Plata, where the population doubles in summer to about one million. Here, the beaches are being exhausted for several reasons:

1. Sand mining has been taking place at the beaches south of Mar del Plata at a rate averaging 250,000 m³/yr during the last 10 years (Colado and Schnack, 1986). Littoral drift has been estimated to be over 500,000 m³/yr northward.
2. Most of the central beaches of Mar del Plata are protected by a system of groynes constructed over the last four decades. Although they solve the problem locally, blockage of sand drifting northward causes erosion in adjacent beaches.
3. The fluvial network is largely insufficient to provide sand to the coastal system, and the cliffs that are exposed on the shoreline do not supply beach materials, except for some pebbles. The lower Paleozoic quartzites are very resistant to erosion and the predominant Plio-Pleistocene cliffs are mostly composed of silts.

In this area, as well as in extensive parts of the Pampas region, beaches are mainly composed of relict sands which are being recycled through fluctuating sea level stands in the Quaternary. North of Mar del Plata, beaches at Mar Chiquita lagoon area are suffering accelerated erosion in the horizontal component (Schnack *et al.*, 1983; Schnack, 1985; Fig. 21.7).

Coastal lowlands such as the Salado Basin depression in the Buenos Aires Pampas

(Fig. 21.4) are exposed to severe impacts from rising sea level. Taking into account the present position of the water-table, drastic changes in flood frequencies would occur as a result of the decreasing ability of the terrain to drain high-intensity rainfall. This is due to the extremely low topographic gradient and the high position of the phreatic surface. Although saltwater intrusion could also be foreseen, the landward migration of the saltwater edge would occur at slow rates (Schnack *et al.*, 1986). Urban and industrial settlements at low-lying areas, such as the Rio de la Plata shore and plain around Buenos Aires capital city, where population density is very high, would be severely affected if rates of sea level rise were to increase.

Concluding remarks

With respect to the east coast of South America, the trends in relative sea level and the physical patterns of coastal development are largely unknown. Although this fact prevents a complete diagnosis of the effects of sea level rise, it can be stated that beach and cliff erosion, as well as wetlands migration or elimination, will be the most critical problems in many areas if the rate of sea level rise increases as a consequence of global warming. If sea level were to rise by 1 m by the end of the next century (a 'high scenario'), the major effects would occur at extremely low-gradient coastal plains. However, if the whole region is considered, such a sea level rise would only exacerbate already existing environmental deterioration. At many locations, human activities (e.g., beach mining, harbour development, urban and industrial settlements) can be regarded as far more deleterious than sea level rise itself in modifying coastal ecosystems. Sea level rise would be an additional factor which must be taken into account if adequate policies for coastal management are to be adopted.

Adjustment strategies should be mainly directed towards the establishment of regulations that consider various levels of environmental, governmental and institutional management. In Argentina most coastal issues are sub-national (provincial) or local (municipal). Usually at these levels there is a lack of resources and expertise to implement the required actions. Only the Province of Buenos Aires, having a major resource and industrial base, can afford to deploy a broad range of management strategies (Brandani and Schnack, 1987).

In all countries, an adequate legislative framework is necessary. The fact that the three countries (Argentina, Brazil and Uruguay) have now established democracies leads one to be cautiously optimistic about the prospects for legislation on the environment in general and on the coastal zone in particular.

It is quite obvious that simple strategies will be employed to solve urgent problems, such as beach nourishment (Isla and Schnack, 1984). Engineering practices (structural approaches) will most likely be followed in lieu of a more comprehensive, systematic approach. Dune conservation will provide better protection for the beach systems. Urbanization in sandy shorelines should also take into account the preservation of dune fields since in many cases they are the only storage bodies for groundwater. In densely populated areas public and governmental concern will influence the practice of management techniques, attention usually being given to local and urgent problems. This applies to areas under urban or industrial development, or in critical beach resorts.

Resources will need to be allocated to build a scientific–technical database on coastal issues. The following information is needed to allow for planning and protection:

a. Environmental Atlases (geological, landforms, hazards);
b. Assessment of possible impacts of SL rise and high-energy events;
c. Inventories of shoreline erosion;
d. Monitoring of coastal processes (waves, currents, mean sea level);
e. Zonation of the coastal zone for multiple uses.

Although shoreline development poses a serious challenge to some critical areas within the region, it is foreseeable that national decisions will give precedence to other natural hazards (e.g., terrestrial floodings, droughts, landslides). Such events have a more direct impact on the economy and population, and are perceived as having higher priority for limited financial resources.

Cooperation between international agencies (UNEP, UNESCO, OAS) and the countries of the region could help in developing technical baseline information and human expertise relevant to coastal systems and their management. This could be accomplished through specific programmes involving research and surveying of particular areas, staff training and comparative analysis of management practices. A good example of such cooperation is the Special Project on Coastal Zone Management which is being carried out in some countries of the Americas under the auspices of the Organization of American States. The Project COMAR (Marine Science Division, UNESCO) is another useful means for the development of technical and human capabilities on a regional basis.

Acknowledgements

The author acknowledges the preparation of illustrations by Vicky Bernasconi, Juliana Bo and Monica Tomas of the Centre for Coastal and Quaternary Geology. Thanks are also due to Nat Rutter and Nestor Lanfredi for review of the paper.

References

Bird, E.C. (1987). *Physiographic Indications of a Rising Sea Level.* IGU Commission on the Coastal Environment, 14 pp.

Brandani, A.A. & Schnack, E.J. (1987). The coastal zone of Argentina: environmental, governmental and institutional features. *J. Shoreline Management*, **3**(3), 191–214.

Brandani, A.A., D'Onofrio, E.E. & Schnack, E.J. (1985). Comparative analysis of historical mean sea level changes along the Argentine coast. *Quaternary of South America and Antarctic Peninsula*, **3**, 187–95.

Bruun, P. (1962). Sea level rise as a cause of shore erosion. *J. Waterways and Harbours Div., Proc. Am. Soc. Civ. Eng.*, **88**, 117–30.

Codignotto, J.O. (1983). Depósitos elevados y/o de acreción Pleistoceno–Holoceno en la costa fueguino–patagónica. *Actas, Simposio Sobre Oscilaciones del Nivel del Mar Durante el Último Hemiciclo Deglacial en la Argentina, Mar del Plata, 6–7 de abril de 1983*, Edit. Univ. de Mar del Plata, pp. 12–26.

Colado, U.R. & Schnack, E.J. (1986). *Criterios para la Fijación de la Linea de Ribera en el Litoral Atlántico de la Provincia de Buenos Aires.* Unpublished Report Fiscalìa de Estado de la Provincia de Buenos Aires, La Plata, 7 pp.

Delaney, P. (1966). *Geology and Geomorphology of the Coastal Plain of Rio Grande do Sul, Brazil and Northern Uruguay, Baton Rouge.* Louisiana State University, South American Coastal Studies Tech. Report 18, 58 pp.

Fasano, J.L., Isla, F.I. & Schnack, E.J. (1983). Un análisis comparativo sobre la evolución de ambientes litorales durante el Pleistoceno tardío-Holoceno: Mar Chiquita (Buenos Aires) – Caleta Valdés (Chubut). *Actas, Simposio sobre Oscilaciones del Nivel del Mar Durante el Último Hemiciclo Deglacial en la Argentina, Mar del Plata, 6–7 de abril de 1983*, Edit. Univ. de Mar del Plata, pp. 24–47.

Feruglio, E. (1950). Descripción geológica de la Patagonia. *Yacimientos Petrolíferos Fiscales, Buenos Aires*, 111, 74–196.

Fidalgo, F., Colado, U.R. & de Francesco, F. (1973). Sobre ingresiones marinas cuaternarias en los partidos de Castelli, Chascomús y Magdalena (Provincia de Buenos Aires). *V Congr. Geol. Argentino (Carlos Paz, Córdoba)*, 111, 227–40.

Frenguelli, J. (1950). *Rasgos Generales de la Morfología y Geología de la Provincia de Buenos Aires*. LEMIT Serie II (53), La Plata, 72 pp.

González, M.A., Weiler, N.E. & Guida, N.G. (1986). Late Pleistocene transgressive deposits from 33°S to 40°S, Republic of Argentina. *J. Coastal Res. Sp. Issue No. 1*, 39–47.

Isla, F.I. & Schnack, E.J. (1984). Repoblamiento artificial de playas, Sus posibilidades de aplicación en la costa marplatense. *IX Congr. Geol. Argentino, S.C. de Bariloche 1984, Actas VI*, 202–17.

Jackson, M. (1984). *Contributions to the Geology and Hydrology of Southeastern Uruguay Based on Visual Satellite Remote Sensing Interpretation*. Münchener Geographische Abhandlungen, Band 31, München, 71 pp.

Leatherman, S.P. (1986). Coastal geomorphic impacts of sea level rise on coasts of South America. *Proceedings, United Nations Environmental Programme Conference, Washington, DC*, 73–82.

Martin, L., Bittencourt, A.C.S.P. & Vilas Boas, G.S. (1982). Primeira ocorrência de corais pleistocènicos da costa brasileira: Dataçao do máximo da penúltima transgressao. *Ciencia da Terra*, 3, 16–17.

Martin, L., Suguio, K., Flexor, J-M., Domínguez, J.M. & Bittencourt, A.C. (1987). Quaternary evolution of the central part of the Brazilian coast: The role of relative sea level variation and of shoreline drift. *UNESCO Reports in Marine Science*, 43, 97–145.

Martins, L.R. & Willwock, J.A. (1987). Eastern South America Quaternary coastal and marine geology: A synthesis. *UNESCO Reports in Marine Science*, 43, 29–96.

Pirazzoli, P.A. (1986). Secular trends of relative sea level changes indicated by tide-gauge record. *J. Coastal Res. Sp. Issue No. 1*, 1–26.

Porter, S.C., Stuiver, M. & Heusser, C.J. (1984). Holocene sea level changes along the Strait of Magellan and Beagle Channel, southernmost South America. *Quat. Res.*, 22, 59–67.

Rutter, N.W., Schnack, E.J., Radke, U., del Río, J.L., Fasano, J.L. & Isla, F.I. (1987). The Pleistocene marine terraces of Patagonia, Argentina. New evidences from amino acid and electron spin resonance dating. *Abstract, International Symp. on the Quaternary of South America, Ushuaia, Tierra del Fuego 1987*, 1 p.

Schnack, E.J. (1985). Argentina. In *The World's Coastlines*, E.C. Bird and M.L. Schwartz (eds.). Van Nostrand Reinhold Co., New York, pp. 69–78.

Schnack, E.J., Alvarez, J.R. & Cionchi, J.L. (1983). El carácter erosivo de la línea de costa de la región marplatense, Provincia de Buenos Aires. *Actas, Simposio sobre Oscilaciones del Nivel del Mar Durante el Último Hemiciclo Deglacial en la Argentina, Mar del Plata, 6–7 de abril de 1983*. Edit. Univ. de Mar del Plata, pp. 118–30.

Schnack, E.J., Bocanegra, E., Fasano, J.L. & Isla, F.I. (1986). Predicting effects of sea level rise on coastal flatlands and related shorelines, southeastern Buenos Aires Province, Argentina. *International Symposium on Sea Level Changes and Applications, Qingdao, China*, Abstract 44.

Schnack, E.J., Fasano, J.L. & Isla, F.I. (1987). Late Quaternary sea levels in the Argentine coast. *International Symposium on Late Quaternary Sea Level Correlation and Applications, Halifax, Canada, July 1987*, Abstracts, Volume 23.

Schnack, E.J., Fasano, J.L. & Isla, F.I., submitted. Fluctuaciones holocenas del nivel del mar en el sudeste de la Provincia de Buenos Aires. *Rev. Asoc. Geol. Arg.*, Buenos Aires.

Suguio, K. & Tessler, M.G. (1984). Planicies de cordoes litoraneos quaternarios do Brasil: origem e nomenclatura. In *Restingas: Origem, Estrutura, Processos*. CEUFF, Niteroi, pp. 15–25.

Suguio, K., Martin, L. & Flexor, J-M. (1980). Sea level fluctuations during the past 6,000 years along the coast of the State of Sao Paulo. In *Earth Rheology, Isostasy and Eustasy*, N.A. Mörner (ed.). John Wiley and Sons, pp. 471–86.

Urien, C.M. & Martins, L.R. (1979). Sedimentación marina en América del Sur Oriental. *Memorias, Seminario sobre Ecología Bentónica y Sedimentación de la Plataforma Continental del Atlántico Sur*. UNESCO-ROSTLAC, 5–28.

22

Future sea level rise in Hong Kong and possible environmental effects

W.W.-S. Yim

Abstract

Sea level has been rising at an average rate of about 0.3 mm/yr in Hong Kong, as estimated from an analysis of tide-gauge data obtained over the period 1962–86 from the North Point station. Because this station is sited on reclaimed land prone to long-term ground settlement, the rate determined is likely to be a maximum rate. Since a moderate positive correlation coefficient of about 0.4 is also found between the annual mean sea level and annual mean precipitation, the local increase in runoff into the sea through rainfall fluctuations and in the low central mean sea level pressure, in addition to global warming, are suggested to be responsible for the rising sea level trend. A review of previous studies on storm surges and tide-gauge observations has confirmed the importance of coastal configurations and cyclone paths in causing high sea level elevations.

For the design and planning of coastal reclamations in Hong Kong, the amount of future sea level rise within an engineering time-scale should be taken into account in the formation level to be adopted. Because this has not been carried out in the past, coastal reclamations affected by large post-constructional ground settlement are facing a high risk of marine inundation. Some examples of possible environmental effects are presented.

Introduction

Although it is difficult to measure and explain recent changes in sea level reliably (Barnett, 1983), sea level has probably risen appreciably over the past century, by 10–20 cm (Warrick and Oerlemans, 1990; Woodworth *et al.*, this volume). In the IPCC scientific assessment (Houghton *et al.*, 1990) it was estimated that global warming due to increasing atmospheric concentrations of greenhouse gases in a 'business-as-usual' (BAU) emissions scenario would lead to a sea level rise of 31–110 cm by the year 2100. Such a rise would have major environmental effects on coastal lowlands including low-lying land reclaimed from the sea. The economic development of these areas will require special planning strategy in order to avoid disasters through marine inundation and erosion associated with future sea level rise. This paper is a regional case study of Hong Kong.

At the end of 1990, Hong Kong had a total population of about 5.8 million and a total land area of about 1,100 km². The climate is tropical monsoonal with pronounced wet summers and dry winters associated with the southwest and northeast monsoons respectively. The coastline is 738 km in length, out of which 246 km is beach fringed (Bird, 1985). It is amongst the most densely populated cities in the world with a population density in some urban areas exceeding 130,000 people/km² (Anon., 1982). The shortage of land for urban development, the difficulty in building on slopes susceptible to landslides, and the rapid population growth have forced the Hong Kong Government to undertake major coastal reclamation during the post-war years. The risks of inundation resulting from a rising sea level need to be given due consideration in future developments. The main objective of this paper is to examine evidence for past sea level rise in Hong Kong through tide-gauge observations and the historical record of storm surges associated with tropical cyclones, known locally as typhoons, and to assess the possible future environmental effects on low-lying coastal areas.

Evidence for recent tectonic movements

Hong Kong is situated just south of the Tropic of Cancer adjacent to the South China Sea near the mouth of the Pearl River Estuary. Sediment discharged by the Pearl River affects only the western Hong Kong waters as a result of the influence of the Coriolis force. The annual sediment discharge into the South China Sea was estimated by Tong and Cheng (1981) to be $85–100 \times 10^6$ tonnes. Minor subsidence through the distribution of mass in the mouth region of the river is therefore likely. The three minor earthquakes with a magnitude of 1.5 on the Richter scale which have been recorded by the Hong Kong Royal Observatory since the establishment of a short-period seismograph network in 1979 (Poon, 1985) support the classification of Lee (1981) that Hong Kong lies within a crustal block with weak seismic activity. However, crustal movement involving both emergence and submergence on the scale reported in adjacent parts of the South China coast by Huang *et al.* (1984) is not known.

Sand bar deposits at Pui O, Lantau Island (Figs. 22.1 and 22.2), which were suggested by previous workers including Berry (1961) to represent remnants of raised beaches, may be used as evidence for Holocene uplift. However, Meacham and Yim (1983) concluded otherwise. The discovery of artefacts dating from the 3rd–6th centuries (Meacham, 1986) within the inner sand bar (Fig. 22.2) provided convincing proof against a higher Holocene sea level origin. Both sand bars were accounted for by a storm beach and/or blown sand origin associated with severe typhoons (Meacham and Yim, 1983).

Sea level measurements

The relative elevation of datums and sea levels in Hong Kong is summarized in Fig. 22.3. Chart Datum (C.D.) is used by the Royal Observatory for all sea level observations while Principal Datum (P.D.) is adopted by the Crown Lands and Survey Office, Hong Kong Government, for the design of all civil engineering works in the Territory. The highest astronomical tide level predicted is only 2.7 m above C.D., but the maximum sea level during typhoons may exceed this elevation by more than 3 m.

The Royal Observatory currently operates the nine tide-gauge stations shown in Fig. 22.4. Data from five of the nine stations are telemetered to the Central Forecasting

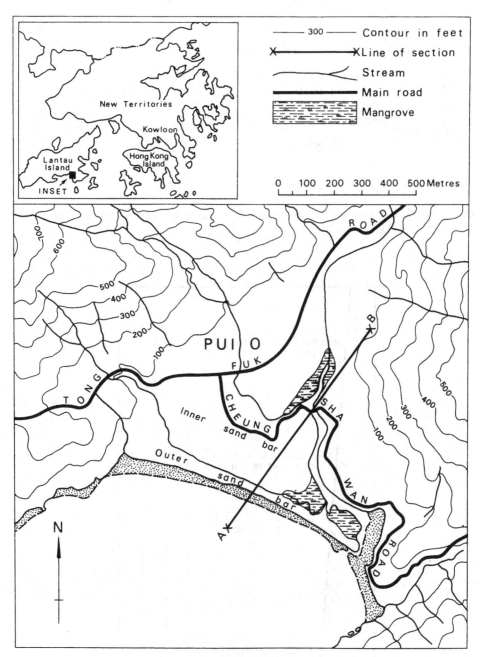

Fig. 22.1 Location map of the Pui O sand bars (1 foot = 0.3m).

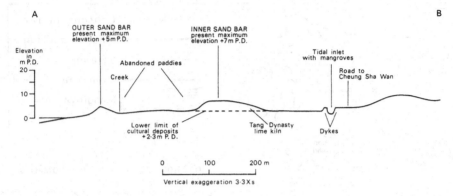

Fig. 22.2 Cross-section of the Pui O sand bars along line of section A–B shown in Fig. 22.1. (P.D. = Principal Datum; see text).

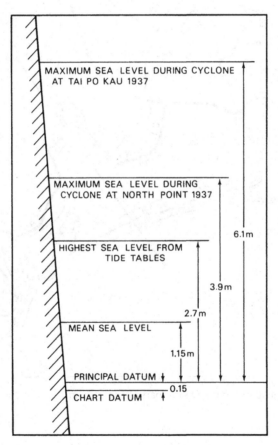

Fig. 22.3 Relative elevation of datums and sea levels in Hong Kong (in metres). Maximum sea level data from Chan (1983). (Reproduced from Yim 1986).

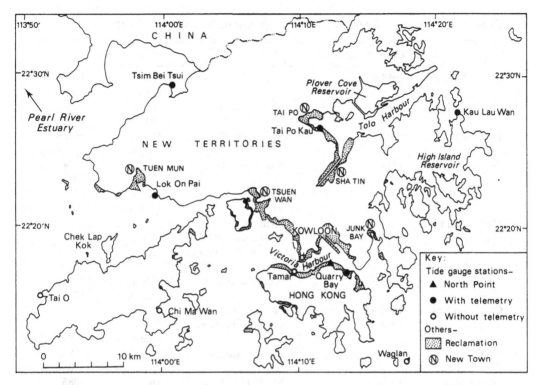

Fig. 22.4 Location map of tide-gauge stations, coastal reclamations and coastal new towns in Hong Kong.

Office on a real-time basis. The longest operational tide-gauge station at North Point was used for 33 years before it was replaced in 1986 by the Quarry Bay tide-gauge station which is located less than 1 km to the east (Fig. 22.4). Both these stations are located on land reclamations and the tide-gauge data obtained from them may be influenced by ground settlement. During the common period between September 1985 and May 1986, tide-gauge results obtained from both stations show only minor differences. Because of this compatibility of results, it is the intention of the Royal Observatory to use tide-gauge observations at the Quarry Bay station from 1986 onwards as a continuation of the North Point station (S.F. Ip, personal communication).

Table 22.1 provides a summary of information on all former and current tide-gauge stations. Because of frequent breakdowns and difficulties in maintenance of tide-gauge stations located in the more remote parts of Hong Kong, discontinuous records of hourly measurements are common in many stations. Out of the nine stations, North Point, Tai Po Kau and to a lesser extent, Chi Ma Wan, have the most complete record of tide-gauge observations.

An account on sea level related activities at the Royal Observatory was given by Cheng (1986). The scope of work includes:

Table 22.1. *Summary of information on tide-gauge stations in Hong Kong*

Station	Operational period	Type	Data available	
North Point	1954–86	Mechanical float	Hourly	1962–86
			Extremes	1961–86
Chi Ma Wan	1960–	Mechanical float	Hourly	1970–
			Extremes	1983–
Tai Po Kau	1962–	Mechanical float	Hourly	1970–
			Extremes	1984–
Tsim Bei Tsui	1974–	Mechanical float	Hourly	1974–
			Extremes	1983–
Kau Lau Wan	1974–	Pneumatic	Hourly	1974–
			Extremes	1983–
Ma On Shan	1974–78	Pneumatic	–	
Waglan	1976–	Pneumatic 1976–83; variable reluctance 1984–	Hourly	1976–
			Extremes	1981–
Lok On Pai	1981–	Mechanical float	Hourly	1981–
			Extremes	1981–
Chep Lap Kok	1982–83	Piezoresistive	–	
Tamar	1984–	Variable reluctance	Hourly	1984–
			Extremes	1984–
Tai O	1985–	Piezoresistive	Hourly	1985–
			Extremes	1985–
Quarry Bay	1985–	Mechanical float	Hourly	1985–
			Extremes	1985–

1. Provision of technical support to the Central Forecasting Office in connection with the issuance of storm surge warnings during tropical cyclone passage and the forecasting of sea state in marine weather bulletins.
2. Provision of a consultative service for coastal engineering development projects.
3. Replying to enquiries related to tide, waves and other aspects of ocean hydrodynamics.

The main focus of attention on sea level studies at the Royal Observatory is on storm surges including the use of empirical forecasting equations to predict peak surges (Lau, 1980), and numerical modelling in a bay or an open coast to assist engineering design studies (Royal Observatory, 1978). However, the use of tide-gauge observations for the prediction of future sea level trends has not been undertaken. In order to fill this need, tide-gauge data were obtained from the Director of the Royal Observatory for analysis in this paper. Previous studies on storm surges are reviewed.

Analysis of tide-gauge observations

Since most tide-gauge stations in Hong Kong have been operational for a period of less than 20 years (Table 22.1), only the three stations with the longest and most continuous record, North Point, Tai Po Kau and Chi Ma Wan, were selected for analysis. The

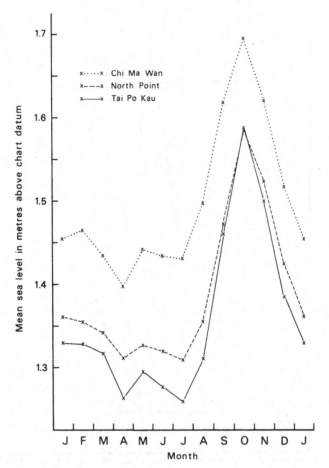

Fig. 22.5 Comparison of 15-year monthly mean sea level during 1970–84 at the North Point, Tai Po Kau and Chi Ma Wan tide-gauge stations.

periods of tide-gauge observations covered by these stations are 1962–86, 1970–85 and 1970–85 respectively.

A comparison of 15-year monthly mean sea level during 1970–84 at the North Point, Tai Po Kau and Chi Ma Wan stations is shown in Fig. 22.5. All stations can be seen to show a similar pattern. The highest monthly mean sea level coincides approximately with the typhoon season during the wet southwest monsoon in the summer, while appreciably lower monthly sea levels between January to July are associated mainly with the dry northeast monsoon in the winter. However, the differences observed between the minimum monthly sea level and the sharpness of the maximum October peak at each station are distinctive, and may be explained by variations in the coastal configuration of the stations. In increasing order, the differences are 0.28, 0.30 and 0.33 m for the North Point, Chi Ma Wan and Tai Po Kau stations respectively. The smallest difference found for North Point is consistent with the shape of Victoria Harbour which is open to both the east and west (Fig. 22.4). This station is the least influenced by coastline configuration and possesses the longest and most continuous tide-gauge

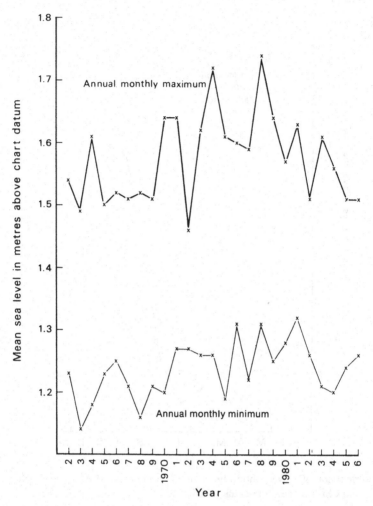

Fig. 22.6 Annual monthly maximum and minimum sea levels of the North Point tide-gauge station during 1962–86.

record. On the other hand, Chi Ma Wan with the highest overall monthly sea level has a funnel-shaped coastline facing east (Fig. 22.4). The predominant easterly wind in Hong Kong which is associated with the northeast monsoon has a maximum effect because of the orientation of the coastline. Tai Po Kau is similar in orientation to Chi Ma Wan but is located within a large body of inshore water, Tolo Harbour, with a narrow northeasterly exit less than 2 km in width into the open sea (Fig. 22.4). At this station, both the large difference between the maximum and minimum monthly sea levels and the sharpness of the October peak (Fig. 22.5) are indicative of shoaling under the influence of the unusual coastline configuration.

The North Point tide-gauge station was chosen for trend analyses because of its relatively long record. The annual monthly maximum and minimum sea levels for this station during a 25–year period from 1962–86 are presented in Fig. 22.6. In terms of

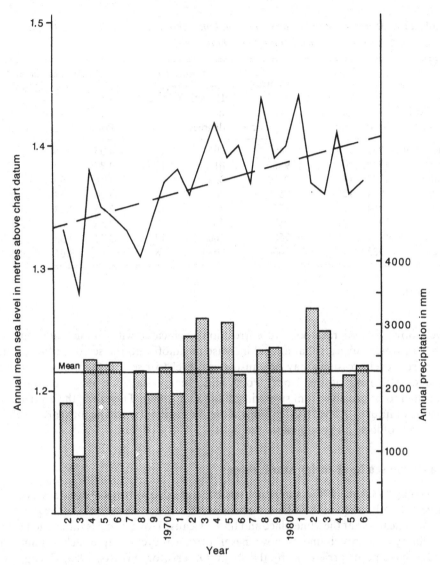

Fig. 22.7 Annual mean sea level trend of the North Point tide-gauge station and annual precipitation of the Royal Observatory station (located in Fig. 22.4 as a cross) during 1962–86. The dashed line shows the mean sea level trend.

variability, the annual monthly maximum is greater than the annual monthly minimum which is best explained by the changing intensity of the monsoonal effect. Fig. 22.7 shows the annual mean sea level trend and the annual precipitation recorded at the Royal Observatory station (located in Fig. 22.4 as a cross). A least-squares regression line fit to the annual means (average of monthly mean values) indicates a rise in sea level over the 25-year period. The mean rate of sea level rise during 1962–86 is about 3 mm/ yr. Since this rate is greater than the estimated global mean eustatic sea level rise of 1–2 mm/yr, regional/local factors may also be contributing to the trend. Annual mean sea

Table 22.2. *Meteorological observations and maximum sea level of selected typhoons in Hong Kong since 1937. Based mainly on Chan (1983)*

Tropical cyclone	Year	Maximum gust in knots	Central MSL pressure in mbar	Movement Direction in degrees	Movement Speed in knots	Maximum sea level (m above C.D.) North Point	Maximum sea level (m above C.D.) Tai Po Kau	Maximum sea level (m above C.D.) Chi Ma Wan
Unnamed	1937	130*	949	296	15	4.05	6.25	—
Mary	1960	103	966*	015	12	2.77	—	—
Wanda	1962	140	944*	325	12	3.96	5.03	—
Ruby	1964	122	954*	303	11	3.14	3.54	3.20
Dot	1964	94	973	360	10	2.65	3.23	2.95
Rose	1971	121	982	360	8	2.56	3.00	2.98
Hope	1979	94	950	260	17	2.78	4.33	2.73
Ellen	1983	128	960	305*	8	—	3.06	3.06

Note:
* Estimated

level shows a moderate positive correlation coefficient with annual precipitation ($r = 0.4$), which suggests that a local increase in runoff into the sea or in low central mean sea level pressures may be causing sea level to rise. Furthermore, long-term ground settlement of the North Point and Quarry Bay tide-gauge stations was indicated to be possible from surveying data (Yim, 1988). Therefore, the estimates of the rate of future sea level rise for Hong Kong from global warming must also account for trends of local and regional origin.

Maximum sea level during storm surges

According to Chan (1983), maximum sea levels are due to tropical cyclones. A storm surge is usually taken as the difference between the observed sea level and the predicted astronomical tide at the same place and time. Because Hong Kong is threatened by the possibility of marine inundation whenever a tropical cyclone approaches, numerous studies have been carried out by the Royal Observatory (Watts, 1959; Cheng 1967; Petersen, 1975; Lau, 1980; Chan, 1983 and others) to assist coastal engineering projects.

The meteorological observations and maximum sea levels of selected tropical cyclones in Hong Kong since 1937 are summarized in Table 22.2. Tai Po Kau shows the highest maximum sea level out of the stations followed by Chi Ma Wan and North Point. However, the difference between the maximum sea level in the two latter stations is small. Generally, the highest maximum sea level found at all stations was associated with maximum gust, the lowest central mean sea level pressure and a west–north-westerly direction of cyclone movement at the greatest speed.

Storm surge levels are influenced by numerous factors such as those presented in Table 22.3 (after Lau, 1980). Out of these factors, coastline configuration and cyclone path appear to be most important in causing the highest maximum sea level. The

Table 22.3. *Factors affecting storm surge levels. After Lau (1980)*

Parameters of storm	Coastal parameters	Local factors
Central pressure	Sea-floor topography	River discharges
Distance of closest approach	Coastline configuration	Seiching
Translational speed		Rainfall runoff
Path		Tidal effects
Size		Wind effects

maximum sea level of 6.25 m above C.D. recorded at the Tai Po Kau in 1937 (Table 22.2) was associated with a west–northwesterly moving tropical cyclone centred about 13 km south of the Royal Observatory at its nearest point (Royal Observatory, 1938).

Both North Point and Chi Ma Wan show appreciably lower maximum sea level during typhoons. In the case of North Point, any increase in surge amplitude caused by shoaling water would be partially offset by horizontal divergence in the westward flow of water (Watts, 1959).

An analysis of past storm surges in Hong Kong was carried out by Watts (1959) to examine the effect of meteorological conditions on tide height. Storm surges were found to be most pronounced when the centre of a typhoon moving westward passes over Hong Kong directly or within 13 km to the south. A map showing typhoon tracks with the greatest surges in Hong Kong is shown in Fig. 22.8. The typhoons typically follow an almost straight west-northwesterly course through the strait between Taiwan and Luzon, maintaining their ferocity and without passing over land. Most of the storm surges were found by Watts (1959) to occur within one hour before the wind maximum and three hours after it. However, no simple relationship was found between surge height and wind speed or the duration of strong winds. Furthermore, the surge is unusually large when a typhoon passing close to Hong Kong is fast moving, when it is accompanied by heavy rain, and when there is an accumulation of water associated with an abrupt change of wind direction.

Storm surge statistics in Hong Kong have been presented by Petersen (1975) and Chan (1983). A statistical method was used by Petersen (1975) to estimate the return periods of sea levels by combining the effects of storm surges and astronomical tides. However, Chan (1983) considered that interactions between storm surges and astronomical tides may lead to uncertainties in the results. Instead, the method of Gumbel (1954) was used by Chan (1983) and applied to data on annual maximum sea levels for the estimation of return periods (Table 22.4).

In the case of Tai Po Kau and Chi Ma Wan, because some of the annual maximum sea levels were extracted from monthly maximum hourly sea levels, it is possible that higher sea levels may have occurred. Therefore, the return periods for various specified sea levels at Tai Po Kau and Chi Ma Wan are likely to be less than predicted.

Predictions of storm surge levels at Tai Po Kau were computed using meteorological parameters measured by the Royal Observatory during storm surges in 1936 and 1954–1964 by Cheng (1967) (Table 22.5). In this study, a storm surge was defined as the difference between the observed water level and that which would have occurred at the same time and place in the absence of the storm. The difference in return periods for the

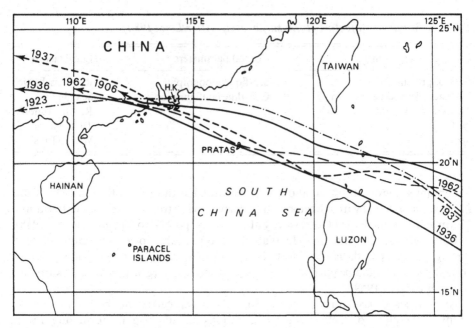

Fig. 22.8 Tracks of typhoons with the greatest surges in Hong Kong. After Petersen (1975). (Redrawn and reproduced with permission from the Royal Observatory, Hong Kong).

same maximum sea level elevation found for Tai Po Kau by Cheng (1967) (Table 22.5) and Chan (1983) (Table 22.4) may be explained by the periods of annual maximum sea level data taken into account. Because of this, it is desirable to make use of the longest record of data available.

In the Tolo Harbour storm surge modelling investigation for the Sha Tin new town, numerical modelling was used by the Royal Observatory (1978) to establish the probability of particular sea levels occurring. The two storm surge models tested were:

1. The open coast 'SPLASH' model (Special Program for Listing Amplitude of Surge Heights of the United States Weather Bureau) applied to about 900 km of the South China coast centred around Hong Kong. Intensity, size, speed and direction of movement of tropical cyclones were used as inputs.
2. The Tolo Harbour 'Bay' model. This program was adapted from the United States operational model for predicting nearshore surges.

An important finding on storm surge phenomena in Tolo Harbour was that it is wave rather than current-flow oriented. Altogether eight post-reclamation coastline configurations shown in Fig. 22.9 were tested. The conclusion was that sizeable reductions of the area of Tolo Harbour would not result in proportional increases in peak surges (Royal Observatory, 1978). However, since the predictions made by the models are based on a relatively short period of meteorological observations of tropical cyclones not exceeding 30 years, the representativeness of the results may be disputed. No evidence for an increase in the frequency of storm surges which may be the result of a rising mean sea level has been found.

Table 22.4. *Return periods of maximum sea level (in m above C.D.) estimated using Gumbel's method. From Chan (1983)*

Return period in years	Period		
	1950–81[a]	1962–81[b]	1963–81[c]
10	3.34	4.13	3.40
50	3.71	5.01	3.74
100	3.86	5.39	3.89
200	4.01	5.76	4.03
500	4.22	6.25	4.22
1000	4.37	6.62	4.37

Station [a]North Point
[b]Tai Po Kau
[c]Chi Ma Wan

Table 22.5. *Prediction of storm surge levels made on several bases at Tai Po Kau. From Cheng (1967)*

Return period in years	Storm surge (m above C.D.)			
	Obtained from extreme gusts	Obtained from extreme 60 minute mean wind	Obtained from extreme hourly minimum near sea level pressure	Average
10	2.56	2.96	2.13	2.56
20	2.96	3.47	2.41	2.96
50	3.47	4.02	2.68	3.38
100	3.87	4.48	2.87	3.75
200	4.27	4.88	3.14	4.08
500	4.82	5.49	3.41	4.57
1000	5.21	5.88	3.60	4.91

Coastal reclamation and development

The area of coastal reclamation in Hong Kong to the end of 1984 is shown in Fig. 22.10. A list of the main types of reclamation is given below:

1. Housing estates
2. Industrial estates
3. Reservoirs with marine sites
4. Airport
5. Container terminals

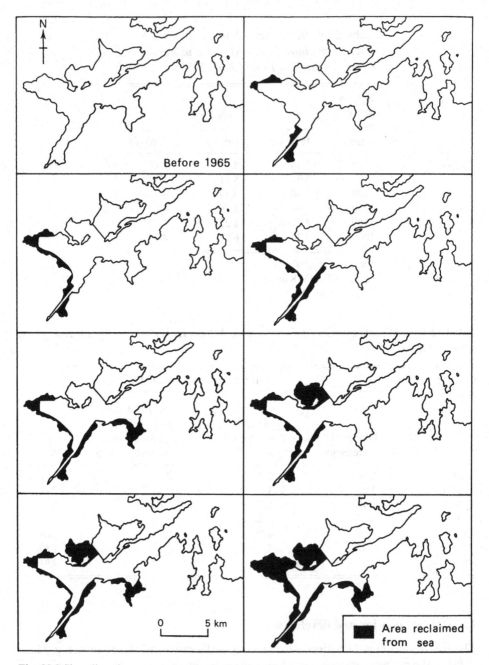

Fig. 22.9 Shoreline shapes tested to determine the effects of varying shoreline on typhoon surge characteristics. (Maunsell Consultants Asia, 1980).

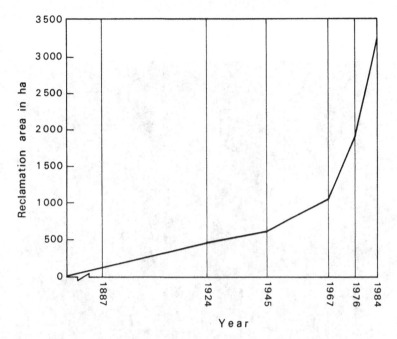

Fig. 22.10 Area of coastal reclamation in Hong Kong to the end of 1984. (Lands Department, 1985).

6. Road and railway network including the Mass Transit Railway
7. Power stations
8. Waste disposal structures including coastal landfills and lagoons created for the disposal of pulverized fuel ash.

The total area of land reclaimed from the sea at the end of 1984 was 33 km² (Lands Department, 1985) (Fig. 22.10). The major housing programme undertaken by the Hong Kong Government in 1972 led to the development of five coastal sites for new towns with a targeted population of over 2 million people. They include Tsuen Wan, Tuen Mun, Junk Bay, Sha Tin and Tai Po (Fig. 22.4) of which the last two are located within Tolo Harbour (Fig. 22.11). Large-scale coastal reclamations are currently in progress for the Port and Airport Development Strategy (PADS).

Land was created by cutting into the hills and dumping the spoils into the sea. In addition to new towns, the container port of Hong Kong at Kwai Chung (Fig. 22.12) which currently ranks second in the world after Singapore in terms of traffic volume, and, the Kai Tak International Airport (Fig. 22.13) are both constructed mainly on land reclaimed from the sea.

Formation level of reclamations

From an engineering viewpoint, it is desirable to keep the formation level of reclamations as low as possible in order to reduce the quantities of fill required and hence the cost of construction. However, low-level reclamations create problems for gravity drainage of surface water. Therefore a compromise has to be reached. The standard of

Fig. 22.11 Oblique air photograph of Tolo Harbour showing the Tai Po and Sha Tin new towns, and the Plover Cove Reservoir. The Tai Po Kau tide-gauge station is located at the end of the straight pier near Tai Po. (Reproduced with permission from the Director of Building and Lands, © Hong Kong Government).

Fig. 22.12 Oblique air photograph of the Tsuen Wan new town and the Kwai Chung Container Terminal. The original coastline is not discernible because of coastal reclamations. (Reproduced with permission from the Director of Building and Lands, © Hong Kong Government).

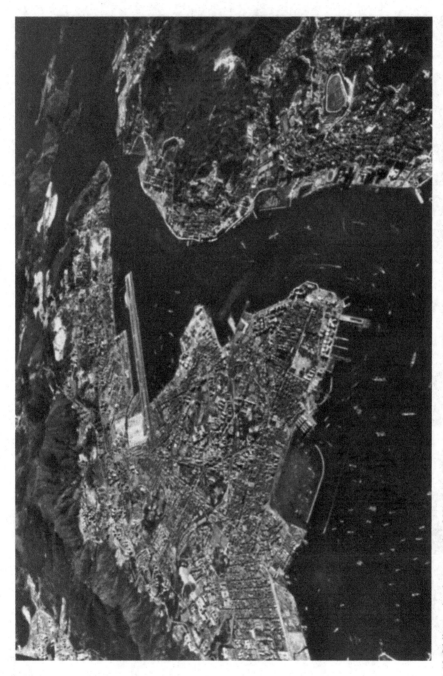

Fig. 22.13 Oblique air photograph of Victoria Harbour showing extensive areas of coastal reclamations including most of Kai Tak International Airport. (Reproduced with permission from the Director of *Building and Lands*, © Hong Kong Government).

Table 22.6. *A comparison of minimum formation levels of selected coastal engineering projects in Hong Kong*

Project	Minimum level (m P.D.)
Tai Po new town	4.5
Sha Tin new town	4.5
Junk Bay new town	4.5
Tsuen Wan new town*	3.9
Tuen Mun new town*	3.9
Wanchai reclamation (Victoria Harbour)	3.5
Mass Transit Railway ground level access points	4.5
Kwai Chung Container Terminal	4.2

Note:
*Refers to most sheltered areas. More exposed sites have increased cope to allow for wave effects. Also all reclamations rise from seawall cope to limit flooding caused by overtopping.

minimum formation level in reclamations is mainly dependent on the planned future usage of the land which in turn is determined by the flood level return period of the site. Future sea level rise is perhaps already adequately covered by the storm surge estimates made by the Royal Observatory for engineers and town planners.

A comparison of minimum formation levels adopted in Hong Kong for selected engineering projects is shown in Table 22.6. Out of the new towns, Tai Po, Sha Tin and Junk Bay have the highest level of 4.5 m above P.D. While Tai Po and Sha Tin are in accordance with the high historical storm surge levels recorded at Tai Po Kau, the level at Junk Bay was fixed after an investigation of expected storm surge and coincident wave heights using mathematical modelling (Maunsell Consultants Asia, 1982). Both Tsuen Wan and Tuen Mun have a lower minimum formation level because there was no history of severe flooding at these localities and storm surges were assumed to be similar to North Point except for corrections on the basis of tidal differences quoted in Admiralty Tide Tables for astronomical tides. In the Wanchai reclamation located on the northern side of Hong Kong island, the formation level is lower because significant flooding problems have not resulted since the completion of the reclamation (New Territories Development Branch, 1981). However, both the Mass Transit Railway ground level access points and the Kwai Chung Container Terminal are considered to be areas of high security and were given minimum flood defence levels of 4.5 and 4.2 m above P.D. respectively.

An example of the provisional standards for minimum formation levels adopted for the Tai Po new town is shown in Table 22.7. However, the decision on using these minimum formation levels was made before carrying out a hydraulic study of the area (Maunsell Consultants Asia, 1980). On completion of the hydraulic study, only the

Table 22.7. *Provisional standards for minimum formation levels proposed for the Tai Po new town. (Maunsell Consultants Asia, 1978)*

Planned usage	Flood return period (yrs)	Minimum level (m P.D.)
Important development areas – all building areas, except in special circumstances	200	5.5
Unimportant development areas – minor roads, car parks, intensive-use open space	50	5.0
Undeveloped areas – including non-intensive-use open space	10	4.0

minimum level for undeveloped areas was raised to 4.5 m above P.D. to provide better protection against lengthy inundation. In the design of major storm water drains, a return period of 200 years is specified by the Civil Engineering Services Department of the Hong Kong Government (Maunsell Consultants Asia, 1978). The same standard is also widely adopted for areas of the highest security such as all building areas and gas plants.

Ground settlement

A major problem affecting development on recent reclamations is the large settlements which take place, particularly when soft marine mud is allowed to remain undisplaced beneath fill material. Ground settlement in reclamations takes place during both the constructional and post-constructional periods. In an account of the Castle Peak reclamation for the Tuen Mun new town, Hunt *et al.* (1982) estimated that a 10 m thickness of marine clay would take about 4 years to achieve 95% consolidation, resulting in a settlement of 1.6 m; even if surface construction were delayed for 2 years after reclamation, 0.4 m settlement would still occur in the succeeding 2 years.

In some reclaimed areas floored by marine mud, a total settlement of about 30% of the marine mud may be expected. Therefore ground settlement is likely to affect the minimum formation level of reclamations resulting in an increase in the risk of marine inundation. An example of the predicted total ground settlement for the Junk Bay new town reclamation is shown in Fig. 22.14. The maximum total settlement is in excess of 5 m and appears to be closely related to the total thickness of marine mud on the sea bed (Maunsell Consultants Asia, 1982). Because the thickness of marine mud increases from nil near the low water mark to a maximum of 16–18 m, the settlement contours generally increase seawards. However, the margin of error is likely to be higher where the thickness of marine mud is the greatest.

In order to keep ground settlement problems under control, three main methods of reclamation are used in Hong Kong. They are:

1. Removal of marine mud by dredging and replacement with fill. However, the cost of dredging and dumping of marine mud, and the cost of fill makes this method uneconomic for general reclamation. It is used only in the construction of sea-

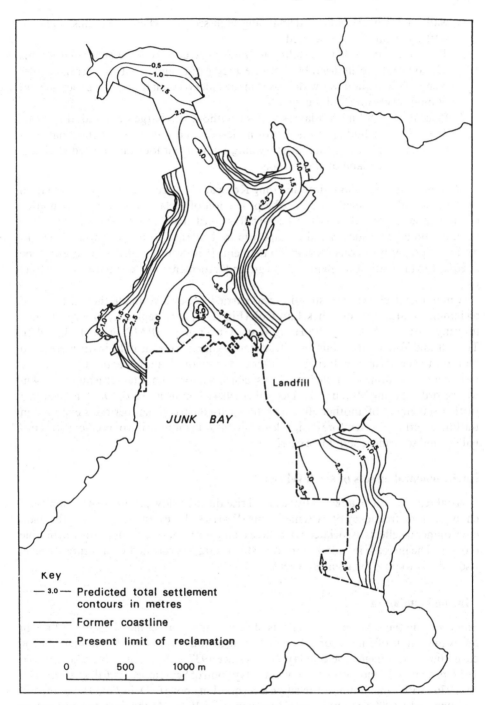

Fig. 22.14 Predicted total ground settlement of the Junk Bay new town (Maunsell Consultants Asia, 1982).

walls and in reclamations demanding high security where post-constructional settlement cannot be tolerated.

2. The use of controlled overfilling techniques. Mud waves are generated by this method and are undesirable because they cause significant differential settlements. This method was widely used in the older reclamations before the problem of mud waves was well-recognised.

3. Placement of fill in thin layers with or without surcharge or foundation treatment. This method prevents the formation of mud waves while consolidation is accelerated by using surcharge loading with or without an inserted drainage system such as sand or wick drains.

Most present day reclamations in Hong Kong are constructed using a combination of the first and third methods. For the protection of reclaimed land against marine erosion either a vertical or inclined sea-wall is constructed at the new coastline. It is common practice to add an additional 0.1 m to the design formation level of the sea-wall to allow for post-constructional settlement. However, this allowance may not be adequate to take into account long-term ground settlement or increased rate of sea level rise.

A major geotechnical investigation was carried out to investigate the feasibility of reclaiming a large area off Chek Lap Kok (Fig. 22.4) for a replacement airport (now forming part of PADS) by the Hong Kong Government (RMP Econ. Ltd., 1982). The fill and foundation soils in a test embankment were heavily instrumented to monitor the performance during and after construction (Handfelt *et al.*, 1987). In one test area with alidrains installed at 1.5 m spacing, a maximum total settlement of 3.4 m was recorded during March 1982–December 1984 (Cheung and Ko, 1986). However, a further settlement of another 10% over the post-constructional period is expected in the longer term. It is therefore desirable that long-term monitoring of post-constructional ground settlement should be carried out.

Environmental effects of sea level rise

Coastal regions will be affected by sea level rise through slow progressive changes and changes in the frequency of extreme events (Warrick, 1986). In Hong Kong, the main environmental effects associated with these changes are marine inundation and marine erosion. The increase in inundation from storm surges associated with future sea level rise is likely to pose the greatest threat.

Marine inundation

Some idea on the environmental effects of future storm surges may be gained through an examination of the historical record of disastrous typhoons. A summary of the casualties and damage of selected typhoons since 1937 is shown in Table 22.8. Prior to 1937, probably the most disastrous was the typhoon of September 1906 documented by Li (1976), with a maximum storm surge level of 6.1 m above C.D. in Tolo Harbour, and an estimated 10,000 persons killed (Hutcheon in Li, 1976). At the time, the population of Hong Kong was 450,000 and the warning of the storm was issued only 16 minutes before the disaster. This resulted in the high number of deaths, mostly of boat people.

Table 22.8. *Casualties and damage of selected typhoons in Hong Kong since 1937.*
Based on Royal Observatory (1983)

Tropical cyclone	Year	Persons killed	Persons injured or missing	Ocean-going vessels in trouble	Small craft sunk or damaged
Unnamed	1937	11000*	Unavailable	28	1855
Mary	1960	45	138	6	814
Wanda	1962	130	Unavailable	36	2053
Ruby	1964	38	306	20	314
Dot	1964	26	95	2	90
Rose	1971	110	291	34	> 303
Hope	1979	12	260	29	374
Ellen	1983	10	345	44	360

Note:
*Estimated

Another disastrous typhoon in September 1937 struck during the night killing an estimated 11,000 persons including a large number in the Tolo Harbour area where the maximum surge level was 6.25 m above C.D. (Chan, 1983). A tidal wave which was estimated as being 30 feet (9.14 m) above mean sea level caused great damage and loss of life in the Tai Po and Sha Tin villages (Royal Observatory, 1938). Such disastrous events will have to be prevented by an improved warning system and better land use planning of all low-lying coastal areas.

Both low-lying coastal areas and reclamations in Hong Kong are facing an increasing risk of marine inundation through storm surges. In low-lying coastal areas, ground settlement through constructional loading and groundwater pumping is likely to further increase the risk. However, the impact of groundwater pumping is unlikely to be great because groundwater is used only in small quantities in areas where a piped water supply from reservoirs is unavailable. Because the minimum formation level in reclamations already takes into account the return period of maximum sea level during storm surges the risk of marine inundation may be acceptable (providing that post-constructional ground surface settlements are minimal).

Given the prospect of future sea level rise, large post-constructional ground surface settlements will greatly increase the risk of marine inundation in reclamations. Since settlement rates are dependent on the ground conditions, the degree of loading and the time factor, old reclamations occurring further inland of new reclamations are likely to have settled more and consequently may face a greater risk of flooding. During urban renewal it is desirable as a remedial measure to raise the formation level. This alternative is preferred over raising the design level of sea-walls, especially since rainstorms in Hong Kong could also give rise to floods in reclamation areas.

An example of a marine inundation hazard map of Tolo Harbour produced for the Tolo Harbour hydraulic study by Maunsell Consultants Asia (1980) is shown in Fig. 22.15. The areas liable to sea water inundation were identified by their low ground elevation. During the construction of the Tai Po and Sha Tin new towns, efforts were

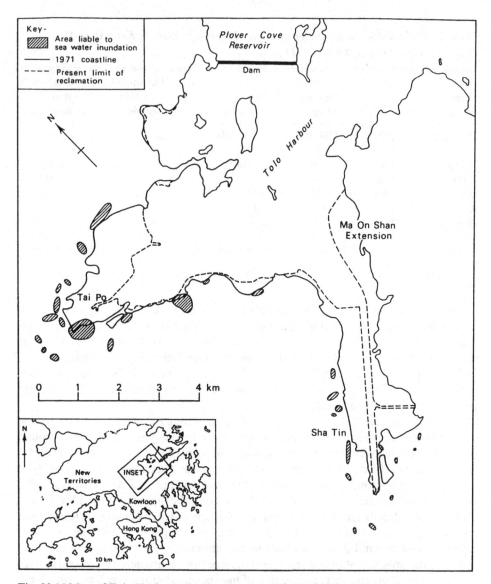

Fig. 22.15 Map of Tolo Harbour showing the present limit of reclamation and areas identified by Maunsell Consultants Asia (1980) to be liable to sea water inundation.

made to raise the formation level of the flood prone areas to reduce the risk of marine inundation.

The efficiency of storm water drains is affected by marine inundation (Titus *et al.*, 1987). Although storm water drains are designed using a 200-year return period in Hong Kong, the efficiency of gravity flow will be reduced and in extreme cases be reversed to become totally ineffective during storm surges. Consequently, the risk of flooding in reclaimed areas may be greatly increased.

Marine erosion

The main preventive measure used for protecting reclamations against wave attack is construction of a sea-wall, either vertical or inclined, topped with a wave-wall. In breakwaters and marine dams, rock and other types of armour are often used for the dissipation of wave energy. Normally, unless high security is required, coastal structures are not designed against overtopping during the severest storms. However, since all coastal structures are affected by post-constructional settlements to some degree, increases in the frequency of extreme events associated with future sea level rise will increase the risk of marine erosion.

The two reservoirs with marine sites in Hong Kong, Plover Cove and High Island (Fig. 22.4) have been constructed at a cost of HK$600 × 10^6 and HK$1.5 × 10^9 respectively. The minimum constructional elevation of the wave-wall on the main dam of Plover Cove varies with chainages from 15.84–16.30 m above P.D. while the constructional elevation of the crest of the east coffer-dam of High Island is 17.5 m above P.D. (K.P. Mok, personal communication). Although post-constructional settlement values are unavailable for the main dam of Plover Cove, it is worth noting that the dam was raised in 1973. In the east coffer-dam of High Island, settlement monitoring was not carried out but the dam structure is designed against overtopping in the severest storms and the inside face is protected against this. Waves up to 12 m in height were reported at the site by Vail *et al.* (1976) and about seven thousand 25-ton units of dolosse providing the best stability factor (Aspden, 1973) were selected to protect the seaward side of the dam. Because both marine dams are permanent features on the coastline and were constructed at great cost, it is desirable to monitor ground settlement and the future rate of sea level rise over the longer term so that remedial measures may be taken whenever necessary to guard against damage caused by marine erosion.

Marine erosion is likely to cause the greatest damage in a depositional coast through the off-shore transport of sediment. A study by Williams *et al.* (1977) in Repulse Bay, Hong Kong indicated that different processes and intensities of these processes were constantly at work. From this, a to-and-fro transfer of sediment within Repulse Bay rather than along it was inferred. However, the depletion of sand on the present beach may be linked to beach erosion induced by sea level rise. The construction of sand-retaining structures and beach nourishment work completed recently would slow down beach erosion only until the next severe tropical cyclone. Because of the greater exposure to oceanic influences in Eastern Hong Kong, marine erosion through a rise in future sea level is expected to cause more damage along the depositional coastline there than in Western Hong Kong.

Conclusions

Hong Kong is facing a continuing, and perhaps increasing risk of marine inundation from storm surges. The minimum formation level of reclamations should be increased to take into account future changes in mean sea level resulting from factors operating at both global- and regional/local-scales. For monitoring of the rate of future sea level rise, the choice of tide-gauge stations located in inshore waters and on reclaimed land requires careful analysis and further studies.

In reclaimed areas, long-term continuous monitoring of post-constructional ground settlement using state-of-the-art surveying methods including satellite altimetry is recommended. This is essential because significantly large ground settlements, in excess of the future rate of sea level rise, are possible within the engineering time-scale. Such affected areas will be particularly prone to marine inundation and should be selected for close monitoring. Furthermore, it is necessary to examine both ground settlement and the rate of future sea level rise together in any environmental impact assessment.

Acknowledgements

This work would not have been possible without the help of Hong Kong Government departments including the Royal Observatory, the Territories Development Department, the Civil Engineering Services Department and the Water Supplies Department; Maunsell Consultants Asia, and Scott Wilson Kirkpatrick and Partners. I am particularly grateful to M.L. Chalmers, Dr M.J. Tooley, R.K.H. Cheung, Professor C.J. Grant, S.F. Ip, Y.M. Lee, C.R. Matson and H. Smith for their assistance. Figures 22.11–22.13 are reproduced by courtesy of the Director of Buildings and Lands, Hong Kong Government. Dr D.R. Workman provided constructive criticisms on an early draft of this paper, while the Croucher Foundation, the University of Hong Kong and the Hui Oi Chow Trust awarded research grants to support this work.

References

Anon. (1982). *Population Map of Hong Kong*. Hong Kong Government.

Aspden, J.A.T. (1973). Some design considerations of the High Island Water Scheme, Paper 4 – Wave protection of the East Dam. *Hong Kong Eng.*, 1/2, 45–6.

Barnett, T.P. (1983). Recent changes in sea level and their possible causes. *Climatic Change*, **5**, 15–38.

Berry, L. (1961). Erosion surfaces and emerged beaches in Hong Kong. *Geol. Soc. Am. Bull.*, **72**, 1383–1394.

Bird, E.C.F. (1985). *Coastal Changes, a Global View*. John Wiley and Sons, Chichester.

Chan, Y.K. (1983). *Statistics of Extreme Sea Levels in Hong Kong*. Royal Observatory, Hong Kong Tech. Note (Local) 35.

Cheng, T.-S. (1986). Sea level related studies at the Royal Observatory, Hong Kong. In *Sea Level Changes in Hong Kong During the Last 40,000 Years*, W.W.-S. Yim (ed.). Abstracts 3, Geological Society of Hong Kong and Department of Geography and Geology, University of Hong Kong, 34–6.

Cheng, T.T. (1967). *Storm Surges in Hong Kong*. Royal Observatory, Hong Kong, Tech. Note 26.

Cheung, R.K.H. & Ko, S.W.K. (1986). *Continued Monitoring of the Test Embankment September 1983–December 1984: Study Report 2B, 1–3, Replacement Airport at Chek Lap Kok*. Civil Engineering Office, Civil Engineering Services Department, Hong Kong Government.

Gumbel, E.J. (1954). *Statistical Theory of Extreme Values and Some Practical Applications*. US Dept. of Commerce, Washington, DC.

Handfelt, L.D., Koutsoftas, D.C. & Foott, R. (1987). Instrumentation for test fill in Hong Kong. *J. Geotech. Engrg.*, **113**, 127–46.

Houghton, J.T., Jenkins, G.J. & Ephraums, J.J. (eds.) (1990). *Climate Change: The IPCC Scientific Assessment*. Cambridge University Press, Cambridge.

Huang, Y., Xia, F., Huang, D. & Lin, H. (1984). Holocene sea level changes and recent crustal movements along the northern coasts of the South China Sea. In *The Evolution of the East*

Asian Environment Vol. 1: Geology and Palaeoclimatology, R.O. Whyte (ed.). Centre of Asian Studies, University of Hong Kong, pp. 271–87.

Hunt, T., Kemp, W.R. & Koo, Y.C. (1982). Castle Peak Bay reclamation, Hong Kong. *Proc. 7th Southeast Asian Geotechnical Conf., Hong Kong*, **1**, 257–69.

Lands Department (1985). *Reclamation and Development in Hong Kong*. Hong Kong Government.

Lau, R. (1980). *Storm Surge Investigations and the Use of Vertically Integrated Hydrodynamical Models*. Royal Observatory, Hong Kong, Tech. Note 53.

Lee, C.M. (1981). Some geological factors and their influence on the seismicity of Hong Kong. *Hong Kong Eng.*, **9/12**, 47–50.

Li, W.Y. (1976). *The Typhoon of 18 September 1906*. Royal Observatory, Hong Kong, Occasional Papers 36.

Maunsell Consultants Asia (1982). *Junk Bay New Town Study – Appendices*. Junk Bay Development Office, Hong Kong Government.

Maunsell Consultants Asia (1980). *Sha Tin New Town Tolo Harbour Investigation Stage 4*. New Territories Development Department, Hong Kong Government.

Maunsell Consultants Asia (1978). *Tai Po Development Study Final Report*. New Territories Development Department, Hong Kong Government.

Meacham, W. (1986). Pui O. *J. Hong Kong Archaeological Soc.*, **11**, 113–18.

Meacham, W. & Yim, W.W.-S. (1983). Coastal sand bar deposits at Pui O. In *Geology of Surficial Deposits in Hong Kong*, W.W.-S. Yim and A.D. Burnett (eds.). Abstracts 1, Geological Society of Hong Kong and Department of Geography and Geology, University of Hong Kong, pp. 44–7.

New Territories Development Branch (1981). *Development Investigation of the Northwestern New Territories*. Public Works Department, Hong Kong Government.

Petersen, P. (1975). *Storm Surge Statistics*. Royal Observatory, Hong Kong, Tech. Note 20.

Poon, H.T. (1985). Seismological measurements in Hong Kong. *Bull. Geol. Soc. Hong Kong*, **2**, 59–65.

RMP Econ. Ltd. (1982). *Replacement Airport at Chep Lap Kok: Civil Engineering Design Studies*. Study Report 1 Site Investigation, Vol. 1 Main Report. Civil Engineering Office, Engineering Development Department, Hong Kong Government.

Royal Observatory (1983). *Part 3 Tropical Cyclone Summaries*. Meteorological Results, Royal Observatory, Hong Kong.

Royal Observatory (1978). *The Tolo Harbour Storm Study Numerical Modelling Investigation*. Unpub. Report Royal Observatory, Hong Kong.

Royal Observatory (1938). *The Typhoons of August 16–17th 1936 and September 1st–2nd 1937, Appendix 3, Meteorological Results*. Royal Observatory, Hong Kong.

Titus, J.G., Chin, Y.K., Gibbs, M.J., Laroche, T.B., Webb, M.K. & Waddell, J.O. (1987). Greenhouse effect, sea level rise, and coastal drainage systems. *J. Water Resources Planning and Management*, **113**, 216–27.

Tong, Q. & Cheng, T.-W. (1981). Runoff. In *Physical Geography of China*, B. Huang (ed.). Science Press, pp. 6–121 (in Chinese).

Vail, A.J., Lee, G.C. & Robertson, I.R.S. (1976). Some aspects of the construction of the High Island Scheme. *Hong Kong Eng.*, **4/8**, 53–63.

Warrick, R.A. (1986). Climatic change and sea level rise. *Climate Monitor*, **15**(2), 39–44.

Warrick, R.A. and Oerlemans, J. (1990). Sea level rise. In *Climate Change: The IPCC Scientific Assessment*, J.T. Houghton, G.J. Jenkins and J.J. Ephraums (eds.). Cambridge University Press, Cambridge.

Watts, I.E.M. (1959). *The Effect of Meteorological Conditions on Tide Height in Hong Kong*. Royal Observatory, Hong Kong.

Williams, A.T., Grant, C.J. & Leatherman, S.P. (1977). Sedimentation patterns in Repulse Bay, Hong Kong. *Proc. Geol. Assoc.* **88**, 183–200.

Yim, W.W.-S. (1988). An analysis of tide-gauge and storm surge data in Hong Kong. In *Future Sea Level Rise and Coastal Development, Abstracts 5*, W.W.-S. Yim (ed.). Geological Society of Hong Kong and Department of Geography and Geology, University of Hong Kong, pp. 18–26.

Yim, W.W.-S. (1986). A sea level curve for Hong Kong during the last 40,000 years. In *Sea Level Changes in Hong Kong During the Last 40,000 years, Abstracts 3*, W.W.-S. Yim (ed.). Geological Society of Hong Kong and Department of Geography and Geology, University of Hong Kong, pp. 23–30.

V

SUMMARIES AND RECOMMENDATIONS

The United Nations Environment Programme/Commission of the European Communities (UNEP/CEC) *Workshop on Climatic Change, Sea Level, Severe Tropical Storms and Associated Impacts*, held at the University of East Anglia in Norwich, 1987, which provided the inspiration for this book, was a unique event. It is doubtful whether any previous invitational workshop had attracted so many experienced scientists from the entire spectrum of research who were given the opportunity to make contributions to a review of the current 'state-of-the-art' understanding of the subject. For this reason, part of the Workshop was devoted to panel discussions in which participants were invited to: identify the major points of the Workshop presentations; reach consensus regarding past changes and future projections, if appropriate; and recommend future areas of research and national or international action.

The following three chapters are the reports of the expert panels. For the reader, the reports have the advantage of summarizing the salient points of a wider range of presentations than contained in this volume, since not all Workshop papers could be included herein. The reader will also find that although other assessments of climate and sea level change, like that of the Intergovernmental Panel on Climate Change, have taken place since the Workshop, the judgements and recommendations of the panels remain relevant and timely, which attests to the combined wisdom and foresight of Workshop participants.

23

Sea level changes

P.L. Woodworth

Panel members: S. Jelgersma (Chairperson), G. Alcock, D.G. Aubrey, W. Carter,
R. Etkins, V. Gornitz, R.B. Heywood, M. Kuhn, D.T. Pugh, P. Rowntree,
M.E. Schlesinger, I. Shennan, T.M.L. Wigley.

Introduction

The purpose of this chapter is to report on the deliberations of the Workshop panel concerned with climate and sea level changes. The relevant portion of the Workshop contained eleven up-to-date expert presentations covering the entire spectrum from paleo-sea levels, through tide-gauge measurement and analysis, to climate modelling and the greenhouse effect (corresponding to Parts I–III of this volume). The Workshop panel was comprised of 14 participants, including experts in most aspects of sea level changes. The presence of experts on past sea level changes was of great benefit to the group, and to the conference in general, in shaping discussion on predictions for the next century. Variations in sea level clearly comprise a set of complicated spatial and temporal functions and cannot simply be extrapolated into the next century, even if there were to be no 'greenhouse effect'. Furthermore, experience of changes in past sea levels and shorelines provide the main source of knowledge for impact predictions for the next century. Consequently, the acquisition and analysis of data on past sea level changes, and their assimilation into models of climate change, comprise an essential complement to and component of discussions of future change.

The group's findings are given in the sections below. Brief summaries of the presentations are given, followed by the actions identified by the group which should be undertaken by national and international agencies or by individual scientists. The actions are divided into 'Technical Recommendations' which should be regarded as comments directed at sea level research workers, and 'Strategic Recommendations' which require the creation and operation of research programmes by groups of scientists and by the various agencies. There will clearly be some overlap between the two sets of recommendations.

Measurement of sea level changes over long time-scales

The paper by Tooley (this volume) emphasizes two crucial points concerning sea level research which are essential to grasp before deliberating future changes in sea level due to climate change.

The first point is that global sea levels have fluctuated by ± 100 m throughout recent Earth history. The results from the Exxon group of workers (Haq *et al.*, 1987), determined from sequence stratigraphy and from the development of depositional models, have demonstrated that variations in sea levels have occurred during the past 250 million years which can be ascribed to episodes of global tectonics and to variations in the size of the ice caps. In particular, and most relevant to the present situation, sea levels have risen by the order of 100 metres over the past 18,000 years since the last ice age. There is therefore a 'natural background' of real sea level variations against which any projected anthropogenic change has to be compared.

The second point concerns the concept of 'eustatic' sea level changes (Suess, 1906) as changes in ocean volume which may be globally uniform. This concept is clearly implicit in many published sea level curves: for example, that of Fairbridge (1962) which attempted to link global mean sea level variations to the ice advance and retreat stages of the last 10,000 years. The work of Mörner (1976) largely replaced 'eustasy' with 'local' or 'regional eustasy' composed of changes in ocean-basin volume ('tectono-eustasy'); ocean-water volume ('glacial eustasy'); and the ocean and earth mass distribution ('geoidal eustasy'), superimposed upon which are smaller terms such as 'sedimento-eustasy' (the filling-in of the ocean basins with sediment) and dynamic sea level changes (or 'ocean climate'). Versions of this picture are nowadays generally accepted by most people although the question of how quickly the geoid can be modified, for example, remains a controversial subject.

The most relevant question to the present discussion is what happens to the concept of eustasy in the event of the removal of an ice-load from Antarctic or Greenland and its redistribution over the global ocean. The work of Clark and Lingle (1977) demonstrated that the magnitude of the resulting rise or fall of sea level in each area would not be the same, and that the time-scales would not necessarily be correlated. The simple picture of 'eustasy' has therefore largely been replaced by conceptually more difficult, although doubtless more correct, computer models of a dynamic earth.

In the last few years a considerable amount of new sea level data from the Late Quaternary, and especially the Holocene, has been acquired from around the world through programmes such as the International Geological Correlation Programme (IGCP) Project 200, (although there is still a scarcity of information from many areas such as China). The measurements have been obtained from different materials (peats, shells etc.) along coastlines with different tidal conditions and have been dated in different ways. Tooley's presentation contained a discussion of the systematic errors which can occur in such diversified sources of information. The inevitable conservative approach one must adopt towards paleo-sea level data makes the use of parallel 'proxy-data' essential. Tooley described a set of his own analyses linking Fairbridge-type sea level curves and ^{18}O Greenland data to calcium carbonate measurements from Northwest Britain; a fair degree of correlation was observed with times of positive and negative sea level rise tendency identified and related to glacier fluctuations in the Northern Hemisphere over the last 12,000 years or so. Notably, while rates of change of sea level have varied by several mm/yr over the past 10,000 years since the Fennoscandian melting (e.g., Tooley's example of the east coast of China), these fluctuations are all at the conservative end of most sea level rise scenarios for the next century, which stresses the historical significance of the climatic situation which may await us.

One conclusion from the body of work on paleo-sea levels and coastlines is that sea level rise does not necessarily imply erosion, and sea level fall does not necessarily imply accretion, although there can be found clear examples of areas where such relationships do apply; for example, for the Holocene beaches and barrier islands along the US East Coast (Dolan and Lins, 1987).

Technical recommendations

1. That sea level movements should be represented not only by a fitted regression line or a single curve but also by an error band which would account for altitudinal, temporal (including ^{14}C to sidereal conversion) and interpretational variations. Extrapolations should also be portrayed as an error band.
2. That there should be applied a unified, standard methodology to areas displaying different vertical motions (i.e., subsidence and uplift) and that correlations should be undertaken to isolate regionally significant tendencies of sea level movement. Periods of positive and negative sea level rise tendency can then be correlated with proxy-data for climatic change, particularly ice advance and retreat.
3. That 'global correlation' should be attempted only when regionally significant tendencies of sea level movement have been established.
4. That ^{14}C dating methods should be replaced where practicable by the new accurate accelerator techniques.
5. That rates of change of regional sea level and associated coastline changes should be calculated and comparisons made with rates of projected changes for the next century.
6. That correlations should be made between variations in ^{14}C isotope and regional eustatic sea level curves and the various transgressive/regressive overlap sequences.

Strategic recommendations

1. That international (preferably regional) databanks be established to store sea level datapoints for the period 15,000–0 BP together with the corresponding ^{14}C (or other) data, material description and other related information. In addition, the databank should contain regularly updated data sets of digitized, published sea level curves for comparison, when applicable, to the raw sea level data points. The databases should be screened in various ways to improve their quality and to increase the number of unequivocal sea level data points.
2. That efforts be made to obtain well-established data for the periods 15,000–9,000 BP and 3,000–0 BP. For the latter period, an overlapping series of sea level data from natural (stratigraphic), archaeological and documentary/instrumental records is urgently required.

Measurement of recent and future sea level changes

Relatively recent changes in the global mean sea level have been estimated from tide-gauge records by several authors (see Gornitz, this volume) with the consensus of opinion in favour of an increase of approximately 15 cm in the global mean level over the past century (see recent reviews by Barnett, 1983, 1985; Revelle, 1983; PRB, 1985 and Robin, 1986). However, the group recognized that there are a number of problems associated with this figure. Firstly, all the analyses have employed essentially the same data set (that available from the PSMSL) so that the general agreement between results cannot be considered as significant (although different analyses have used different subsets of the data set). Secondly, the poor geographical distribution of the historical tide-gauge data set and the inherent ambiguity in the data between land measurements and real sea level changes preclude definitive statements regarding the climate signal in the sea level records.

These difficulties have been discussed previously in the literature (e.g., Pirazzoli, 1986) and were emphasized in the presentations of Gornitz and of Aubrey and Emery (this volume). Nevertheless, the Workshop panel felt confident that the acquisition of sea level data over longer time-scales (both from tide-gauges and from paleo-sea level studies), the development of geodynamic models of the Earth, and the independent measurement of vertical land movements by advanced geodetic techniques should eventually allow the unambiguous determination of long-term global climatic sea level change. As examples, Gornitz and Aubrey both presented applications of extrapolated paleo-sea level curves and of the Peltier geodynamic model to US East Coast tide-gauge data. In general, the scatter in the tide-gauge secular trends was reduced with a 'corrected' mean US trend of approximately 1.3 mm/yr, not dissimilar to the analyses of global mean trends.

Aside from the question of 'eustatic' sea level changes, tide-gauge records provide a useful means of investigating current relative vertical land movements for comparison to movements obtained over geological and archaeological time-scales or inferred from geodynamic models; Gornitz (this volume) and Aubrey and Emery (this volume) provide good examples of this application. (Also see papers by Yim and by Schnack, this volume).

The presentations of Pugh and of Carter *et al.* describe what sort of global tide-gauge network there may be in the next decade. The full GLOSS ('Global Level of the Sea Surface') plan (Pugh, this volume) is initially based on a proposal for a global network of about 300 gauges, approximately 150 of which already exist. These criteria may well be modified subject to the requirements of satellite altimetry and as experience is gained during the World Ocean Circulation Experiment (WOCE). Carter *et al.* (this volume) explained the methods and the present accuracies of linking gauges into regional and global geodetic reference systems, so that independent measurements of vertical land movements in a geocentric frame can be made, thereby removing the land-ocean ambiguity from the tide-gauge records. Accuracies of 1 cm over 200 km are currently being obtained from mobile VLBI (Very Long Baseline Interferometry) and from differential GPS (Global Positioning System) with decimetric accuracy for VLBI over 10,000 km. Advanced geodetic programmes are underway in several countries for different applications; for example, in oceanography the development of accurate geoids together with the geocentric positioning of the gauges should yield absolute geostrophic ocean surface currents, rather than simply current-variability as at present. The enhanced global tide-gauge network and the new geodetic methods (VLBI, GPS, absolute gravity, lasers, Doppler etc.) should, if all goes well, completely revolutionize the topic of sea level measurements within the next few years.

In principle, it is still possible to envisage changes in global or regional mean sea level which, through gyral spin-up, are not interpreted correctly at the tide gauges, most of which are on continental coastlines. The answer is to routinely measure the level of the entire ocean by means of satellite radar altimetry: that is, replace 'mean sea level' at a finite number of locations with 'mean sea surfaces'. The Workshop panel noted that the precision of several centimetres anticipated from altimeters (such as that to be carried on TOPEX/POSEIDON) is not very different from the accuracy of a single tide gauge.

The panel felt that in no general sense could data on coastline changes during the past century be used as a proxy-variable for sea level changes. The IGU Commission on the Coastline Environment (Bird, 1985) completed a semi-quantitative review of global coastline changes over the past century and compiled a list of 14 major influences on coastlines, only one of which is sea level change. The major conclusion from that study (Bird, 1987) is that sandy beaches appear everywhere to be in general retreat. The consequences to the coastlines of a large rise in sea level as a result of the greenhouse effect forms an important area for ongoing research.

The recommendations below benefited from the direct experience of one-half of the panel members with analysis of the PSMSL data set or with the installation and operation of tide gauges. To some extent the recommendations can be considered an update and reinforcement of those of previous working groups, such as IAPSO (1985).

Technical recommendations

1. That in analyses of secular trends in global mean sea level, attempts should be made to correct the tide-gauge data for the effects of local vertical land movements. This can be done through the use of extrapolated Holocene sea level curves (e.g., Gornitz *et al.*, 1982), although it was realised that such curves contain a real historical eustatic contribution which will inevitably modify the current climate signal estimate. Alternatively, the tide-gauge data may be corrected by means of geodynamic models of the Earth such as those of Peltier (1982).

2. That the geodynamic Earth models be further developed and improved.

3. That the large amount of interannual variability (annual and decadal time-scales) in tide-gauge records should be modelled, and understood, thereby improving estimates of low-frequency sea level secular trends. This can be done by means of regression analysis of the atmospheric and oceanographic forcings (winds, air pressures, air and ocean temperatures, salinities, runoff etc.), such as employed in Thompson (1986) or Woodworth (1987), or through more sophisticated ocean numerical modelling.

4. That very short tide-gauge records (less than approximately 10–15 years) should not, in general, be used in analyses of secular trends in mean sea level in view of the large amount of interannual variability present in most tide-gauge records. This recommendation clearly depends on the details of the particular analysis; short records of approximately 10 years can possibly be used in the study of relative secular changes in sea level between nearby stations as much of the interannual variability will be coherent.

5. That analyses of the historical tide-gauge data set should continue for all areas for which there are sufficient data with more sophisticated statistical averaging techniques employed than hitherto.

6. That tide-gauge records should be continually examined for evidence of 'accelerations' or 'inflexion points' in sea levels. This may take the form of a test of change in slope (Barnett, 1984) or incorporation into a general climate 'early warning system' (Wigley *et al.*, 1985).

7. That, as an input to discussions of local impacts, mean sea level may be a less useful variable than mean high waters. For example, de Ronde (1983), Führböter and Jensen (1985) and Rossiter (1969) report larger increases in mean high waters than mean low waters in the last century for the Dutch and German coasts and the Thames estuary respectively. These increases in tidal range are not explicable simply as a result of the increased water depth shown by the mean sea level records, and presumably are related in some way to the changing geomorphology of the area.

Strategic recommendations

1. That the currently geographically limited global tide-gauge network should be improved and enhanced thereby eventually improving the estimates of global sea level trends. The group, therefore, recognised and endorsed the establishment of the GLOSS sea level network (IOC, 1986: Pugh, this volume) with its emphasis on the quantity and quality of the tide-gauge data: the uniformity of recording methods; the storage of hourly height data for oceanographic purposes; on geodetic connections between benchmarks; on data exchange mechanisms; on training; and on the interaction of tide-gauge measurements with those by satellite radar altimetry. The group strongly recommended that the appropriate national and international agencies be encouraged to participate in the GLOSS programme.

2. That tide-gauges used for global mean sea level analysis must be connected to an appropriate geodetic reference system and regularly (annually or more frequently) re-surveyed. The minimum control should be through a 'local' levelling network of benchmarks extending a few kilometres about the tide-gauge, one or more of which would be located at sites where GPS (Global Positioning System) observations can be made (i.e., sky visibility down to 10–15 degrees elevation, low radio-frequency interference). Where possible, the tide-gauges should be tied to a 'regional' geodetic network using differential GPS observations, and ultimately the regional network should be tied to the global VLBI (Very Long Baseline Interferometry) network. As an independent check on the vertical stability of the gauges, high accuracy (± 1–3 microgal, where one microgal is essentially equivalent to 5 mm) absolute gravity surveys should be made in the vicinity of as many of the gauges as resources permit, and in particular in regions of large rates of submergence or uplift.

3. That long-term support and encouragement be given to the various national and international programmes of satellite radar altimeter measurements of the oceans which should eventually result in truly global sea level measurements within a global geodetic system.

4. That further working groups be established to study the relationships between sea level changes and coastlines as a guide to potential impacts in the next century.

Forcings responsible for sea level changes

Five presentations were given during the Sea Level Session directed at understanding the forcings responsible for the observed sea level changes of the past century. Two papers, by Wigley and Raper (this volume) and by Schlesinger, were concerned with the contribution to sea level rise from thermal expansion of the oceans; Kuhn (this volume) described the contributions from glacier ice wastage; Oerlemans (this volume) presented a review of studies of changes in the Greenland and Antarctic ice sheets; and Aubrey and Emery (this volume) discussed the processes responsible for regional variations to global sea level change. Two further papers, by Etkins and by Schlesinger, were also concerned with the forcings responsible for changes in global sea levels. Etkins presented a picture of the oceanic deep and surface circulation, in particular insofar as it relates to high-latitude ice melting, salinity decrease, and sea surface temperature and precipitation changes (see below); Schlesinger reviewed the ability of five coupled ocean–atmosphere General Circulation Models (GCMs) to reproduce patterns of regional climatic change for a doubling in greenhouse gas forcing.

It is interesting to consider the opinions of previous workshops on the forcings responsible for the apparent rise of order 15 cm in global sea levels during the past century. The estimates of the relative importance of thermal expansion to ice wastage vary widely. The PRB (1985) and Barnett (1985) de-emphasize the role of thermal expansion, while Hoffman et al. (1983), Gornitz et al. (1982) and others believe that the ratio is approximately 1:1. Robin (1986) assumes thermal expansion to have contributed approximately one-third to sea level rise since 1900. These previous estimates can be compared to those obtained from this workshop described below.

Thermal expansion

The simulations of the thermal expansion contribution to sea level rise over the past century presented in Wigley and Raper (this volume) and by Schlesinger are far more sophisticated than those made previously.

The Wigley and Raper model is a one-dimensional (1-D) mixed-layer scheme with diffusion and with upwelling balanced by high-latitude downwelling. The authors show that it is possible to treat the Northern and Southern Hemispheres separately as their temperature histories are quite different. Either the individual forcings or the upwelling parameters in each hemisphere can be changed separately; but different mechanisms which yield similar temperature responses in fact give different sea level results. In this way an optimum set of parameters can be arrived at by comparison to the global temperature and sea level data sets.

There are three interesting results from the Wigley and Raper model. Firstly, while the surface temperature change for a specified greenhouse gas forcing is found to be not too sensitive to the choice of vertical diffusivity parameter 'K', the change in mean sea level is sensitive to it because sea level is an integral throughout the water column. If the greenhouse gas forcing in the model is stabilized (i.e., set to a constant after some specified period) the surface temperatures reach equilibrium within a relatively short time. Mean sea level, however, continues to rise for many decades (or even a century) afterwards as the heat propagates down to the lower ocean layers; and the long time-scales involved clearly have consequences for those involved in long-term impact decision making processes.

Secondly, the model suggests a sea level rise due to thermal expansion over the past century of order 2–5 cm. This is approximately one-eighth to one-third of the total sea level rise (see Table 2.1 in Gornitz, this volume) and is slightly less than the Villach conclusions (Robin, 1986). Thirdly, the authors predict a thermal expansion induced sea level rise over the next 40 years within the 'outside range' of 3–13 cm and an 'inside-range' of 4–8 cm which are less than some previous predictions (for example, those of Hoffman et al., 1983).

Schlesinger used the results of the Oregon State University three-dimensional (3-D) coupled ocean–atmosphere GCM (two-layer atmosphere, six-layer ocean with a seasonal cycle) to simulate the effect of an instant CO_2 doubling. The model clearly shows heat penetrating to the lower ocean levels, particularly at high latitudes, and sea level increasing by about 9 cm after 20 years. The results of the 3-D model were used to calibrate a much simpler 1-D model (superficially similar to the Wigley and Raper model) which was then employed in a simulation of past and future sea level rise due to thermal expansion with realistic greenhouse gas forcing. This 1-D model, with parameters optimally tuned to reproduce the 3-D results, suggests a rise in sea level due to thermal expansion of approximately 7 cm over the past century and a further rise of about 8 cm over the next 40 years. These estimates are slightly larger than those of Wigley and Raper, but the differences are hardly significant and may be a feature of the slightly different greenhouse gas forcings employed in the two modelling exercises.

The agreement between the two independent model studies represents an important contribution from this conference to the understanding of the thermal expansion component of past and future sea level changes.

Schlesinger's second presentation focused on the ability of five GCMs (the GFDL, GISS, NCAR, OSU and UK Met. Office models) to reproduce patterns of regional climate change for a greenhouse gas doubling. The sensitivity of the models to greenhouse gas forcing was found to depend critically on the control climate, and therefore in practice on the presence of factors not yet included in the model; these include soil moisture processes and, in particular, the cryosphere. The conclusions from the model comparisons are that a greenhouse gas doubling would result in global temperature increases in the range 2.8–5.2 °C and 7–15% increase in global precipitation, although the geographical distributions of the change in each parameter are only in qualitative agreement. It is clearly important for sea level estimates that regional climate change over mountain glaciers and the poles is modelled as well as possible (see Schlesinger, this volume). Further work is required to simulate the present (i.e., control) climate better and to increase model resolution.

Glacier ice

Kuhn (this volume) presented a detailed review of the few-tenths of a percent of the world's ice which is stored in mountain glaciers and on Arctic islands 'on the borders of existence'. The Meier (1984) analysis, which suggested an average sea level rise of 0.46 ± 0.26 mm/yr over the period 1900–61, is still the best estimate available and would represent approximately one-third of the total signal of Gornitz's Table 2.1 (this volume). Meier uses a formula for estimating the ice balance of glaciers unsurveyed for long-term changes, which essentially has the balance of an unsurveyed glacier set equal to that of a known glacier scaled by the difference between the summer and winter balance. Kuhn explained that this is an acceptable procedure as long as the area of the glacier system does not change significantly. In many cases the total volume of a glacier can be estimated from its length alone.

For the near future (say the next 40 years), the 'present' rate of order 0.5 ± 0.3 mm/yr sea level rise equivalent would be a reasonable extrapolation. For a greenhouse gas climate change scenario over 100 years (including an air temperature rise of 4 degrees), a rise of 9–31 cm in global sea level due to glacier melting would result. In the long term, however, the wastage of glacier ice is surely not a major problem given the uncertainties in the contributions from other forcings (particularly polar ice melting) and given that there is only 33–61 cm of sea level equivalent in the global store of glacier ice anyway.

Polar land ice

The presentation of Oerlemans (this volume) summarized present knowledge of processes concerning the polar ice sheets and ice shelves relevant to sea level changes. At present, it is unknown whether polar ice is a major net contributor, either positively or negatively, to global sea level rise. Due to uncertainties in the measurement of mass balance, it is possible that the Antarctic ice sheet could have increased or decreased sea level by as much as 5 cm over the last 100 years. For a one degree warming, the contributions to sea level from Greenland and Antarctica (including West Antarctic instabilities) are $+0.5 \pm 0.25$ and -0.5 ± 0.85 mm/yr, respectively. Potentially 7.5 m and 65 m rises of global sea level would result from the total melting of the Greenland and Antarctic ice sheets, respectively.

Oerlemans explained the importance of the ice shelves to the sea level question, even though they themselves do not contribute directly to sea level variations. In one scenario of climate warming, the basal melt rates would increase, which would increase the retreat of grounded ice, although the increased accumulation from higher precipitation would dominate for Antarctica as a whole. Meanwhile, the Greenland ice sheet would shrink as a temperature increase would, overall, cause more melting and runoff than snowfall. Oerlemans' previous work (Oerlemans and van der Veen, 1984) suggests that the effect of the instabilities of the West Antarctic ice sheet on sea level will be negligible during the next century, but thereafter the disintegration may cause a sea level rise of several metres following the retreat of the rims of the ice shelves.

Presumably the only 'surprise' in scenarios of sea level change during the next century would arise through surges in the Antarctic ice sheets. One recent study (DOE, 1987), however, concluded that this is unlikely primarily because, unlike the situation for mountain glaciers, most snowfall in Antarctica occurs near the coast at the low end of the ice streams.

Melting of Greenland ice is also important in the context of freshening of the North Atlantic discussed by Etkins. In his thermo-haline circulation model, the Greenland melting influences the production of North Atlantic deep water, polar heat transport and temperatures and salinities through the water column, and modifies the relative sea temperatures of the Northern and Southern Hemispheres. Etkins showed a list of

North Atlantic oceanographic time series of temperatures and salinities for the middle twentieth century, and discussed how they may relate to the two degree warming over Greenland in the 1930s and the probable consequent release of fresh water.

Other factors

The panel did not discuss the contributions from other, particularly human activity related, factors on mean sea level, such as ground water pumping or the storage of water behind dams (Newman and Fairbridge, 1986). It was considered that, as sea level rises because of the greenhouse effect, these other factors will become less important.

Technical (modelling) recommendations

1. That the utility of simple linear regressions of global air temperatures against global sea levels as a basis for future sea level prediction is not supported by the models described above.
2. That the limits of the utility of the box-upwelling-diffusion climate models should be further explored by means of comparison to the results of the more sophisticated 3-D GCMs.
3. That the difficulties currently encountered with the GCMs should be further investigated. In particular:
    (a) The reasons for the disagreements between the results of the several GCMs should be identified and research carried out to resolve the differences.
    (b) The present climate should be simulated better thereby improving confidence in the sensitivity of the models to changes in greenhouse gas forcing.
    (c) The resolution of the models should be improved both horizontally and vertically.
    (d) The various positive and negative feedbacks be better understood.
    (e) The cryosphere should be added to the coupled models. In particular, ice sheets and glacier models should be employed with the coupled ocean-atmosphere-model results as input.
    (f) Detailed ocean models, which are in the course of construction as part of the Tropical Ocean and Global Atmosphere (TOGA) and World Ocean Circulation Experiment (WOCE) programmes, as well as shelf models in important areas, should be incorporated into the coupled models.
4. That modelling of polar ice sheets and ice shelves and precipitation rates over these areas should continue in view of current uncertainties in polar accumulation and wastage rates and in view of the potential crucial role to be played by the ice sheets in long-term sea level change.
5. That modelling of the mass balance of glaciers should also continue along the lines explained by Kuhn (this volume).

Strategic recommendations

1. That measurements of the area, topography and volume of low-latitude glaciers by conventional surveying techniques should be continued and expanded in view of the estimated importance of this forcing to current sea level rise. The group

noted and encouraged the further activities of the World Glacier Inventory (Zürich).

2. That monitoring of ice caps and glaciers by means of satellite radar altimetry should be undertaken on a routine basis as a complement to ground surveys and polar deep drilling.

3. That many other forms of remote sensing are, or will, soon be in existence using devices from aircraft (e.g., laser altimetry) and from space (for a review of which see Thomas, 1986) which will be essential to the acquisition of a comprehensive data set of climate change. These should be incorporated into combined analyses of polar and ocean changes of which sea level variations will form one component. The intensive monitoring of the atmosphere and oceans (repeated CTD transects, surface flux measurements, transient tracers, wind stress measurements, deep circulation studies etc.) via ground stations and research vessels as part of the World Ocean Circulation Experiment should be continued afterwards, as then considered appropriate, in order to provide the most effective climate monitoring system.

4. That length-of-day and polar motion measurements should also be continued in order to provide long-term data sets free from systematic errors which can be included in future studies of mean sea level secular changes.

Consensus on future global sea level changes

The panel sought to make a conservative statement of consensus on sea level rise for the 40-year period 1987–2027. Over this period the greenhouse gas forcing, and therefore the thermal expansion component of sea level rise, can be reasonably estimated. In addition, the ice wastage rates from polar regions and from glaciers can be extrapolated from the present rates and from estimates of the sensitivities to temperature changes. It is emphasized that this conservatism is not necessarily a criticism of previous predictions of future sea levels, nor is it an attempt to minimize the potential importance of long-term sea level rise; rather, it is simply the view of this panel that this is the most useful statement which can be made at the present time.

For a 'baseline' rate of global sea level secular trend we take an estimate of 1.5 ± 0.5 mm/yr (Gornitz, this volume). This already includes current ice wastage rates which we assume to be similar to those which will occur over the next 40 years. To the 6 ± 2 cm for the 'baseline' over the next 40 years we add 7 ± 2 cm for thermal expansion, based on the analyses by Wigley and Raper (this volume) and by Schlesinger. (Strictly, the 'baseline' also includes 1–2 cm of thermal expansion over 40 years which we ignore). This yields a total sea level rise of 13 ± 4 cm over the next 40 years. This estimate is about double the rise that would be anticipated from simple extrapolation of current trends, but is rather less than most previous estimates.

A second exercise involves the calculation of sea level rise during the next 100 years. Greenhouse gas forcing is assumed to increase only up to 2050 and to be constant thereafter; we take this to be a reasonable low-estimate of the potential forcing. The sea level rise due to thermal expansion, measured from the present-day, would be around 31 cm, based on model projections by Schlesinger. (Again the relatively small thermal expansion contribution in the 'baseline' is ignored). Kuhn (this volume) provides an estimate of 20 cm for glacier melting from which we subtract the 5 cm already included in the 'baseline'. For polar ice we assume no overall change. Added to the 'baseline' 15 cm, this gives a total rise of 61 cm which is approximately four times the current rate of sea level rise.

The uncertainties in the above calculations are large (and to a great extent unquanti-fiable) but it does appear that, based upon our limited current knowledge, we must anticipate global sea levels to increase at least by several decimetres during the next century. The estimates from this workshop are similar to those of other recent studies (as shown in Table 1.1, this volume). However, this similarity cannot be taken as evidence that the sea level research community is in overall agreement as to the potential rise. Each study has invoked a different mixture of climatic forcings in order to arrive at its predictions which detracts from confidence in the overall similarity. Clearly many of the forcings are interrelated through climatic feedback mechanisms and it cannot be claimed that any of them are modelled adequately at present. We conclude that more research efforts in measurement and modelling are required and that the results we have quoted are the best that are given at the present time.

References

Barnett, T.P. (1983). Recent changes in sea level and their possible causes. *Clim. Change*, **5**, 15–38.

Barnett, T.P. (1984). The estimation of 'global' sea level change: a problem of uniqueness. *J. Geophys. Res.*, **89**, 7980–8.

Barnett, T.P. (1985). On long-term climatic changes in observed physical properties of the oceans. In *Detecting the Climatic Effects of Increased Carbon Dioxide*, M.C. MacCracken and F.M. Luther (eds.). US Department of Energy, DOE/ER-0235.

Bird, E.C.F. (1985). *Coastline Changes*. Wiley.

Bird, E.C.F. (1987). The modern prevalence of beach erosion. *Marine Pollution Bulletin*, **18**, 151–7.

Clark, J.A. & Lingle, C.S. (1977). Future sea level changes due to West Antarctic ice sheet fluctuations. *Nature*, **269**, 206–9.

De Ronde, J.G. (1983). Changes of relative mean sea level and of mean tidal amplitude along the Dutch coast. In *Seismicity and Seismic Risk in the Off-shore North Sea Area*, A.R. Ritsema and A. Gürpinar (eds.). D. Reidel Publishing Co., pp. 131–41.

DOE (1987). *On the Surging Potential of Polar Ice Streams*. US Department of Energy Report DOE/ER/60197–H1.

Dolan, R. & Lins, H. (1987). Beaches and barrier islands. *Sci. Am.*, **257**, 52–9.

Fairbridge, R.W. (1962). World sea level and climatic changes. *Quaternaria*, **6**, 111–34.

Führböter, A. & Jensen, J. (1985). Long-term changes of tidal regime in the German Bight (North Sea). *Proceedings of the Fourth Symposium on Coastal and Ocean Management*, Baltimore.

Gornitz, V., Lebedeff, S. & Hansen, J. (1982). Global sea level trend in the past century. *Science*, **215**, 1611–14.

Haq, B.U., Hardenbol, J. & Vail, P.R. (1987). Chronology of fluctuating sea levels since the Triassic. *Science*, **235**, 1156–67.

Hoffman, J.S., Keyes, D. & Titus, J.G. (1983). *Projecting Future Sea Level Rise: Methodology, Estimates to the Year 2000, and Research Needs*. US GPO#055-000-0236-3, GPO, Washington, DC.

IAPSO (1985). Changes in relative mean sea level. Working party report of the International Association for the Physical Sciences of the Ocean (Chairman: D.E. Cartwright). *EOS Trans. Am. Geophys. Union*, **66**, 754–6.

IOC (1986). *Global Sea Level Observing System (GLOSS) Implementation Plan.* Intergovernmental Oceanographic Commission of UNESCO, IOC/INF-663.

Meier, M.F. (1984). Contribution of small glaciers to global sea level. *Science,* **226**, 1418–21.

Mörner, N.A. (1976). Eustasy and geoid change. *J. Geol.,* **84**, 123–51.

Newman, W.S. & Fairbridge, R.W. (1986). The management of sea level rise. *Nature,* **320**, 319–20.

Oerlemans, J. & van der Veen, C.J. (1984). *Ice Sheets and Climate.* Reidel Publishing Company, Dordrecht.

Peltier, W.R. (1982). Dynamics of the ice age earth. *Adv. Geophys.,* **24**, 1–146.

Pirazzoli, P.A. (1986). Secular trends of relative sea level (RSL) changes indicated by tide-gauge records. *J. Coastal Res.,* **SI1**, 1–26.

Polar Research Board (PRB) (1985). *Glaciers, Ice Sheets and Sea Level: Effect of a CO_2-induced Climatic Change.* Report of a workshop held in Seattle, Washington, September 13–15, 1984. DOE/ER/60235-1, US Department of Energy, Washington, D.C.

Revelle, R. (1983). Probable future changes in sea level resulting from increased atmospheric carbon dioxide. In *Changing Climate, Report of the Carbon Dioxide Assessment Committee.* National Academy Press, Washington, DC, pp. 433–48.

Robin, G. de Q. (1986). Changing the sea level. In *The Greenhouse Effect, Climatic Change and Ecosystems, SCOPE 29,* B. Bolin, B. Döös, J. Jäger and R.A. Warrick (eds.). John Wiley and Sons, Chichester, pp. 323–59.

Rossiter, J.R. (1969). Tidal regime in the Thames. *The Dock and Harbour Authority,* **49**, 461–2.

Suess, E. (1906). *The Face of the Earth.* Clarendon, Oxford.

Thomas, R.H. (1986). Future sea level rise and its early detection by satellite remote sensing. In *Effects of Changes in Stratospheric Ozone and Global Climate, Volume 4: Sea Level Rise,* J.G. Titus (ed.). UNEP/EPA pp. 1–18

Thompson, K.R. (1986). North Atlantic sea level and circulation. *Geophys. J. R. Astron. Soc.,* **87**, 15–32.

Wigley, T.M.L., Kukla, G.J., Kelly, P.M. & MacCracken, M.C. (1985). Recommendations for monitoring and analysis to detect climate change induced by increasing carbon dioxide. In *Detecting the Climate Effects of Increasing Carbon Dioxide,* M.C. MacCracken and F.M. Luther (eds.). US Department of Energy Report DOE/ER-0235.

Woodworth, P.L. (1987). Trends in U.K. mean sea level. *Marine Geod.,* **11**, 57–87.

24

Severe tropical storms and storm surges

A.B. Pittock and R.A. Flather

Panel members: Anwar Ali, S.K. Dube (Chair), B. Johns, S.C.B. Raper.

Papers

The relationship between tropical cyclones (TCs) and sea surface temperature (SST) was examined by Raper (this volume). SST was selected as the independent variable because it should change with the greenhouse effect, and because it is better described by data than other possible causal factors. There is a long but inhomogeneous record of TC occurrence. The short record of TC occurrence from satellites is insufficient for independent validation of relationships with SST.

The data suggest different relationships in different areas. For example, there is a positive correlation between SST and the number of TCs in the Bay of Bengal, but a negative correlation in the Western North Atlantic. Interpretation is difficult since many relevant questions are not fully answered: where and when has a TC formed? Is the SST more critical in the period and region of precursor formation or during the growth to a full TC? To what extent do TCs change the SST (by enhanced vertical mixing) and so affect the simultaneous correlations? How important are upper-air conditions such as vertical and horizontal wind shear? What is the appropriate time-scale for statistical studies, given that TC development is influenced by short-term anomalies in the mean flow? How are perturbations in the flow related to the local mean?

A better understanding of the conditions which lead to the intensification of tropical depressions into tropical cyclones is critical for predicting the possible consequences of the greenhouse effect. Both the intensity and area of occurrence of TCs would be affected by SST changes if SST were the limiting factor, but it is not always limiting.

During the Workshop, M.E. Schlesinger reviewed the theory of TC formation and, in particular, the hypothesis (Emanuel, 1987) that the maximum intensity and frequency of occurrence of TCs would increase significantly due to SST increases in the tropics. Experiments with existing general circulation models (GCMs), with the exception of Manabe *et al.*, (1970), have not shown the development of TCs, probably because of poor horizontal resolution. However, Manabe *et al.* (1970), with better horizontal resolution, did show TC development, and the European Centre for Medium-Range Weather Forecasting (ECMWF) operational model predicts the occurrence of some, but not all, TCs. Improved resolution, and perhaps better physics, will be necessary for GCM studies of the effect of the greenhouse warming on TC occurrence. A horizontal resolution better than 2 degrees in latitude and longitude seems to be essential, and improved performance in predicting the track and intensification of a cyclone was evident with the ECMWF model at a resolution of 0.5 degrees. There is considerable potential for the use of high-resolution nested grids in GCM studies.

Regarding storm surges, Flather and Khandker (this volume) and B. Johns each presented papers which reviewed various aspects of the problem and described results from tide surge models for the Bay of Bengal. The meteorological forces generating storm surges are wind stress and atmospheric pressure, heavily conditioned by water depth profiles. Tides are generated in the deep oceans by gravitational forces. Given accurate data on wind and pressure forcing, a knowledge of tidal variations at the shelf edge, and detailed bottom and coastal zone topography, water levels due to tides and storm surges can be predicted with good accuracy. On the basis of comparisons with observations in well-observed regions (e.g., the North Sea), typical errors are of the order of 10%. Validation for the Bay of Bengal is, however, very limited due to a serious lack of water level observations during cyclones. In regions where large areas are subject to inundation, models with moving coastlines are essential. Such models should, therefore, be able to predict tides and storm surges with specified coastlines, bathymetry and storm characteristics as changed by the greenhouse effect.

Results of model experiments to determine the effect of an increase in mean sea level on tidal, storm surge and combined tide and storm surge levels were described by Flather and Khandker (this volume). The changes were small compared with the change in mean sea level producing them. Thus the main impact of an increase in mean sea level on the maximum water level during a storm will result from the change in mean sea level itself: consequent changes in tide and surge elevations relative to mean sea level will be of secondary importance. Changes in tidal regimes may, however, be significant in particular areas such as coastal wetlands and some embayments and estuaries.

Assessment of future effects

It is not yet possible to give exact predictions of the changes in TC frequency, intensity or location of occurrence due to the greenhouse effect. However, it is expected that, in general, an increase in SST would lead to the occurrence of TCs at somewhat higher latitudes.

Changes in maximum water levels will arise predominantly from any increase in mean sea level. With some possible exceptions, changes in the tides and storm surges relative to mean sea level will be secondary effects. In some regions, other factors such as changes in river flow, sedimentation rates and cyclone occurrence will be important.

Possible 'surprise'

The effects of changes in atmospheric circulation on ocean circulation and its consequent feedbacks need to be considered. For example, interregional differences may change the relative difference in SST across the tropical Pacific Ocean which is related to the El Niño–Southern Oscillation system. Such changes could affect regional mean sea level and SST, and thus the frequency of occurrence and preferred location of tropical and extra-tropical storms. A full understanding of these effects will require experimentation with fully coupled dynamic ocean–atmosphere models. Verification of such models with modern and paleo-climatic data is needed. In lieu of fully verified coupled ocean–atmosphere models, it may be useful to conduct carefully designed sensitivity studies with simpler models.

Recommendations

1. Despite uncertainties introduced by probable redistribution of sediments and changes in storm intensity, studies of the socio-economic impacts of predicted storm surges, with higher mean sea level and consequent coastline changes, should be carried out in selected areas.
2. The resolution of coupled atmosphere–ocean GCMs should be improved to enable them to resolve TCs accurately. Studies of TC occurrence under conditions of $1 \times CO_2$ and $2 \times CO_2$ should then be carried out.
3. Further global-scale statistical studies of TCs in relation to various atmospheric and oceanic parameters should be carried out. These studies should be designed to test theoretical models of TC development and dependence on SST and other climatic parameters.
4. If estimates of sea level change are refined, investigations of the interaction between river flow, sedimentation and tide/surge motion should be carried out.
5. Studies of the potential effect of changes in the frequency of tropical and extra-tropical cyclones on regional mean sea level should be carried out.
6. Better access to data on coastal topography and bathymetry, meteorological data, and observed storm surges is required to facilitate the development and validation of models for sensitive areas such as the Bay of Bengal.
7. Off-shore tidal measurements should be carried out to facilitate the development and validation of storm surge-tide models for vulnerable areas such as the Bay of Bengal.
8. The impact of short waves superimposed on tides and storm surges should be studied.
9. The frequency, location and intensity of extra-tropical cyclones should also be studied in relation to climate change and coastal effects.

References

Emanuel, K.A. (1987). The dependence of hurricane intensity on climate. *Nature*, **326**, 483–5.
Manabe, S., Holloway, J.L. Jr., & Stone, H.M. (1970). Tropical circulation in a time-integration of a global model of the atmosphere. *J. Atmos. Sci.*, **27**, 580–613.

25

Regional effects of sea level rise

J.G. Titus

Panel members: H. Brammer, E.B. Peerbolte, J.M. Broadus, J.W. Day,
J.G. de Ronde, B.R. Döös, S.P. Leatherman, Y.S. Qin, E.J. Schnack,
J.C. Stevenson, S. Sudara, P. Usher, H.G. Wind, W.S. Yim.

Introduction

This chapter summarizes the discussion of the Workshop Panel on the regional effects of sea level rise. The Panel identified the principal physical effects and socio-economic impacts of sea level rise (Table 25.1) and then discussed them in light of specific regional contexts. The Panel then put forth a set of research recommendations and concluding remarks.

Physical effects

The major effects of a rise in sea level are the loss of land due to inundation and erosion, increased flooding during storm surges and rainstorms and the intrusion of saltwater into aquifers, estuaries and wetlands. In addition, sea level rise raises water tables, and can change both currents and sedimentation rates.

During the Workshop, Qin, Yim (this volume) and Broadus (this volume) provided examples of areas of potential inundation in China, Hong Kong, and the Bangladesh and Nile Deltas, respectively. In the Panel discussion, it was also noted that the Republic of the Maldives is perhaps even more vulnerable to inundation, with most of the country situated less than 2 m above sea level.

Although inundation is the most obvious impact of a rise in sea level, erosion of land above sea level could be just as important. During the Workshop, Leatherman explained simple methods of estimating erosion rates (also see Clayton, this volume). Schnack (this volume) showed areas where current erosion rates are very high on the east coast of South America. Some of the most highly valued properties in the United States are the coastal beach resorts that are currently eroding at a rate of about 0.5 m/ yr. However, in the Panel discussion, it was noted that, in many cases, severe erosion is not necessarily caused by sea level rise. This does not diminish the importance of global warming and sea level rise; to a large extent, the ability to undertake impact studies rests upon knowledge gained from the study of erosion in general.

Sea level rise can increase flooding due both to storm surges and to rainstorms. A higher mean sea level allows storm surges to build on a higher base, with resulting

Table 25.1. *Issues related to sea level rise*

I Physical effects	III Responses
A Land loss	*A Curtail sea level rise*
1 Inundation	1 Control greenhouse gases
2 Erosion	2 Impound water
	3 Curtail extraction, loading, tidal range
B Flooding	changes
1 Higher storm surges	
2 Backwater effect during rainstorms	*B Responses to land loss*
	1 Defence (walls)
C Saltwater intrusion	2 Raise land surface
1 Groundwater	3 Abandon low areas
2 Surface water	4 Accept and adjust to inundation
3 Wetlands	
	C Responses to flooding
II Environmental and socio-economic	1 Responses to land loss, above
impacts	2 Flood walls
	3 Flood proofing
A Lost productive capacity of land	4 Tidal barriers
1 Agricultural	5 Warning and evacuation
2 Residential	
3 Tourism	*D Saltwater intrusion responses*
4 Recreation	1 Supply freshwater
5 Industrial	2 Barriers
	3 Shift to alternative supplies
B Capital losses	
1 Houses	
2 Roads	
3 Factories	
4 Infrastructure	
C Ecological	
1 Wetlands	
2 Dunes	
3 Estuaries	
4 Fish and wildlife	
D Water supply	
1 Salinity	
2 Dunes lost	
3 Pollution	
E Flood control and coastal defence works	

increases in damages. Higher water levels also have a 'backwater effect'. During rainstorms, areas drain more slowly because the higher water levels cause drain pipes, creeks and canals to back up more easily. Bangladesh offers extreme examples of both types of flooding. In 1970, the storm surge from a tropical storm killed hundreds of thousands of people, and in September, 1987 riverine flooding affected nearly one-half of the country. Hong Kong is also subject to surge inundation, but has a state-of-the-art system for evacuating people in the event of a major storm.

Both erosion and flooding raise the following issue: 'Can't we ever learn?' Brammer (this volume) pointed out that due to a shortage of available land and resources in Bangladesh, people move onto land where the last occupant was killed by a flood. Schnack (this volume) pointed out that in Brazil and elsewhere in South America, people build high-rise buildings 'too' close to the beach, repeating the discredited practices of the US East and Gulf Coasts. Clearly, more needs to be learned about individual decision making in creating situations of hazard.

Relatively little was presented at the Workshop concerning impacts due to saltwater intrusion, sedimentation or changes in currents due to climatic variation. However, studies conducted in the United States suggest that Florida and Philadelphia could have water supply problems (Hull and Titus, 1986). Saltwater intrusion into wetlands is causing rapid loss of cypress swamps in Louisiana, where land is sinking about one metre per century (Day *et al.*, this volume).

Environmental and socio-economic impacts

Studies of the impacts of future sea level rise on the United States and other developed nations have emphasized the potential implications for environmental quality and the loss of beach houses (e.g., Smith and Tirpak, 1989). For most of the developing world, however, the Panel concluded that other factors had to be considered

Perhaps most importantly, sea level rise could lead to human casualties. Although the slow rate of rise does not, by itself, threaten to harm anyone, it could be too slow to alarm anyone. When the infrequent severe storm occurs, however, more people could be at risk because the surge would be higher. Moreover, larger areas of cropland could be flooded with saltwater than in previous storms, possibly resulting in larger crop failures and food shortages in nations like Bangladesh.

The importance of losing land would vary from country-to-country. The minor loss of area is of little importance for a country with plenty of land like the United States. However, for a densely populated country like Bangladesh, loss of land would be a serious disruption. Indonesia's transmigration programme involves moving people from crowded locations to low-lying areas which could be subject to future flooding. Finally, countries such as the Maldives that are faced with the possibility of total inundation must take the prospect of land loss seriously whether or not they are crowded today.

Capital losses could be serious, but the Panel generally felt that, in absolute terms, the developing regions of the world would have less to lose in this regard. On the other hand, the disruption of beach resorts could threaten foreign exchange earnings in some cases.

Sea level rise could also further endanger coastal ecosystems and water supplies. Coastal wetlands are important to commercial fisheries in many areas, particularly near major river deltas. In the United States, 50–80% of coastal wetlands could be lost in the event of a 1.5–2 m rise in sea level (Armentano *et al.*, 1988). Studies of the potential importance of sea level rise relative to other threats to coastal ecosystems and water supplies in developing nations are fewer in number.

Sea level rise could threaten the operation of existing flood works. In countries with extensive dyke and pumping systems, such as The Netherlands (Wind and Peerbolte, this volume), this could be one of the most important impacts of sea level rise.

Societal responses

The responses to sea level rise can be classified broadly into two categories: keep the sea from rising, and adapt to the rise. The most obvious way to curtail the rise in sea level due to global warming would be to curtail the emissions of greenhouse gases. The international agreement reached by the Montreal Protocol on Substances that Deplete the Ozone Layer (and subsequent agreements to strengthen the Protocol) to cut chlorofluorocarbon (CFC) production is a major step in the right direction. Nevertheless, curtailing emissions of CO_2 and the other greenhouse gases enough to keep sea level rise from accelerating is a formidable task. Due to lags in the climate system, the changes in radiative forcing that have already occurred may make some future global warming and sea level rise inevitable.

It has been suggested that the sea could be prevented from rising by impounding water somewhere besides the oceans. Newman and Fairbridge (1986) investigated such a possibility but concluded that even the most ambitious set of impoundments would only lower sea level about 7 cm.

There are many established engineering and planning solutions by which nations could adapt to a rise in sea level. Densely developed areas could be protected by the use of walls to keep the sea out supplemented by pumping systems to remove excess water. However, this option may result in the loss of the valuable coastal resourcess such as wetlands and beaches. As Schnack (this volume) and Leatherman pointed out during the Workshop, a common procedure followed by beach communities is to raise the land surface with fill. However, both erecting walls and raising land surfaces can be expensive. In developed nations such as the United States, it is often argued that, in many cases, it is more economic to abandon areas and allow the sea to encroach than to employ expensive engineering solutions. In less-developed nations, there may be little choice. Finally, in a few cases, it may be preferable to allow the sea to encroach without abandoning the area, converting streets to canals as in Venice, or converting rice paddies to fish ponds.

Established responses to saltwater intrusion include injection of freshwater, construction of barriers, and abandonment in favour of alternative water supplies.

Research recommendations

The Panel offered five main recommendations for research to international and national bodies.

1. *Identify nations most vulnerable to sea level rise*. This task has the highest priority and should also include the identification of decisions sensitive to sea level rise.

These tasks would make it possible to identify which people should be concerned about sea level rise.

2. *Make rough estimates of the potential implications of sea level rise.* Such estimates would be useful for two reasons. From the standpoint of environmental organizations weighing the merits of limiting a global warming, one needs estimates of the costs of sea level rise to compare with the costs of preventing a significant rise. From the standpoint of adapting to the rise in sea level, it often takes an estimate of the potential stakes to convince decision-makers that they should devote time and resources to a new problem.

3. *Evaluate adjustment strategies.* If convinced that the potential stakes are sufficiently large, nations should investigate how they might respond to sea level rise. Particularly important is the timing of the response: should preparations begin now, or when sea level rise reaches the critical stage

4. *Improve the environmental data base.* For most areas, good environmental maps showing coastal wetlands and elevations within one metre do not exist (Shennan, this volume). Moreover, even where data exist, it is not always made available for research purposes due to national security considerations. In general, shoreline change and sediment transport rates are poorly known, and models have yet to be developed.

5. *Specify probability distributions of future sea level rise.* The development of high and low scenarios of future sea level rise was a useful first step to permit impact studies to go forward. However, engineers, planners and policy-makers need to make assumptions regarding the likelihood of particular scenarios. There has been an understandable reluctance to assign probabilities to such estimates given the crude methods that have been used. But the danger is that in the absence of explicit probability statements, undue weight may be attached to the extreme high and extreme low scenarios.

Conclusions

The Panel also reached five major conclusions.

1. For many, if not most, of the world's coasts, *local subsidence and emergence of land is currently far more important than global sea level rise.* This assertion does not impugn the importance of future global sea level rise; rather, it helps to explain the intense interest that the issue has generated. Current relative sea level rise has provided researchers and policy makers with an immediate reason for understanding the consequences of sea level rise.

2. *Many developing nations have a flexibility that developed nations do not share.* Because they have not yet invested huge amounts of fixed capital in the coastal zone, such nations may be able to plan for orderly retreat from the shore as sea level rises. However, many developing nations are intensifying the development of their shorelines. Such countries should consider whether they can learn from the experience of developed nations, or must simply repeat it.

3. *Decision-makers should consider a wide range of future sea level rise scenarios.* Precise estimates of future sea level rise are simply not possible. More harm than good can result from overstating the accuracy of future sea level estimates. If

there is a small chance of a very large rise in sea level, for example, decision-makers should decide for themselves whether that small chance warrants a response; such information should not be withheld on the grounds that decision makers will overreact.

4. *Deltaic and small coral reef nations appear to be the most vulnerable to sea level rise.* In the absence of increased sedimentation or other compensating factors, a 2 m rise in sea level would flood 20% of Bangladesh and a similar fraction of the Nile Delta (see Broadus, this volume). But such a rise might completely inundate small coral reef nations such as the Maldives, particularly if the health of the coral that protects the islands is allowed to degenerate.

5. *The highest priority for international bodies should be to communicate existing information to appropriate officials around the world, by translating articles into the various languages and by holding regional and national workshops. Strategic assessments should have a higher priority than regional case studies.* The Panel felt that developing nations do not need large comprehensive analyses that consider all the implications of sea level rise. Rather, they need smaller, strategic assessments that identify opportunities to incorporate sea level rise into ongoing projects and development policies.

The Panel closed with a consensus that even though there are substantial uncertainties, much is known about possible future climate change and sea level rise. It was agreed that, at least for the sea level rise issue, it is most appropriate to alert decision-makers of the potential risks now, rather than waiting until future research succeeds in narrowing the range of possible outcomes. It is just as important for policy makers to provide scientists with priorities as it is for scientists to provide the policy makers with assessments of what to expect. There must be a co-evolution of science and public policy.

References

Armentano, T.V., Park, R.A. & Cloonan, C.L. (1988). Impacts on coastal wetlands throughout the United States. In *Greenhouse Effect, Sea Level Rise, and Coastal Wetlands*, Titus, J.G. (ed.). US Environmental Protection Agency Washington, DC.

Hull, C.H.J. & Titus, J.G. (1986). *Greenhouse Effect, Sea Level Rise, and Salinity in the Delaware Estuary*. US Environmental Protection Agency and Delaware River Basin Commission, Washington, DC.

Newman, W.S. & Fairbridge, R.W. (1986). The management of sea level rise. *Nature*, **320**, 319–21.

Smith, J.B. & Tirpak, D. (1989). *The Potential Effects of Global Climate Change on the United States*. US Environmental Protection Agency, Washington, DC.

ANNEX:

Global mean temperature and sea level projections under the 1992 IPCC emissions scenarios

T.M.L. Wigley and S.C.B. Raper

After the publication in 1990 of the Intergovernmental Panel on Climate Change's report on the greenhouse problem (Houghton *et al.*, 1990, henceforth IPCC90), IPCC was charged with the task of producing an update as background for the June, 1992 United Nations Conference on Climate and Development (UNCED). The new Working Group 1 report (Houghton *et al.*, 1992, henceforth IPCC92) contains information on significant new scientific developments and, in addition, a new set of future emissions scenarios for greenhouse gases and related compounds. The IPCC update does not, however, include any detailed discussion of the implications of these new scenarios in terms of climate and sea level. This information is given by Wigley and Raper (1992) in an independent paper published at the same time as the IPCC update. We give here a brief summary of these new results.

IPCC90 gave four 'policy scenarios' for future emissions. Scenario A (otherwise referred to as the Business as Usual scenario – BAU – or as SA90) was given as an 'existing policies' reference scenario: it represented a possible future under the assumption of no new policy responses to the greenhouse problem. The other three IPCC90 scenarios, B, C and D, were developed using a 'top down' approach. In other words, certain radiative forcing goals were specified (specifically, stabilization at an equivalent $2 \times CO_2$ level by a certain year), and future emissions scenarios were determined so as to achieve these goals.

In IPCC92, six new scenarios were given (IS92a–f), *all* of which correspond to an existing policies situation. Even with no radically new policies to control greenhouse gas emissions, future emissions become increasingly uncertain with time due to uncertainties in population growth, economic growth, etc. The IPCC92 scenarios account for these uncertainties. The new scenarios also account for the recent changes in emissions control policies and the major political changes that have occurred since 1990. The most important of these changes is the strengthening of the Montreal Protocol to reduce halocarbon production. As a consequence, future emissions of halocarbons in the IPCC92 scenarios, and the consequent radiative forcing effects, are much less than in the IPCC90 BAU scenario.

In addition to the new emissions scenarios, new scientific results are included in the IPCC update. The most important of these are

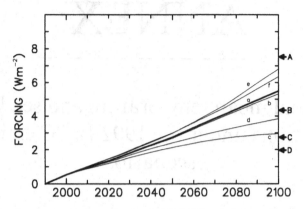

Fig. A.1 Radiative forcing projections for the six IPCC92 emissions scenarios (IS92a–f). The results shown are best guess values from Wigley and Raper (1992). They include the effects of CO_2 fertilization feedback in calculating CO_2 concentration changes, stratospheric-ozone-depletion feedback in calculating the radiative forcing changes due to halocarbons, and the negative forcing due to SO_2-derived sulphate aerosols. IPCC90 scenario results are shown on the right of the Figure (A, B, C and D).

* the negative forcing (i.e., cooling effect) of anthropogenic sulphate aerosols, due to sulphur dioxide emissions arising from fossil fuel combustion. IPCC92 includes scenarios for SO_2 emissions.
* a negative feedback effect with halocarbons arising through stratospheric ozone depletion. Reduced ozone causes cooling in the lower stratosphere which, for many halocarbons, leads to a substantial reduction in the net radiative forcing effect. The overall warming effect of these gases is thereby reduced.
* a recognition of problems with regard to balancing the carbon budget, and the possible consequences of the so-called CO_2 fertilization effect. This process is thought to sequester carbon in the biomass and so reduce the rate of increase of CO_2 concentrations in the atmosphere.

Wigley and Raper (1992) have attempted to account for all of these factors in their assessments of future global mean temperature and sea level changes under the new scenarios. The overall consequence of these factors is to reduce the radiative forcing changes for any given emissions scenario. In other words, if the implications of the IPCC90 BAU emissions scenario were re-evaluated, then one would have to . . .

* reduce the halocarbon emissions to account for a strengthened Protocol,
* reduce the halocarbon forcing to account for ozone-depletion feedback,
* reduce future CO_2 concentrations to account for CO_2 fertilization,
* add in a negative forcing term to account for aerosol effects.

A minor increase in forcing would also occur because of an upward revision of the lifetime of methane.

The following figures show the range of possible future forcings for the new IPCC92 scenarios (Fig. A.1), temperature and sea level changes for the IPCC92 'a' scenario (IS92a) accounting for model uncertainties (Fig. A.2), and the range of possible temperature and sea level changes spanning all six scenarios, but with best guess model

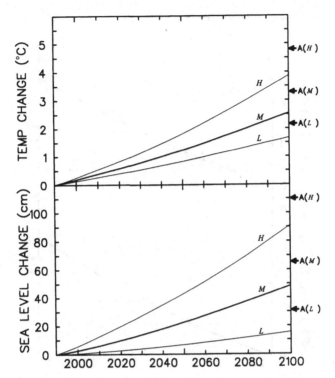

Fig. A.2 Global mean temperature and sea level projections for IPCC92 scenario 'a'. This is the scenario judged most nearly equivalent in its underlying assumptions to the IPCC90 BAU scenario. The projections correspond to the forcing shown in Fig. A.1. Temperature values given are for climate sensitivities, $\Delta T_{2\times}$, of 1.5, 2.5 and 4.5 °C (denoted L, M and H) with other climate model parameters kept at their best guess values (see Wigley and Raper, this volume). For sea level, corresponding thermal expansion values are combined with ice-melt contributions based on low, middle and high model parameter values (see Wigley and Raper, this volume). Equivalent low, middle and high (A(L), A(M) and A(H)) results for the IPCC90 BAU scenario are shown on the right of the figure for comparison.

parameters only (Fig. A.3). IPCC90 results for the year 2100 are also shown on these figures for comparison. The temperature and sea level models used are those described by us elsewhere in this book. For further details of these new projections, see Wigley and Raper (1992).

References

Houghton, J.T. Callander, B.A. and Varney, S.K. (eds.) (1992). *Climate Change 1992. The Supplementary Report to the IPCC Scientific Assessment.* Cambridge University Press, Cambridge, 200pp.

Houghton, J.T., Jenkins, G.J. and Ephraums, J.J. (eds.) (1990). *Climate Change: The IPCC Scientific Assessment.* Cambridge University Press, Cambridge, 365pp.

Wigley, T.M.L. and Raper, S.C.B. (1992). Implications for climate and sea level of revised IPCC emissions scenarios. *Nature*, **357**, pp. 293–300.

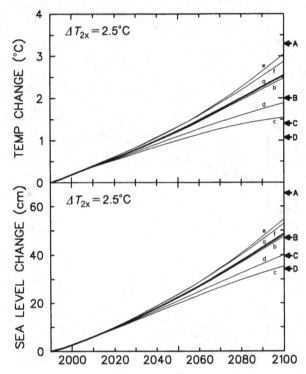

Fig. A.3 Global mean temperature and sea level projections for the six IPCC92 scenarios based on best guess model parameter values (the most important of which is the climate sensitivity, for which a value $\Delta T_{2x} = 2.5\ °C$ is used). Results for the IPCC90 scenarios (A, B, C and D) are shown on the right of the figure.

Index